High Performance Chelation Ion Chromatography

RSC Chromatography Monographs

Series Editor:
R.M. Smith, *Loughborough University of Technology*, UK

Advisory Panel:
J.C. Berridge, *Sandwich, UK*, G.B. Cox, *Indianapolis, USA*, I.S. Lurie, *Virginia, USA*, P.J. Schoenmakers, *Eindhoven, The Netherlands*, C.F. Simpson, *London, UK*, G.G. Wallace, *Wollongong, Australia*

Titles in this Series:
 1: Chromatographic Integration Methods
 2: Packed Column SFC
 3: Chromatographic Integration Methods, Second Edition
 4: Separation of Fullerenes by Liquid Chromatography
 5: Applications of Solid Phase Microextraction
 6: HPLC: A Practical Guide
 7: Capillary Electrochromatography
 8: Hyphenated Techniques in Speciation Analysis
 9: Cyclodextrins in Chromatography
10: Electrochemical Detection in the HPLC of Drugs and Poisons
11: Validation of Chromatography Data Systems: Meeting Business and Regulatory Requirements
12: Thin-layer Chromatography: A Modern Practical Approach
13: High Temperature Liquid Chromatography: A User's Guide for Method Development
14: High Performance Chelation Ion Chromatography

How to obtain future titles on publication:
A standing order plan is available for this series. A standing order will bring delivery of each new volume immediately on publication.

For further information please contact:
Book Sales Department, Royal Society of Chemistry, Thomas Graham House, Science Park, Milton Road, Cambridge, CB4 0WF, UK
Telephone: +44 (0)1223 420066, Fax: +44 (0)1223 420247, Email: books@rsc.org
Visit our website at http://www.rsc.org/Shop/Books/

High Performance Chelation Ion Chromatography

Pavel N. Nesterenko
Australian Centre for Research on Separation Science (ACROSS), University of Tasmania, Hobart, Tasmania, Australia

Phil Jones
School of Geography, Earth and Environmental Sciences, University of Plymouth, Plymouth Devon, UK

Brett Paull
Irish Separation Science Cluster (ISSC), National Centre for Sensor Research, Dublin City University, Dublin, Ireland

RSCPublishing

RSC Chromatography Monographs No. 14

ISBN: 978-1-84973-041-9
ISSN: 1757-7055

A catalogue record for this book is available from the British Library

© The Royal Society of Chemistry 2011

All rights reserved

Apart from fair dealing for the purposes of research for non-commercial purposes or for private study, criticism or review, as permitted under the Copyright, Designs and Patents Act 1988 and the Copyright and Related Rights Regulations 2003, this publication may not be reproduced, stored or transmitted, in any form or by any means, without the prior permission in writing of The Royal Society of Chemistry or the copyright owner, or in the case of reproduction in accordance with the terms of licences issued by the Copyright Licensing Agency in the UK, or in accordance with the terms of the licences issued by the appropriate Reproduction Rights Organization outside the UK. Enquiries concerning reproduction outside the terms stated here should be sent to The Royal Society of Chemistry at the address printed on this page.

The RSC is not responsible for individual opinions expressed in this work.

Published by The Royal Society of Chemistry,
Thomas Graham House, Science Park, Milton Road,
Cambridge CB4 0WF, UK

Registered Charity Number 207890

For further information see our web site at www.rsc.org

Printed in Great Britain by CPI Antony Rowe, Chippenham and Eastbourne

Preface

Ion exchange is a fundamentally important mode of liquid chromatography, forming the basis of many modern analytical methods, from solid phase extraction and sample clean-up, to modern high-performance ion chromatography. A number of significant books and monographs exist on each of these ion exchange based technologies, detailing fundamental aspects of the exchange process, methodology and their analytical and industrial applications. Traditionally ion exchange is viewed as a technique based upon the exchange of ions between phases, where retention on a stationary phase is predominantly based upon simple electrostatic attraction. However, in cation exchange chromatography a significant variation from simple cation exchange exists, where the retention of cations is no longer dominated by electrostatic attraction, but instead is controlled through the formation of metal–ligand bonds with immobilised ligands, forming metal–ligand complexes, or chelates, on the stationary phase surface. Chelating stationary phases are those which are capable of forming such complexes, and are available in many variations for both classical and analytical chromatographic application.

In this monograph we selectively detail the wide and varied application of chelating ion exchange materials to the 'high-performance' separation of metal ions. As 'ion chromatography' has naturally evolved as the descriptive name to describe high-performance separations achieved based upon all modes of ion exchange, here we collectively describe as 'chelating ion chromatography' all those methods based upon chelating ion exchange phases for the high-performance liquid chromatographic separation of metal ions. This descriptive definition specifically excludes the significant body of work on chelating phases for extraction, batch separation, concentration and classical chromatographic applications, which is also well documented elsewhere. The monograph provides the reader with a detailed description of the fundamental theory of the chelating exchange mechanism, the impact of thermodynamics and kinetics on the chelating exchange process, the physical and chemical requirements for efficient chromatographic separations of metal ions and the selectivity offered by the wide and varied range of chelating ligands utilised in this significant mode of ion chromatography.

The authors of this monograph have worked both together and individually on chelating stationary phase materials and their chromatographic applications for over two decades, having published a significant number of joint papers on the subject. They have now come together again to publish this definitive and unique resource for all those analytical scientists, bio-analysts, chemical engineers and chromatographers working on the analysis of metal ions in complex and challenging matrices.

The authors would like to acknowledge the great support and advice from the Royal Society of Chemistry book publishing group in the preparation of this monograph. In addition the authors would like to thank Dr. Ekaterina (Katya) Nesterenko for her much appreciated administrative support helping in the preparation of this work.

Pavel Nesterenko
Phil Jones
Brett Paull

Contents

Chapter 1 Chelation and its Role in Contemporary Liquid Chromatography   1

    1.1 Basic Chromatographic Principles   1
    1.2 Chelation as a Mechanism for Obtaining Separation Selectivity   3
    1.3 Terminology and Definitions   4
    1.4 Historical Developments   6
    1.5 The Evolution of Chelating Stationary Phases   9
    1.6 The Current State of HPCIC and Context   11
    References   13

Chapter 2 Retention Mechanism and Chelation Theory   16

    2.1 Introduction   16
    2.2 Surface and Solution Complexation   16
        2.2.1 Complexation Kinetics   17
    2.3 Equilibria Within the Chelating Ion Exchanger   21
        2.3.1 Distribution Ratio Within a Mixed Mode Mechanism   23
        2.3.2 Selectivity in Non-complexing Mobile Phases   24
        2.3.3 Secondary Equilibria Within the Mobile Phase   26
    2.4 Temperature Effects in HPCIC   27
        2.4.1 Thermodynamic Effects   27
        2.4.2 Kinetic Effects   28
    2.5 Mobile Phase pH   29
    2.6 Organic Solvent Additives   31
    References   32

Chapter 3 Chelating Stationary Phases — 35

- 3.1 Introduction — 35
- 3.2 Types and Properties of Chelating Ion Exchange Ligands — 36
  - 3.2.1 Adsorption of Metal Ions — 37
  - 3.2.2 Ligand Chemical Stability — 37
  - 3.2.3 Suitable Coordinating Sites — 37
  - 3.2.4 Functional Selectivity — 37
  - 3.2.5 Suitable Acid–Base Properties and Surface Charge — 39
- 3.3 Chelating Ion Exchangers with Covalently Bonded Chelating Groups — 42
  - 3.3.1 Types of Chelating Ion Exchangers and Synthetic Methods — 42
  - 3.3.2 Polymer Based Chelating Ion Exchangers — 46
  - 3.3.3 Silica Based Chelating Ion Exchangers — 54
  - 3.3.4 Commercially Available Chelating Ion Exchange Phases — 64
- 3.4 Stationary Phase Matrix Effects — 67
  - 3.4.1 Surface Distribution of Covalently Bound Ligands — 70
  - 3.4.2 Effects of Phase and Particle Porosity — 73
  - 3.4.3 Effect of Stationary Phase Structure upon Separation Selectivity — 74
  - 3.4.4 Monolithic Chelating Ion Exchangers — 77
- 3.5 Dynamically Modified and Impregnated Stationary Phases — 79
  - 3.5.1 Impregnated (Pre-coated) Phases Using Metallochromic Ligands — 80
  - 3.5.2 Dynamically Modified Phases — 95
- References — 108

Chapter 4 Elution — 116

- 4.1 Mobile Phase Parameters Influencing Separation Performance in HPCIC — 116
  - 4.1.1 Ionic Strength — 116
  - 4.1.2 Temperature Effects — 124
  - 4.1.3 pH of the Mobile Phase — 130
  - 4.1.4 Organic Solvents in HPCIC — 132
  - 4.1.5 Addition of Oxidising and Reducing Agents — 134
- 4.2 Elution Modes — 135
  - 4.2.1 Isocratic Elution — 135
  - 4.2.2 Gradient Elution — 147
- References — 154

Contents ix

Chapter 5 Liquid–Liquid Chromatographic Methods 158

    5.1 High Speed Counter-current Chromatography of
        Metal Ions 158
        5.1.1 Extracting Ligands and Metal Ion Selectivity 161
        5.1.2 Efficiency of Metal Separations using HSCCC 167
        5.1.3 Applications 169
    5.2 High-performance Extraction Chromatography 170
        5.2.1 Stationary Phase Supports 175
        5.2.2 Ligand Loading Stability and Retention
            Mechanism 177
        5.2.3 Type of Extracting Reagent 178
        5.2.4 Applications 180
    References 191

Chapter 6 Detection 194

    6.1 Background 194
    6.2 Post-column Reactions for the Photometric Detection
        of Metal Ions 195
        6.2.1 Construction of Post-column Reactors 196
        6.2.2 Reagents for Post-column Derivitisation 203
        6.2.3 Recent Developments in High-sensitivity
            Reagents for Post-column Reactions 219
    6.3 Practical Methods for Improving Limits of Detection
        in Liquid Chromatography 222
        6.3.1 The Detection Limit 222
        6.3.2 Types of Baseline Noise 224
        6.3.3 Noise Reduction Methods 226
    6.4 Other Types of Post-column Reaction Detection 230
        6.4.1 Displacement Reactions 230
        6.4.2 Fluorescence Detection 232
        6.4.3 Chemiluminescence Post-column Reaction
            Detection 233
    6.5 Hyphenated Techniques 234
    References 236

Chapter 7 Practical Applications 242

    7.1 Potential HPCIC Applications 242
    7.2 Relative Advantages of HPCIC 243
    7.3 Fresh and Potable Waters 250
    7.4 Saline Samples 254
        7.4.1 Seawater and Estuarine Water 255
        7.4.2 Highly Saline Waters 261
        7.4.3 Commercial Products and Fine Chemicals 262

7.5	Solid Samples		269
	7.5.1	Sediments, Soils and Minerals	270
	7.5.2	Biological Materials and Foodstuffs	277
	7.5.3	Miscellaneous	280
References			282

Subject Index   **284**

For
Ekaterina and Dima
Eliza and Fleur
Aidan and Benjamin

CHAPTER 1
Chelation and its Role in Contemporary Liquid Chromatography

1.1 Basic Chromatographic Principles

The performance of any separation system depends on two main factors, namely, separation efficiency and separation selectivity. As a rule, high separation efficiency can be achieved through the correct optimisation of the physical and chemical properties of the separation media (in the case of column liquid chromatography – column length and diameter, particle size and porous structure of the stationary phase), and through design and control of the separation conditions (column temperature, pressure, viscosity of the mobile phase *etc*). Obtaining greater or alternative separation selectivity is a significantly more difficult challenge, as it is predominantly associated with the precise adsorption mechanism, which is dependent upon the chemistry of either the adsorbate or adsorbent, and in most practical circumstances, both of them.

Contemporary *high-performance liquid chromatography* (HPLC) includes all possible combinations of non-specific and specific interactions between separated analytes (molecules or ions) and different adsorbents or stationary phases to achieve maximum separation selectivity. Chromatographic methods based on non-specific, usually weak interactions (van der Waals forces, induction and dispersion), have been reported for many years as reversed-phase (RP-HPLC) and classical normal-phase (NP-HPLC) modes of HPLC (Table 1.1). However, more specific and higher energy interactions (hydrogen bonding, π–π interactions, coordinate bonding and others) can provide a significantly higher degree of separation selectivity, the ultimate examples of which could be a β-cyclodextrin bonded phase for the use in inclusion chromatography or a highly specific bioaffinity phase. Consequently, the long-established trend in

Table 1.1 Typical bonding and nonbonding (intermolecular) forces exploited in liquid chromatography.

Type of interaction	Energy ($kJ\,mol^{-1}$)	Description	NP-HPLC	RP-HPLC	IEC	HPCIC
Van der Waals forces including:		**Non-specific (weak) interactions**				
Dispersion (London) forces	0.05–40	Instantaneous induced dipole-induced dipole	x	x		
Induction (polarisation) forces	2–10	Permanent dipole and a corresponding induced dipole	x	x		
		Specific (strong) interactions				
Hydrogen bonding	10–40	Sharing of an electron from electronegative atom with a hydrogen atom	x[a]			
Aromatic or π–π interaction	0–50	Intermolecular overlapping of p-orbitals in π-conjugated systems		x[a]		
Electrostatic interactions including:						
Ion-induced dipole	3–15	Ion-charge–polarisable electron cloud			x	x
Dipole–dipole	5–25	Between permanent dipoles	x			
Ion–dipole	50–200	Ion charge–dipole charge			x	x
Ion–ion	100–350	Attraction of oppositely charged and repulsion of similarly charged ions			x	x
		Bonding				
Coordinate bonding (dative covalent)	150–1100	Weak reversible covalent bond with two electrons coming from only one of the atoms				x

[a]For specific types of solutes and phases.

chromatographic research and development is the search for new and highly selective stationary phase materials, and their subsequent exploitation in innovative and emerging modes of HPLC, such as chiral phase chromatography, zwitterionic chromatography and others.

Complexation and chelation represent another category of specific interactions, here between metal ions and ligands, which have long been used within the mobile phase in liquid chromatography, through the addition of reagents for greater control and optimisation of separation selectivity. However, achieving desired separations via complexation at the surface of the adsorbent, with such complexation being the sole or dominant separation mechanism, has not been demonstrated too frequently in liquid chromatography. Historically, this has been due to practical and synthetic difficulties in the preparation of high efficiency chelating substrates, and additional problems resulting from the slow kinetics associated with the reversible bonding of ions within these chelating stationary phases. However, such difficulties are now well understood and documented, with considerable improvements having been demonstrated in this area, which collectively have resulted in the emergence of new high-performance modes of liquid chromatography based upon pure stationary phase chelation (or chelating ion exchange) interactions. This current monograph details these developments, in particular detailing the various methods of stationary phase preparation, exhibited and exploited stationary phase properties and selectivity, discussion of the theoretical and experimentally determined retention mechanisms, and finally the various applications of such chelating stationary phases for the separation and determination of metal ions.

1.2 Chelation as a Mechanism for Obtaining Separation Selectivity

As alluded to previously, separation selectivity in a chromatographic system can be increased, or modified, through the exploitation of multi-point or multi-bond interactions between the adsorbate molecules and the adsorbent, in combination with molecule specific effects, such as molecular size/weight *etc*. Chelation is an example of such an interaction, here being defined as the formation of two or more simultaneous and spatially separate covalent binding events between a single polydentate ligand and a central metal ion. The corresponding thermodynamically based 'chelation effect' results in a dramatic increase in the observed affinity of any such polydentate ligand towards specific metal cations. Thus stationary phase chelation has been utilised to provide the all important mechanism for achieving enhanced separation selectivity in many different modes of liquid chromatography, such as for example, ligand-exchange chromatography, or so-called *immobilised metal ion affinity chromatography* (IMAC).

For the chromatographic separation of metal ions, methods showing chelation taking place simultaneously in both the mobile and stationary phases have been developed. However, the chromatographic system exploiting chelation in only the mobile phase is much simpler, both theoretically and experimentally. The addition of various chelating reagents to the mobile phase has become a common way to regulate the separation selectivity of metal ions in ion exchange chromatography, and also within RP-HPLC, where stable metal–ligand complexes have been separated based upon differences in overall complex charge or hydrophobicity.

Ion exchange chromatography (IEC) is based on electrostatic interactions between ion exchangers and solvated ions. Such electrostatic interactions comprise ion–ion, ion–dipole and dipole–dipole interactions, with interaction energies of $100–350\,kJ\,mol^{-1}$, $50–200\,kJ\,mol^{-1}$ and $5–25\,kJ\,mol^{-1}$, respectively (Table 1.1). In some cases, cation–π interactions with energies of $5–80\,kJ\,mol^{-1}$ can also take place in certain instances/examples of IEC. However, when considering the chelating effect, the formation energy for chelates on the surface of a chelating stationary phase, exhibiting coordinate bonding or dative covalent bonding, should be considerably higher under optimum conditions. Additionally, as chelating stationary phases may be neutral (*e.g.* β-diketone functionalised phase), positively charged (*e.g.* 8-hydroxyquinolinol bonded phases) or negatively charged (*e.g.* iminodiacetate resins), it is often probable, and in fact inevitable, that such phases exhibit some degree of mixed mode retention, including both chelation and ion exchange interactions occurring simultaneously, but of different relative strengths, dependent upon the nature of the solvated metal ion. Where both interactions are known to occur simultaneously without inhibition of the other, the term 'chelating ion exchange' may be a more fitting description of the retention mechanism than simply chelation. In most modes of liquid chromatography it is well known that multiple retention mechanisms interacting simultaneously are unlikely to result in achieving high chromatographic efficiency. Therefore, to obtain efficient chromatographic separations using a chelating stationary phase, it is important to have chelation as the dominant retention mechanism. In chelation chromatography this can be achieved by suppression of any electrostatic interactions occurring between the solvated metal ions and the charged functional groups of the stationary phase. The simplest way to do this is through increasing the ionic strength of the mobile phase, typically using an alkali metal salt.

1.3 Terminology and Definitions

Historically, most modes of liquid chromatography are named systematically according to the dominant nature of the adsorbate–adsorbent interaction. Exceptions to this rule have occurred where names have been adopted according to the stationary phase used, such as *chiral phase chromatography* (CPC) and *immobilised artificial membrane chromatography* (IAMC), or according to the kind of separated analytes, as in standard *ion chromatography* (IC), which actually can also be found described under the traditional rules as *high-performance ion exchange chromatography* (HPIEC). Where chelation in the stationary phase is the dominant mechanism providing for the retention and separation of ions, clearly the technique cannot be described using the term 'ion exchange', as this clearly contradicts the recommendations of IUPAC,[1] who define ion exchange as the

> equivalent exchange of ions between two and more ionised species located in different phases, at least one of which is an ion exchanger, without the formation of new types of chemical bonds.

For similar reasons, the use of the term 'adsorption-complexing chromatography' is also not well suited to describe the chelation based separation mechanism under consideration. One of the earliest papers, published in 1979 by Jezorek and Freiser,[2] utilising a chelating stationary phase (silica immobilised 8-hydroxyquinoline) for the chromatographic separation of metal ions, described the new approach as *metal-ion chelation chromatography* (MICC). However, this term does not properly distinguish the technique from low efficiency classical chromatographic methods used for metal ion extraction using chelating resins and so was not adopted universally. Thus, with the lack of suitably accurate accepted terminology available, a new name for chromatographic methods utilising stationary phase chelation as the dominant retention mechanism for the efficient (high-performance) chromatography of metal ions was presented, based upon a combination of the above nomenclaturial traditions. The name *high-performance chelation ion chromatography* (HPCIC) places a correct emphasis on the exact adsorbate–adsorbent interaction and the nature of the adsorbate itself. The name also correctly describes the technique as specifically referring to high-performance applications of chelating phases and as a sub-discipline of the more widely applicable term *ion chromatography*, which now is accepted as a term describing any modern high-performance liquid chromatographic method for the separation of ions.

A similar term, *complexation ion chromatography*, was suggested by Timerbaev and Bonn.[3] However, according to them, this definition should include

> all ion chromatographic modes in which complexation is exploited for the separation and detection of metals in different ways.

Obviously, in this interpretation complexation ion chromatography would cover too broad an area, including many with many different approaches in various fields of liquid chromatography, so it would be difficult to consider it as a unique separation technique.

A possibly confusing factor connected with the terminology prescribed above may have arisen from the emergence in the early 1990s of the so-called *chelation ion chromatography* (CIC) system from the Dionex Corporation and the associated trade name. The automated technique developed utilised switching valves within a complex ion chromatographic system, incorporating short chelating and ion exchange columns for on-line metal ion extraction and preconcentration prior to their separation on coupled ion exchange columns.[4] Obviously, the correct description/terminology for this method should be simply 'ion chromatography of metal ions with on-line preconcentration of trace metals', as no actual separation of metal ions is achieved on the short chelating column. This example does however further justify the inclusion of the 'high-performance' preface to the name for the work described herein, to distinguish it from all other methods using chelating resins for simple preconcentration or extraction of metal ions (of which there is a considerable amount of literature references available, describing low-pressure columns packed with large particle low efficiency chelating resins, typically of 0.2–0.3 mm diameter).

1.4 Historical Developments

The basic principles of coordination chemistry were developed more than 120 years ago by the Swiss chemist Alfred Werner[5] who was awarded the first Nobel Prize in chemistry. The term *chelate* was introduced in 1920 by Gilbert Morgan and Harry Drew[6] who gave a very clear description for the specific type of complexes formed by polydentate ligands:

> The adjective chelate, derived from the great claw or chela of the lobster or other crustaceans, is suggested for the caliper like groups, which function as two associating units and fasten to the central atom so as produced heterocyclic rings.

This description was strongly supported by later discoveries, such as the important 'rule of rings' described by Lev Chugaev, who found that chelates containing five- to six-membered rings are usually the most stable.[7] These very early breakthroughs paved the way for the extensive exploitation of the extraordinary selectivity of the chelate effect in many varied applications of analytical chemistry, including chromatography.

One element of chelation based chromatographic methods which has historically received considerable attention is the nature of the chelating phase itself, and in particular the approach to the immobilisation of what are often relative complex structures (as compared to simple ion exchange functional groups). One of the first chromatographic techniques reported which utilised chelation as an element of the separation mechanism was given the name, *precipitation chromatography*, as described by Erlenmeyer and Dahn in 1939.[8] In this very early work a micro-crystalline powder of a pure chelating ligand was used as the column packing, thus completely eliminating the need for any immobilisation strategies. The authors explored the differences in solubility of metal chelates formed at the surface of 8-hydroxyquinoline crystals as the basis for their separation. There is a well-defined correlation between solubility and stability constants of certain metal chelates, so this work could be considered as the first example of a form of chelation ion chromatography. The principles and further development of precipitation chromatography have been documented in an early monograph[9] and a more recent review.[10] However, clearly, the use of chromatographic columns packed with pure organic reagents as demonstrated by Erlenmeyer and Dahn[8] was not the most cost effective or practical approach, so the use of various support materials, having developed chelating surface chemistries, including alumina, titania, silica, calcium carbonate, and various ion exchangers, became more popular phases in the technique, which continued into the late 1980s to be reported under the name 'precipitation chromatography'. The use of paper impregnated with organic reagents such as thiocyanates, 8-hydroxyquinolinol, dimethylglyoxime and others, quickly gained popularity in precipitation chromatography because of simplicity and improved sensitivity, resulting from distinctive colourful chromatographic bands (Figure 1.1).[11]

Most precipitates produced during the separation of metals in precipitation chromatography are formed with complexing ligands which can themselves be retained to some extent on the support. With this in mind, a different separation mechanism known as *adsorption–complexing chromatography* was suggested in 1954 by Gurvich and Gapon.[12] In adsorption–complexing chromatography, the adsorbent retains both complexing reagent and its associated metal complexes, such that the difference in stability of these metal complexes was used as the basis for their separation, and hence the separation of the metals themselves. Technically the proposed method is very reminiscent of precipitation chromatography, but here no new solid phase (precipitate) is formed. However, it should be noted that the complexing stationary phase was often prepared by adsorption or immobilisation of complexing reagents on the surface of the

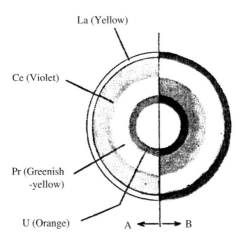

Figure 1.1 Separation of lanthanides and uranium by paper precipitation chromatography with 8-hydroxyquinoline (A) and its radioautogram (B) obtained after irradiation with thermal neutrons followed by exposition to the photographic plate. Reproduced, with permission, from Nagai.[11] Copyright (1964) The Chemical Society of Japan.

support material prior to separation. This extremely useful approach is still widely used today for the separation or isolation of specific metals, particularly if the selective reagent itself cannot be easily covalently attached to the surface of the support.

The slow (or in some cases rapid) bleeding of the chelating ligand from the column is a significant drawback of adsorption–complexing chromatography, such that the idea of using partition chromatography with two immiscible liquid phases was developed as an alternative approach. *Extraction chromatography*, based on this principle, received quick recognition following its first description by Siekierski and Fidelis, detailing the separation of lanthanides (Figure 1.2) in 1960.[13] (Also see Dietz *et al.*[14]) In extraction chromatography, a hydrophobic reagent, often a bulky organic chelating ligand, is strongly retained in the immobilised liquid stationary phase, providing the desired separation selectivity. The technique is still regularly reported for application in the isolation and separation of radionuclides and heavy metals.[14,15] A similar mechanism exists in *counter-current chromatography* (CCC) when applied to metal ion separations.[16–18] As a rule, both of these separation techniques explore chelation in the liquid stationary phase, in a similar fashion to multi-step liquid–liquid extraction, to achieve selective separations from complex mixtures of metal ions. Of course, the large volume of experimental data obtained on extraction of metal ions with various organic reagents provided a

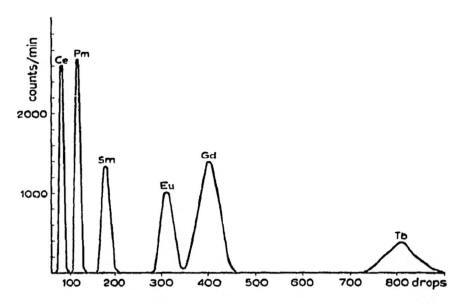

Figure 1.2 Separation of lanthanides by extraction chromatography with tributyl-phosphate (TBP) loaded 110 × 3 mm i.d. column packed with 80 μm particles of Kieselguhr treated by dimethyldichlorosilane. Mobile phase: 15.1 M nitric acid. Reproduced, with permission, from Siekierski and Fidelis.[13] Copyright (1960) Elsevier.

good basis for the fast development and application of both extraction chromatography and CCC to the separation of metal ions.

The appearance of a variety of covalently modified chelating resins in the 1960s intensified the development of chelation based separation methods, presenting new and unique separation selectivity. These new materials were also applied not only to the separation of metal ions or metal complexes, but significantly to the separation of large biomolecules. For example, *ligand exchange chromatography* was first reported in 1961 by Helfferich.[19] This type of highly selective chromatography was further developed by Porath *et al.* in 1975[20] and is now more commonly known as IMAC. In both methods the separation takes place on a stationary phase pre-loaded with various metal ions, with chelating ion exchangers providing the best option for the strong retention of the metal ions required, in a form suitable for further coordination with the target biomolecules, thus providing their retention and subsequent separation. Today this technique is widely used for the selective separation of polar organic molecules such as amines, amino alcohols and amino acids, phenols, carbohydrates, peptides and proteins. Indeed some of the commercially produced chelating phases and columns, when in free ligand form, can also be effectively used in chelation ion chromatography of metal ions.

As discussed previously, the terminology used to describe chromatographic methods should ideally reflect the actual separation mechanism. However, in some cases it is difficult to identify a single dominant type of interaction to name a new method in a suitably descriptive manor. The following method, often described as either simple *complexation chromatography* or more correctly *silver ion* or *argentation chromatography*, being a case in point. This method was first introduced by Morris in 1962[21] and is based on the highly specific interaction between a silver ion loaded adsorbent and the electron-rich double bonds in organic molecules such as fatty acids and lipids. Technically, complexation itself occurs in the stationary phase; however, here there is no need to use complex chelating ion exchange resins to retain the silver cations, as in this type of chromatography the use of a non-polar mobile phase ensures that a simple silver loaded sulfonated cation exchanger, or indeed silica or alumina gels loaded with a suitable silver salt, are stable enough to be used as the complexing stationary phase.

1.5 The Evolution of Chelating Stationary Phases

In general terms, any selective chromatographic separation of a large set of metal ions is not possible without increasing the selectivity of the system by using complexing or chelating ligands. However, for many years, simple addition of various ligands to the mobile phase in either IEC or RP-HPLC was applied to this problem. Significantly, much less attention was paid to the use of chelating resins as the actual stationary phases. The emergence of HPCIC as a technique for metal ion separations was a new direction, which rapidly gained in popularity, having since established itself as an important ion chromatographic method. Indeed, it is the principal goal of this monograph to document this significant body of work, and demonstrate the power and versatility of chelation and complexation in the stationary phase and its application to the selective separation of metal ions.

The development of HPCIC would not have been possible without intensive research into the area of stationary phase synthesis and surface chemistry, and the subsequent investigation of the properties of various types of chelating substrates. Early significant advances in this area were connected with the invention of simple ion exchange resins and with the attempts to improve their affinity towards metal groups or single metal species. The primary and secondary amino groups utilised in weak anion exchangers and carboxylic groups found in weak cation exchangers can exhibit distinct complexing properties (see more discussion in Chapter 2). Therefore, unsurprisingly, soon after the discovery of the first ion exchange resin by Adams and Holmes in 1935,[22] descriptions of their complexing properties were immediately reported, notably by Griessbach in 1939,[23] in this case for the weak cation exchange column, Wofatit C. This resin was prepared by the condensation of formaldehyde with α-resorcylic or 3,5-dihydroxybenzoic acid and contained phenolic and carboxylic groups in adjacent aromatic rings divided by five carbon atoms (Figure 1.3A). Obviously, these groups could form multiple bonds with metals and a chelate effect could be observed via the formation of an eight-membered cyclic complex.

A few years later, in 1948, the preparation of the first metal specific polystyrene based chelating resin, containing dipicrylamine functional groups (Figure 1.3B), with enhanced selectivity to potassium was reported by Skogseid.[24] However, it was not until the 1950s when the real boom in publications on the preparation and application of chelating ion exchange resins began, with Gregor et al. first reporting the synthesis of a m-phenylenediglycine resin demonstrating unique selectivity towards heavy metals.[25] It should be noted that the authors of this particular publication specifically named the prepared substrate as a chelating ion exchange resin, indicating an enlightened awareness of the dual nature of its interactions with metal ions. Two years later, the synthesis of one of the most successful iminodiacetic acid (IDA) type

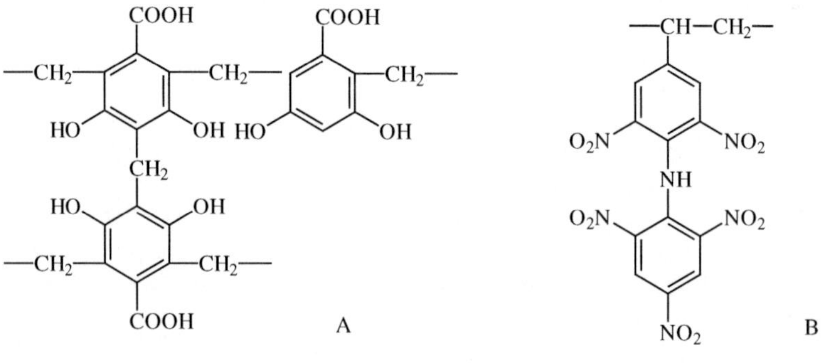

Figure 1.3 The structures of carboxylic cation exchange resin (A) Wofatit C and (B) Skogseid's potassium selective resin.

chelating ion exchangers produced to date was reported, produced via the reaction of a chloromethylated poly(styrene-divinylbenzene) matrix (PS-DVB) with iminodiacetonitrile, followed by hydrolysis of the nitrile groups.[26] Such pioneering work led to IDA functionalised resins becoming commercially available in the late 1950s, and hence to subsequent widespread analytical application to this day. The enormous success of this type of resin intensified research into the synthesis of resins with other types of chelating functional groups, such as phosphonic, arsenic, polyamine, hydroxyquinoline, amidoxime, aminocarboxylic and many others. An excellent historical overview of the development of chelating resins together with the different applications was given by Hering in an early monograph[27] and later reviewed by Sahni and Reedijk.[28] It was Hering who also outlined resins with aminoacidic and iminodiacidic functional groups for the first time as separate classes of chelating ion exchangers. The synthesis, properties and applications of a wide variety of chelating ion exchangers produced mainly in USSR have been documented in a monograph by Kopylova and Saldadze,[29] and most recently in a subsequent review paper.[30] In these works, the authors also considered the different mechanisms of chromatographic separations of metals on columns packed with chelating ion exchangers. The application of these various chelating resins to the selective preconcentration of various trace metals has been described in the reviews by Myasoedova and Savvin[31] and Bilba et al.,[32] and finally, some important information on the characteristics and applications in analytical chemistry of ion exchangers exhibiting complexing properties can be found within the monograph on the subject by Marhol.[33]

Classical ion exchange and chelating resins have a three-dimensional distribution of ligands within the polymer particles, which seriously limited their application in column chromatography, because of the problems of swelling/shrinking in different solutions and of the low diffusion rates for ions within the stationary phase. The use of substrates with chelating groups immobilised only at the surface of inert and structurally rigid supporting substrates could be seen as an obvious solution to these problems. Such materials based upon physically adsorbed organic reagents, on various neutral and charged substrates, are of course readily available, as they have been developed successfully for use in the previously described methods of precipitation chromatography,[34] adsorption–complexing chromatography[12] and in some variants of extraction chromatography.

1.6 The Current State of HPCIC and Context

The earliest separations on chelating substrates were obtained with gravity flow columns with very limited possibility for achieving efficient separations. So, as with standard forms of ion exchange based chromatography (and indeed all modes of liquid chromatography), the research effort over the past three decades in chelation based methods, has focused itself on development of high efficiency phases for high-performance separations. The introduction of IC in

1975 induced intensive work on the synthesis of highly cross-linked microparticulated polymer materials and chemically modified silica gels, with covalently bonded chelating functional groups. For example, in 1977, Fritz and Moyers[35] reported several separations of metals on a short column packed with a 44–56 μm PS-DVB based resin grafted with propylenediaminetetraacetic acid. An example from these early separations is shown in Figure 1.4. Importantly, the authors used a high-pressure pump to deliver the mobile phase and a post-column reagent containing either arsenazo III, arsenazo I or 4-(2-pyridylazo resorcinol) (PAR) for photometric detection of the eluting metal ions. Thus, this work can probably be truly considered as the first example of HPCIC.

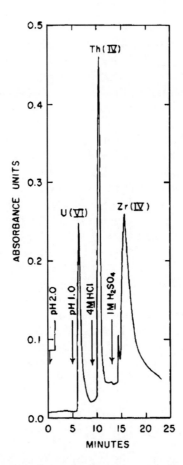

Figure 1.4 Separation of model mixture of metals on 28 × 6 mm i.d. column packed with PS-DVB resin (44–56 μm) with propylenediaminetetraacetic acid functional groups. Flow rate 2 mL min^{-1}. Sample volume 214 μL of 1.2 mM U(VI), 0.4 mM Th(IV) and 4 mM Zr(IV). Photometric detection after PCR with arsenazo III at 635 nm. Reproduced, with permission, from Moyers and Fritz.[35] Copyright (1977) American Chemical Society.

Of course, like simple ion exchange chromatography, HPCIC requires sensitive detection methods for the separated metal ions and this has been achieved predominantly from the very beginning,[35] through the introduction of post-column reagents. The use of metallochromic ligands for such sensitive photometric detection of metals, as first reported by Sickafoose in 1971,[36] has been critical in the development of HPCIC. Indeed, according to the review by Dasgupta on post-column detection techniques,[37] the importance of the detection of metal ions with the help of post-column reactions (PCRs), is second only to the development of the ninhydrin reaction for detection of amino acids. With this in mind, within this monograph the significant progress achieved through the years in the post-column reaction of metal ions, following their chromatographic separation, has been given special attention.

At the current juncture in the long development of HPCIC, research continues to be focused upon improvements in both efficiency and selectivity, in tandem with a steady but continuing amount of applications, showing how HPCIC is now a popular solution to the quantification of metal ions in very complex samples, such as seawater or industrial brines. Current research publications on the improvement of the efficiency of chelating ion exchangers through using long capillary columns, monolithic columns or columns packed with 2–3 µm particles continue to appear, showing the technique has a significant following, which continues to develop new possibilities and advances in the field. Interestingly, improvements in separation selectivity have recently been more focused on the use of complexing elution systems and less on the selection of chelating surface groups. Obviously the use of complexing mobile and stationary phases together gives an extra dimension to manipulating selectivity for specific applications, but can add an unwanted level of complexity to the system. Finally, there remain continuing developments in the PCR detection of metals, which are becoming more sensitive with the common application of fluorescent and chemiluminescent reactions, together with the development of electronic noise reduction systems,[38] or through coupling of HPCIC with elemental analysers, such as inductively coupled plasma mass spectrometry (ICP-MS).

References

1. R. Harjula and J. Lehto, *React. Funct. Polym.*, 1995, **27**, 147–153.
2. J. R. Jezorek and H. Freiser, *Anal. Chem.*, 1979, **51**, 366–373.
3. A. R. Timerbaev and G. K. Bonn, *J. Chromatogr.*, 1993, **640**, 195–206.
4. A. Siriraks, H. M. Kingston and J. M. Riviello, *Anal. Chem.*, 1990, **62**, 1185–1193.
5. A. Werner, *Z. Anorg. Chem.*, 1893, **3**, 267–330.
6. G. T. Morgan and H. D. K. Drew, *J. Chem. Soc. Trans.*, 1920, **117**, 1456–1465.
7. L. Tschugaeff, *Ber. Dtsch. Chem. Ges.*, 1906, **39**, 3190–3209.

8. H. Erlenmeyer and H. Dahn, *Helv. Chim. Acta*, 1939, **22**, 1369–1371.
9. K. M. Ol'shanova, V. D. Kopylova and N. M. Morozova, *Osadochnaya khromatografiya*, Izd. Akad. Nauk SSSR, Moscow, 1963.
10. A. A. Lur'e, *Russ. Chem. Rev.*, 1968, **37**, 39–53.
11. H. Nagai, *Bull. Chem. Soc. Japan*, 1964, **37**, 1076–1078.
12. A. M. Gurvich and T. B. Gapon, *Zavod. Lab. SSSR*, 1957, **23**, 1037–1042.
13. S. Siekierski and I. Fidelis, *J. Chromatogr.*, 1960, **4**, 60–64.
14. M. L. Dietz, E. P. Horwitz and A. H. Bond, Extraction Chromatography: Progress and Opportunities, in *Metal-ion Separation and Preconcentration. Progress and Opportunities*, ed. A. H. Bond, M. L. Dietz and R. D. Rogers, American Chemical Society, Washington, 1999, Chapter 16.
15. *Extraction Chromatography*, ed. T. Braun and G. Ghersini, Elsevier Scientific, Amsterdam, New York 1975.
16. A. Berthod, T. Maryutina, B. Spivakov, O. Shpigun and I. A. Sutherland, *Pure Appl. Chem.*, 2009, **81**, 355–387.
17. B. Y. Spivakov, T. A. Maryutina, P. S. Fedotov and S. N. Ignatova, Different two-phase liquid systems for inorganic separations by countercurrent chromatography, in *Metal-ion Separation and Preconcentration. Progress and Opportunities*, ed. A. H. Bond, M. L. Dietz and R. D. Rogers, American Chemical Society, Washington, 1999, Chapter 21.
18. S. Muralidharan and H. Freiser, Fundamental aspects of metal-ion separations by centrifugal partition chromatography, in *Metal-ion Separation and Preconcentration. Progress and Opportunity*, ed. A. H. Bond, M. L. Dietz and R. D. Rogers, American Chemical Society, Washington, 1999, Chapter 22.
19. F. Helfferich, *Nature*, 1961, **189**, 1001–1002.
20. J. Porath, J. Carlsson, I. Olsson and G. Belfrage, *Nature*, 1975, **258**, 598–599.
21. L. J. Morris, *Chem. Ind.*, 1962, 1238–1240.
22. B. A. Adams and E. L. Holmes, *J. Soc. Chem. Ind.*, 1935, 1–6T.
23. R. Griessbach, *Angew. Chem.*, 1939, **52**, 215–219.
24. A. Skogseid, *Norges Tekniske Hopkole. Trondheim*, 1948.
25. H. P. Gregor, M. Taifer, L. Citarel and E. I. Becker, *Ind. Eng. Chem.*, 1952, **44**, 2834–2839.
26. S. L. S. Thomas, *Chem. Eng. News*, 1954, **32**, 1896.
27. R. Hering, *Chelatbildende Ionenaustausher*, Akademie Verlag, Berlin, 1967.
28. S. K. Sahni and J. Reedijk, *Coord. Chem. Rev.*, 1984, **59**, 1–139.
29. K. M. Saldadze and V. D. Kopylova, *Complex-forming Ion-exchangers. (Complexites)*, Khimiya, Moscow, 1980.
30. V. E. Kopylova, *Solvent Extr. Ion Exch.*, 1998, **16**, 267–343.
31. G. V. Myasoedova and S. B. Savvin, *CRC Crit. Rev. Anal. Chem.*, 1986, **17**, 1–63.
32. D. Bilba, D. Bejan and L. Tofan, *Croat. Chem. Acta*, 1998, **71**, 155–178.
33. M. Marhol, *Ion Exchangers in Analytical Chemistry: Their Properties and Use in Inorganic Chemistry*, Elsevier, Amsterdam, 1982.

34. T. B. Gapon and E. N. Gapon, *Dokl. Akad. Nauk SSSR*, 1948, **60**, 401–404.
35. E. M. Moyers and J. S. Fritz, *Anal Chem.*, 1977, **49**, 418–423.
36. J. P. Sickafoose, Inorganic separation and analysis by high speed liquid chromatography, PhD thesis, Iowa State University, Ames, IA, 1971.
37. P. K. Dasgupta, *J. Chromatogr. Sci.*, 1989, **27**, 422–448.
38. P. Jones, *Analyst*, 2000, **125**, 803–806.

CHAPTER 2
Retention Mechanism and Chelation Theory

2.1 Introduction

The ion chromatographic separation of metal ions is generally based upon electrostatic interactions between metal cations, in either free or partially complexed form, and the negatively charged functional groups within the cation exchange resin. Alternatively, similar interactions can be exploited between negatively charged metal complexes formed within the mobile phase and positively charged functional groups within an anion exchange resin.[1–3] A further possibility exists, whereby the separation is based upon the ability of chelating ion exchangers to form kinetically labile surface complexes with metal ions, with relative retention dependent upon the stability of the corresponding complexes. This last method, the high-performance mode of which is HPCIC, which is discussed herein, exhibits an alternative and often orthogonal selectivity toward the majority of divalent and trivalent metal cations.

2.2 Surface and Solution Complexation

Differences in the process of metal chelate formation in solution and upon the surface of a chelating ion exchanger are related to the restricted mobility of the immobilised ligands along that surface. The stoichiometry of such surface complexes is generally assumed to be 1:1, with the exception of phases modified with relatively small ligands, exhibiting a high degree of surface coverage and ligand density (see Chapter 3). A typical example of the coordination of a single metal cation with two immobilised ligands is the interaction of Cu(II) with two aminopropyl groups at the surface of 3-aminopropylsilica,[4] which is noted for a dense surface coverage. Of course, surface complex stoichiometry may also depend on the length of the linker between the attachment point and the

---

RSC Chromatography Monographs No. 14
High Performance Chelation Ion Chromatography
By Pavel N. Nesterenko, Phil Jones and Brett Paull
© The Royal Society of Chemistry 2011
Published by the Royal Society of Chemistry, www.rsc.org

chelating functional group, although in most cases this is usually relatively short. However, when considering the complexation equilibria governing the retention mechanism in chelation based chromatography methods, the assumption that the dominance of surface complexes is of 1:1 type stoichiometry, greatly simplifies the description of complexation equilibria in heterogeneous systems.

2.2.1 Complexation Kinetics

According to Budarin and Yatsimirskii,[5] the nature of the ligand has minimal effect upon the rate of complexation with metal cations in solution. For example, rate constants for the complex formation of Co(II) with imidazole, glycine, 2-*p*-dimethylaminophenylazopyridine and glycyldiglycine are 1.3×10^5, 4.6×10^5, 4.0×10^5 and $4.6 \times 10^5 \, M^{-1} s^{-1}$, respectively. However, the kinetics of interaction between the functional groups of a chelating ion exchanger and metal cations is expected to be rather different.

The grafting of ligands upon a surface within a chelating ion exchanger provides an excessive concentration of the ligand within a relatively small volume of the bonded layer, which metal ions from the bulk solution or mobile phase need to reach for any interaction to occur. According to Walcarius *et al.*[6] diffusion processes are faster in macro-porous adsorbents with lower amounts of grafted groups, and smaller particle size. For example, with 3-aminopropyl and 3-mercaptopropyl modified silica the apparent diffusion coefficients of Co(II) and Hg(II) were found to be much lower (1000–10 000 times) than those obtained in homogeneous solutions. The effect of pore size on column efficiency is also discussed in Chapter 3 on chelating stationary phases.

The apparent diffusion coefficients of metal cations migrating to the surface and within the stationary phase is somewhat different for chelating substrates having either only basic or acidic functional groups. This is due to long distance type electrostatic interactions between charged groups on the surface and the metal cations (Scheme 2.1). Chelation or coordination of the metal by the immobilised ligand can be viewed as a rather short-distance interaction, which only occurs when the metal ion is already in close proximity to the binding site.

In HPCIC the kinetics of complexation at the surface of a chelating ion exchanger should be fast enough to provide the possibility of obtaining sufficiently efficient separations. Undoubtedly, the kinetics of complexation depends upon the chemical nature of both the bonded ligands and of the metal species involved. In most instances, for fast complexation reactions[5] complex formation can be considered as a multi-step process, in which the key step is associated with substitution reactions. For example, in the case of a negatively charged ligand (L^-) in solution, the initial formation of weakly associated ionic pairs with the hydrated metal cation takes place due to electrostatic interactions:

$$M^{y+}aq + L^-aq = [M^{y+}aq][L^-aq] \qquad (2.1)$$

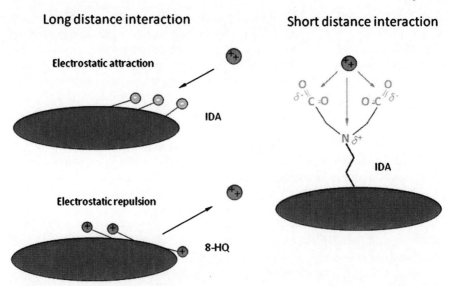

Scheme 2.1 Two-step formation of complexes at the surface of chelating ion exchangers.

The above associated pair then converts into a stronger formally recognised ion-pair as follows:

$$[M^{y+}aq][L^-aq] = [M^{y+}aq\, L^-] \qquad (2.2)$$

Complex formation then follows, with the release of a number of water molecules as the final step:

$$[M^{y+}aq\, L^-] = [M\, L^{(y-1)+}] + aqua \qquad (2.3)$$

The above basic reaction scheme describes complexation within a homogeneous system, when both reagent and cation are in solution, but is also applicable to the adsorption of metal ions within a heterogeneous chelation based system, such as within HPCIC. As mentioned above, the main difference between complexation in homogeneous and heterogeneous systems, is related to the limited conformational mobility of immobilised ligands, in most cases preventing the formation of complexes with stoichiometry other than 1:1.

The formation of weakly associated ionic pairs or outer-sphere complexes (eqn (2.2)) is usually limited by the diffusion of interacting ions. In the case of HPCIC, the diffusion of metal cations towards and within the bonded layer of the stationary phase plays a very important role. Possible effects of porous structure, bonding chemistry, and ligand density on the diffusion of metal cations have been considered in Chapter 3. The rate constant for the formation of the initial weakly associated ionic pair (eqn (2.1)), and subsequently the more

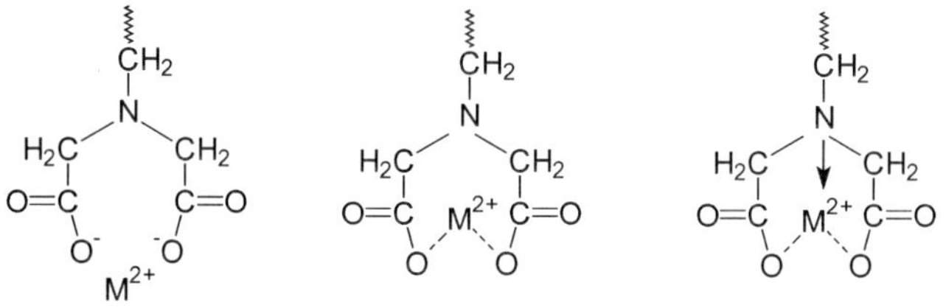

Scheme 2.2   The stepwise interaction of IDA functional groups with divalent metal cations.

strongly held ion-pair (eqn (2.2)), strongly depends on the charge of the immobilised functional groups, as electrostatic interactions are responsible for the long-distance coordination between the metal cation and the ligand. In this way it can be considered as a simple ion exchange interaction. This is illustrated in Scheme 2.2, for the formation of ion-pairs (structure **I**) between iminodiacetic acid (IDA) and a divalent metal cation.

The initial formation of ion-pairs within the surface complexation process, explains the very low chromatographic efficiency of separations obtained during the early stages of the development of HPCIC, when basic species such as 8-hydroxyquinoline[7,8] were used as bonded ligands. Clearly, the repulsion of metal cations from the protonated nitrogen within the attached 8-hydroxyquinoline ligand, would significantly hinder the formation of any corresponding ionic associates. Kuo and Mottolla[9] actually used 8-hydroxyquinoline as a cationic counter-ion for the separation of inorganic anions, so the repulsion of cations from a surface containing this protonated ligand would clearly not provide the favourable complexation kinetics required for the high-performance separation of metals within HPCIC.

The rate-determining process in the formation of a complex is the conversion of an outer-sphere to an inner-sphere arrangement (structure **III** in Scheme 2.2), with release of associated water molecules, according to eqn (2.3). Depending upon the individual characteristics of the metal ion, the corresponding rate constants may vary over a wide range,[10] as shown here in Figure 2.1, which ultimately results in poor chromatographic efficiency for metals ions such as Al(III), Cr(III), Be(II) and V(II). However, an increase in mobile phase acidity, ionic strength and column temperature can effectively disrupt the hydration sphere of a metal cation, which acts to increase such rate constants and provide improved peak efficiency in HPCIC.

The role of the ligand itself upon complexation kinetics can be illustrated by the reaction of Fe(III) with a group of aminocarboxylic chelating reagents, including IDA, NTA, EDTA and DTPA.[11] In the reaction between $[Fe(H_2O)_5OH]^{2+}$ and each of the above ligands, complexation was studied under acidic conditions (pH = 1.3–2.0) and ionic strength $I = 1.0$, with the determined rate constants increasing in the order, IDA < NTA < EDTA < DTPA. Clearly, in this example, rate constants were higher for the anionic form of the

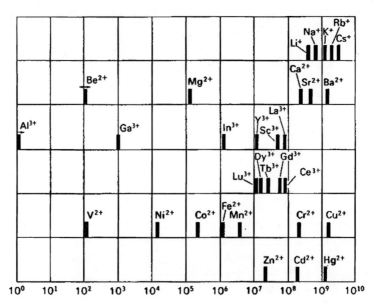

Figure 2.1 Characteristic rate constants (s^{-1}) for substitution of inner sphere H_2O from various hydrated cations. Reproduced, with permission, from Diebler et al.[10]

ligand, with the authors concluding that the rate-determining step of the reaction is the replacement of a water molecule from the coordination sphere of Fe(III), but not the formation of the chelate ring.

However, factors limiting separation efficiency in HPCIC are not only connected with kinetics of complex formation, but also upon the impact of complex dissociation kinetics (eqn (2.4)), which in practice can be equally, if not more significant.

$$[\overline{MR_n}] = [M^{y+}] + n[\overline{R}^-] \quad (2.4)$$

where $(\overline{R})$ represents the surface immobilised ligand. Here complex dissociation rate constants are strongly dependent on the nature of the immobilised ligand, but are always smaller than the corresponding complex formation rate constants for the same ligands. This important relationship was reported by Ahmed and Wilkins[12] who identified a significant correlation between thermodynamics and kinetics for the dissociation of complexes having the same central atom but a differing number of ligands within the complex. If the elimination of water is the rate-determining factor in complex formation (eqn (2.3)), then the following correlation between the stepwise stability constant (β_1) and the dissociation rate constants (k_d) is true:

$$\Delta \log k_d \sim \Delta \log 1/\beta_1 \quad (2.5)$$

The above represents a significant observation, as it concludes that stationary phases with bonded ligands which form very stable complexes with metals ions are unlikely to provide efficient separations within HPCIC.

Desorption of metal ions in HPCIC can be achieved by competitive complexation with additional ligands (L$^-$) added to the mobile phase:

$$[\overline{MR_n}] + [L^-] = [ML^{(y-1)+}] + n[\bar{R}^-] \qquad (2.6)$$

However, the kinetics of ligand substitution within metal complexes is too complex for detailed consideration here. For example, for the reaction described within eqn (2.6), faster kinetics would be expected for outer-sphere complexes formed within the mobile phase, rather than for inner-sphere complexes. Outer-sphere complexes are weaker than inner-sphere complexes, so their rates of dissociation are more rapid.[13,14] Specific interactions between functional groups within the immobilised ligand and the mobile phase ligand could also play a role, particularly as the concentration of the ligand at the surface of the adsorbent and in bulk solution would be expected to be different. Thus, a combination of many poorly defined interactions make the accurate description of such equilibria at the surface of chelating ion exchangers difficult.

2.3 Equilibria within the Chelating Ion Exchanger

For kinetic reasons considered earlier in Section 2.2, the most efficient separations of metal ions on chelating ion exchangers is generally only achieved with those exchangers having negatively charged or acidic functional groups. In such cases, both electrostatic interactions (simple cation exchange) and chelation at the surface can take place, under mobile phase conditions of low ionic strength. The corresponding equilibria for the above scenario, for a metal cation, and immobilised chelating groups, H_nR, can be described as follows:

$$[M^{y+}] + [\bar{H}_n\bar{R}] = [\bar{M}_n^{y+}\bar{R}^{n-}] + nH^+ \qquad (2.7)$$

$$n[B^+] + [\bar{H}_n\bar{R}] = [\bar{B}_n^+\bar{R}^{n-}] + nH^+ \qquad (2.8)$$

$$[M^{y+}] + [\bar{B}_n^+\bar{R}^{n-}] = [\overline{MR}^{(y-n)+}] + nB^+ \qquad (2.9)$$

$$[M^{y+}] + [\bar{R}^{n-}] = [\overline{MR}^{(y-n)+}] \qquad (2.10)$$

where M^{y+} is an analyte metal cation interacting through both an ion exchange (eqn (2.7)) and complexation based mechanism (eqn (2.10)) with the chelating ion exchanger. H^+ and B^+ represent the hydronium and alkali metal cations introduced in the mobile phase, either from the addition of a buffer, or from the addition of an ionic strength modifier salt, respectively. The latter, in sufficiently high concentration, effectively suppresses these electrostatic interactions (eqn (2.7)) between the analyte metal ion and the negatively charged functional groups within the immobilised chelating ligand, $\bar{R}$. However, unlike these suppressed electrostatic interactions, comparative data for stability constants of complexes measured at different values of ionic strength, show they remain practically unchanged, meaning the impact of increased ionic strength upon the retention of metals due to surface complexation alone, is relatively small.

An excellent illustration of the suppression of electrostatic interactions by increasing the concentration of a strong electrolyte within the mobile phase is provided by Kopylova and Saldadze.[13,15] Figure 2.2 shows plots of Cu(II)) sorption capacity from solutions of differing ionic strength, for a number of cation and chelating ion exchangers. For phosphorylated PS-DVB (KF-1) and carboxylic methacrylate-DVB (KB-4) cation exchangers, the capacity decreases with ionic strength. For the complexing vinylpyridine anion exchangers (AN-25, AN-40, AN-43, AN-47) and the anion exchanger AN-31, prepared by condensation of epichlorohydrine, polyethyleneimine and ammonia, an increase in the adsorption of Cu(II)) was noted. Polyampholyte ion exchangers were prepared through the oxidation of methyl groups within a co-polymer of 2-methylvinylpyridine and DVB (AN-25) resulting in the resin ANKB-2, or through the treatment of anion exchanger AN-31 with chloroacetic acid, producing resin ANKB-1. The effects of ionic strength were less prominent on these ampholytic exchangers than for their corresponding original anion exchangers. Due to incomplete oxidation of the methyl group and

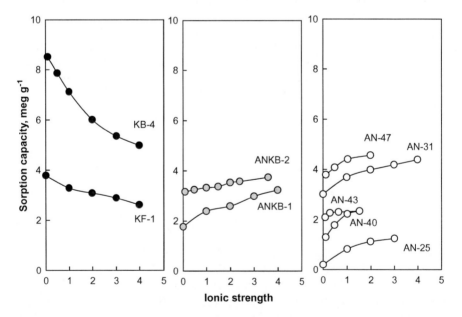

Figure 2.2 Influence of solution ionic strength on the sorption of Cu(II) by complexing cation (left), anion (right) and ampholytic (middle) ion exchangers. Plotted in accordance with data published in Saldadze and Kopylova[13] and Kopylova.[15]

carboxymethylation of amino groups within the AN-31 material, anion exchange properties were still prevailing for these ion exchangers, so some increase in adsorption of Cu(II) from solutions with higher ionic strength occurs. In all cases suppression of the repulsion of Cu(II) from either complexing positively charged protonated groups, or excessive attraction to the dissociated acidic type negatively charged complexing groups, is responsible for the observed effects. Theoretically, the variations in ionic strength should not have an effect on the adsorption of Cu(II)) for the polyampholite ion exchanger, which exhibits equal amounts of negatively and positively charged groups.

2.3.1 Distribution Ratio Within a Mixed Mode Mechanism

Pure sorption mechanisms are very rare in chromatography, and chelation ion exchange, as the previous discussion of equilibria suggests, is no exception. The distribution ratio, D_M, of a metal cation, M^{y+}, between the negatively charged chelating ion exchanger and the mobile phase can be described as follows:

$$D_M = \frac{[\overline{MR}^{(y-n)+}]_e + [\overline{MR}^{(y-n)+}]_c}{[M^{y+}]} \quad (2.11)$$

Where $[\overline{MR}^{(y-n)+}]_e$ and $[\overline{MR}^{(y-n)+}]_c$ are equilibrium concentrations of the cation retained by the stationary phase due to electrostatic interactions (conventional ion exchange) and chelation ion exchange, respectively. $[M^{y+}]$ is the concentration of the cation in the mobile phase. The retention factor k can be expressed as:

$$k = D_M \cdot \varphi \quad (2.12)$$

where φ is a characteristic constant for a given chromatographic column, expressed as the ratio of the volumes of stationary phase, V_s, and mobile phase, V_m:

$$\varphi = V_s/V_m \quad (2.13)$$

At constant mobile phase pH, in the presence of an excess of non-complexing electrolyte or ionic strength modifier, the equilibrium described within eqn (2.8) is shifted to the right and the ion exchange process of cation M^{y+} and alkali metal cation B^+ as the competing cation, on the chelating ion exchanger can be described directly by eqn (2.9). In this case the corresponding selectivity ratio is given by:

$$K_{M,B} = \frac{[\overline{MR}^{(y-n)+}]_e [B^+]^n}{[M^{y+}][\overline{B_n^+ R^{n-}}]} \quad (2.14)$$

Since the formation of complexes at the surface with more than one immobilised chelating ligand is unlikely due to sterical restrictions, then in accordance with eqn (2.10), the following expression can be presented:

$$[\overline{\mathrm{MR}}^{(y-n)+}]_c = \beta_1 [\mathrm{M}^{y+}][\overline{\mathrm{R}}^{n-}] \tag{2.15}$$

where $[\overline{\mathrm{R}}^{n-}]$ is the concentration of functional groups and β_1 is the formation constant of complex, $\overline{\mathrm{MR}}^{(y-n)+}$, formed at the surface.

2.3.2 Selectivity in Non-complexing Mobile Phases

Taking into account the two possible types of interactions described within eqns (2.11), (2.14) and (2.15), the retention factor of a metal cation can also be expressed as follows:

$$k = \left(K_{M,B} \frac{[\overline{\mathrm{B}}_n^+ \overline{\mathrm{R}}^{n-}]}{[\mathrm{B}^+]^n} + \beta_1 [\overline{\mathrm{R}}^{n-}] \right) \varphi \tag{2.16}$$

The expression shows both the combined impact of conventional ion exchange interactions and chelation on retention. As mentioned above, electrostatic interactions are strongly suppressed by the addition of an electrolyte to the mobile phase, but the effect of ionic strength on chelation remains small.

In general, the effect of ionic strength on stability constants is rather complex and is associated with changes in the activity coefficients of ions in electrolyte solutions. For the estimation of activity coefficients in relatively concentrated solutions of electrolytes Brönsted proposed Specific Ion Interaction Theory (SIT),[16] which was further developed by Guggenheim and Turgeon,[17] Scatchard[18] and Ciavatta.[19] Taking into consideration activity coefficients the equilibrium formation constant can be expressed as follows:

$$K_1 = \frac{[\mathrm{ML}]}{[\mathrm{M}][\mathrm{L}]} \frac{\gamma_{\mathrm{ML}}}{\gamma_M \gamma_L} \tag{2.17}$$

while activity coefficients depend on ionic strength, I, according to the following equations:

$$\log \gamma_\pm = -\frac{A|z_+ \cdot z_-|\sqrt{I}}{1 + B\sqrt{I}} \tag{2.18}$$

in which

$$I = \frac{1}{2} \sum_i m_i z_i^2$$

and where z_+ is the cation charge, z_- is anion charge, m is molal concentration and A and B are constants. Alternative descriptions of the impact of ionic strength on equilibrium constants have been suggested by Pitzer.[20,21]

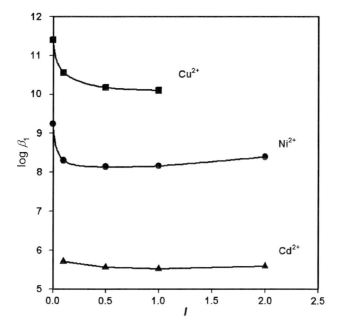

Figure 2.3 Dependence of experimentally measured and calculated stability constants for formation of metal complexes with IDA on ionic strength (NaNO$_3$). Plotted according to NIST data. Source: 'Critically selected stability constants of metal complexes'.[22]

The typical dependences of equilibrium stability constants on ionic strength are shown in Figure 2.3 for complexes of metals with IDA. It is clear that the equilibrium stability constants decrease with increasing ionic strength from 0 to 0.5, but remains approximately the same in the range $0.5 < I < 1.0$ and increases at $I > 1.0$. For example, critical values of the stability constant, β_1, for the complex of Cd(II) with IDA are 5.71, 5.56, 5.52 and 5.59, at ionic strength equal to 0.1, 0.5, 1.0 and 2.0, respectively.[22] Of course, all above mentioned equilibria are derived for homogeneous systems and are not necessarily directly transferably to HPCIC.

Therefore, assuming little change in β_1 with high ionic strength (>0.5), the retention of a metal ion due to chelation should not depend upon the concentration of alkali metal cations [B$^+$], but solely on the concentration of functional groups in the chelating ion exchanger. In practice, this means electrostatic interactions need to be suppressed for chelation to be the dominant sorption mechanism and once domination by chelation is established, as described in eqn (2.19):

$$\beta_1[\bar{R}^{n-}] \gg K_{M,B} \frac{[\bar{B}_n^+ \bar{R}^{n-}]}{[B^+]^n} \qquad (2.19)$$

then the retention of the metal ions depends solely on formation constants, β_1^M, column capacity, $[\bar{R}^{n-}]$ and the phase ratio. The separation selectivity ratio, α, for a pair of metal ions is then defined by the ratio of the corresponding formation constants:

$$\alpha = k_2/k_1 = \beta_1^{M2}/\beta_1^{M1} \tag{2.20}$$

Strictly, this should be the ratio of conditional stability constants, but for a given ligand with no hydrolysis of the metal ions, the ratio is essentially the same as the thermodynamic stability constants. The possible selectivity changes that can occur with an increase in the concentration of an indifferent (non-complexing) electrolyte will be discussed in Chapter 4.

The above expressions describe the retention of a metal cation on a chelating ion exchanger with negatively charged groups. Similar expressions can be derived for chelating ion exchangers with positively charged functional groups ($\bar{R}^{n+}$) where suppression of electrostatic repulsion of metal cations from the surface depends on the adsorption of anions (X^-) from a strong electrolyte added to the mobile phase (eqn (2.21)):

$$k = \left(\beta_1 [\bar{R}^{n+}] - K \frac{[\bar{X}_n^- \bar{R}^{n+}]}{[X^-]^n} \right) \varphi \tag{2.21}$$

For neutral immobilised chelating ligands the retention of metals depends purely on the formation constant, as follows:

$$k = \beta_1 [\bar{R}] \varphi \tag{2.22}$$

2.3.3 Secondary Equilibria Within the Mobile Phase

Additional equilibria may have an effect upon the retention and the separation selectivity of metal ions in HPCIC. According to eqn (2.22), retention under conditions dominated by HPCIC can be defined in bi-logarithmic form as:

$$\log k = \log \beta_1 + \log [\bar{R}] + \log \varphi \tag{2.23}$$

In the case of secondary competing equilibrium within the mobile phase, the concentration of the free metal cation $[M^{y+}]$ must be corrected by application of the complex formation coefficient, $\alpha_{M(L)}$, according to Weiss:[23]

$$\alpha_{M(L)} = 1 + [L]\beta_{ML} + [L]^2 \beta_{ML2} + \ldots \tag{2.24}$$

where [L] is the concentration of the competing ligand in the mobile phase and β_{ML}, β_{ML2} are formation constants for complexes ML, ML$_2$, respectively. Therefore, the retention of metal ions within a HPCIC system, in the presence of secondary equilibria in the mobile phase can be expressed as:

$$\log k = \log \beta_1 - \log \alpha_{M(L)} + \log[\bar{R}] + \log \varphi \qquad (2.25)$$

Regulation of the separation selectivity of metal ions on a chelating ion exchange column is possible if the values of β_1 for the complex at the surface are comparable with $\alpha_{M(L)}$ or, for simplicity, with $[L]\beta_{ML}$. In other words, the separation selectivity may be changed in the presence of small concentration of strong complexing agents or of significantly higher concentration of weak complexing agents in the mobile phase (see Chapter 4).

In the case of an IDA-silica column, changes in separation selectivity were noted with additives of not only relatively strong complexing agents such as oxalic acid,[24,25] dipicolinic acid,[25-27] picolinic acid,[25,28] sulfosalicylic acid,[27] and with tartaric, maleic, malonic, citric and various other carboxylic acids,[24,26,29] but also for relatively weak complexing agents for transition metals such as chloride.[25,30,31] More information on the manipulation of separation selectivity for IDA functionalised resins through the addition of complexing reagents to the mobile phase can be found in a recent review.[27]

2.4 Temperature Effects in HPCIC

There are several effects of column temperature changes on chromatographic performance in HPCIC, which can be divided into two groups related to thermodynamic and kinetic properties.

2.4.1 Thermodynamic Effects

The general thermodynamic effect of column temperature upon retention (ln k) of metal cations in IC and HPCIC can be expressed by the van't Hoff equation:

$$\ln k = -\Delta H/RT + \Delta S/R + \ln \varphi \qquad (2.26)$$

where ΔH and ΔS are sorption enthalpy and entropy, respectively; R is the molar gas constant and φ is the phase volume ratio as described by eqn (2.13). There exists a clear difference between heats of adsorption of metal cations due to pure electrostatic interactions and due to the formation of surface complexes. In the case of conventional ion exchange, the temperature effects are exothermic (negative values of ΔH) and heats of adsorption do not exceed 8–13 kJ mol^{-1}. In chromatographic systems with a dominant chelation mechanism, values of ΔH are usually much higher and both exothermic and endothermic effects can be observed. According to micro-calorimetry measurements, the formation of surface metal complexes with weak base functional groups is an exothermic process ($\Delta H < 0$), while the complexation of metals with cationic and amphoteric functional groups is endothermic ($\Delta H > 0$).[15] The reported sorption enthalpy ($-\Delta H$) values for Cu(II) on pyridyl and primary/secondary amino-functionalised PS-DVB resins with 8% degree of cross-linking are between 19.6 and 32.3 kJ mol^{-1}. Similar enthalpy values were measured for various IDA and picolinic acid types of ampholytic ion exchangers based on the same matrix. In the case of cation exchangers, sorption enthalpy values depend strongly on the type of functional groups, showing more profound thermal effects for carboxylic (-15.9 kJ mol^{-1}) and phosphonic (-10.5 kJ mol^{-1}) cation exchangers as compared with -5.6 kJ mol^{-1} obtained for sulfonated PS-DVB resin.[15]

Chelating ion exchangers have different responses for different metals, so changes of column temperature can be used for the variation of the separation selectivity. So-called temperature-responsive selectivity has been used by Muraviev et al. in traditional ion exchange chromatography for the separation of transition metal ions on IDA and aminomethylphosphonic acid functionalised resins.[32]

The sorption enthalpy comprises the enthalpies of different interactions including metal coordination within the sorbent, the changes in the solvation of metal ions and functional groups, possible thermal deformation of the structure of the bonded layer or polymer matrix, and various additional processes. For example, increased column temperature has been shown to cause a relatively sharp change in the retention of cations due to bonded layer conformational changes for poly(maleic acid) functionalised PS-DVB substrates.[33]

The entropy of metal cation–chelation group/groups interaction may also impact on retention, especially in the case of multi-dentate ligands serving as chelating groups. Thus, the thermodynamic aspect of temperature can produce significant effects on resultant separation selectivity.

2.4.2 Kinetic Effects

The kinetic theory of chromatography describes peak broadening events during the separation process as a function of flow rate and the physical/chemical properties of the adsorbent and chromatographic column. The most widely accepted kinetic based model for chromatography (eqn (2.27)) was originally developed by van Deemter and has been subsequently modified by Knox for application to HPLC:

$$H = Au^{0.33} + B/u + C_M u + C_S u \qquad (2.27)$$

where H is the plate height, and u is the linear velocity of the mobile phase. The A term takes into account the contribution of eddy diffusion and type of flow to band broadening, the B term represents band broadening due to axial or longitudinal diffusion of the analyte, and the terms C_M and C_S describe the broadening caused by kinetic effects in the mobile and stationary phases, respectively. The expanded van Deemter equation (eqn (2.28)) for adsorption chromatography with a solid stationary phase can be written as:

$$H = Ad_P \left(\frac{ud_p}{D}\right)^{0.33} + \frac{2D}{(1+\varepsilon_i/\varepsilon_e)u} + \frac{f(k)d_p^2 u}{D} + \frac{2t_d ku}{(1+k)^2} \quad (2.28)$$

where D is the analyte diffusion coefficient in the mobile phase, d_p is equal to the average size of the phase particles, ε_i is the intra-particle porosity, ε_e is the inter-particle porosity, t_d is the desorption rate of the solute from the stationary phase, k is the retention factor, and $f(k)$ is a function of k, which increases with k. Under normal circumstances, an increase in column temperature produces an increase in the diffusion coefficients of the solutes, according to the equation of Wilke and Chang (eqn (2.29)):[34]

$$D = \frac{7.4 \times 10^{-12} T\sqrt{\Psi M}}{\eta V_s} \quad (2.29)$$

The diffusion coefficient, D, is directly proportional to the absolute temperature, T, and inversely proportional to the viscosity of mobile phase, η, which reduces with temperature. The other parameters in eqn (2.29), are the solvent constant, Ψ, the molar mass of the solvent, M, and the molar volume of the solute, V_s. Clearly then, an increase in the column temperature should improve column efficiency and have a subsequent effect upon resultant peak resolution.

There has traditionally been a widespread concern that the surface complexation kinetics are a limiting factor governing efficiency in HPCIC. Generally, improved kinetics of mass transfer between mobile and stationary phases should occur at higher temperatures, due to changes in diffusion coefficients and to a reduction in viscosity of the mobile phase. However, an additional important parameter effecting separation efficiency is the desorption rate, t_d in eqn (2.28), which also increases with T. In HPCIC the desorption rate corresponds to dissociation rate of metal complexes formed at the surface of the chelating ion exchanger, and here any improvements through increased temperature may be limited. This is because in some instances, particularly with chelating ligands attached to supporting substrates *via* flexible linker groups, the increased conformational mobility resulting from increased temperature, could lead to the formation of complexes with a higher denticity, and thus a slower dissociation rate. The latter factor can in fact cause a decrease in separation efficiency.

2.5 Mobile Phase pH

In general, the pH of the mobile phase controls the degree of dissociation/protonation of both functional groups within the chelating ion exchanger and of some complexing agents within the mobile phase. In the absence of any secondary equilibria (complexation, hydrolysis *etc.*) in the mobile phase the reaction of a chelating resin with a metal ion can be expressed according to eqn (2.30) as follows:

$$[M^{y+}] + \overline{[H_nR]} = \overline{[\tilde{M}_n^{y+}\bar{R}^{n-}]} + nH^+ \qquad (2.30)$$

The equilibrium constant for this surface reaction (predominantly complexation) is given by:

$$K = \frac{[H^+]^n \overline{[MR^{(y-n)+}]}}{[M_{aq}^{y+}]\overline{[H_nR]}} \qquad (2.31)$$

The ratio of metal concentrations between the adsorbent or stationary phase and in solution or mobile phase, $[\tilde{M}\bar{R}^{(y-n)+}]/M_{aq}^{y+}$, is equal to the distribution ratio of metal between two phases, D_M, so the retention factor in accordance with eqn (2.31) can be expressed as follows:

$$\log k = \log D_M + \log \varphi = \log K + \log \overline{[H_nR]} - n\log[H^+] + \log \varphi \qquad (2.32)$$

The phase ratio, φ, for the same column used is constant. Also, under chromatographic conditions the column loading is negligible, as compared with the amount of the functional groups in the stationary phase or column capacity. Thus the value of $\overline{[H_nR]}$ is also assumed to be constant under isocratic elution conditions or at a constant concentration of a mineral acid within the mobile phase in HPCIC. Two significant conclusions can be drawn from eqn (2.32). Firstly, the linear dependence of log k versus pH with slope n should take place for the retention of a selected metal ion. Secondly, at constant concentration of an acid in the mobile phase, or at constant pH value, a linear correlation between log k and logarithms of stability constants should be obtained for the set of equally charged metal cations displaying the same surface complex stoichiometry.

The majority of functional groups of immobilised chelating ligands utilised within HPCIC include weak acid or weak base groups (see Chapter 3). This means that the pH of the mobile phase is reflected within the conditional stability constants for their complexes with each metal ion. However, at the same time, these groups will dissociate or become protonated with an increase or decrease in pH, thus also changing the electrostatic interactions with separated metal ions. However, as discussed previously (eqn (2.19)) this effect is not significant if operating under suppressed ion exchange conditions.

If for the attached chelating groups $\overline{[H_nR]}$ the dissociation constant is expressed as:

$$K_{diss}^R = \frac{[H^+]^n[\bar{R}^{n-}]}{\overline{[H_nR]}} \qquad (2.33)$$

So the final expression is:

$$\log k = \log \beta_1 + \log [\overline{H_n R}] + \log K_{diss}^R - n\log[H^+] + \log \varphi \qquad (2.34)$$

An analysis of the equation shows that once the separation selectivity is established a change in pH should not change the selectivity (see Chapter 4). So, assuming no changes in retention mechanism take place due to hydrolysis of some metal ions or other secondary equilibria, an increase in the mobile phase pH will lead to an increase in retention times of the metal cations. Usually these dependences are linear using bi-logarithmic axes, while the slopes can provide some information about the stoichiometry of interaction between the metal ion and chelating group. Such dependences have been observed for poly(itaconic)acid functionalised resin,[35] mixed carboxylic and phosphonic ion-exchangers,[36] and IDA silica.[37,38]

For ligand impregnated phases (see Chapter 3), any change in the mobile phase pH may also affect the actual concentration of the adsorbed ligand itself. The protonation/dissociation of ionogenic functional groups within such ligands changes the total charge of ligand and hence their adsorption, which in turn, may alter the ion exchange capacity of adsorbents.[39,40]

2.6 Organic Solvent Additives

The presence of organic solvents within the mobile phase may also affect the dissociation/protonation of functional groups within immobilised ligands and their stability constants for metal complexes formed on the surface. There is no general theory explaining all the effects of the presence of organic solvents on chelation, but some examples of the complexity of such dependences for some metal complexes with IDA have been described by Field and McBryde.[41] These authors reported an increase in metal complex stability with an increase of methanol content in water–methanol solutions, as shown in Figure 2.4.

Additionally in HPCIC, the presence of organic solvent in the mobile phase can under certain circumstances cause swelling of organopolymer stationary phases. Less important effects within HPCIC could be connected with changes in hydrated ion radii, possible substitution of water molecules for organic solvent molecules in the hydration sphere and in dielectric constants of the mobile phase. Organic solvent additives to the mobile phase may improve the conformational mobility of chelating groups attached to hydrophobic matrices and change their accessibility for interaction with cations. In the case of hydrophilic amino acid type silica based chelating ion exchangers the addition of ACN, methanol or isopropanol did not produce any significant changes in separation selectivity, but reduced the retention of alkaline earth and transition metal cation by 10–15%.[42,43] Hatsis and Lucy[44] investigated the effect of addition of ACN on the separation of alkaline earth metal cations on an Ionpac

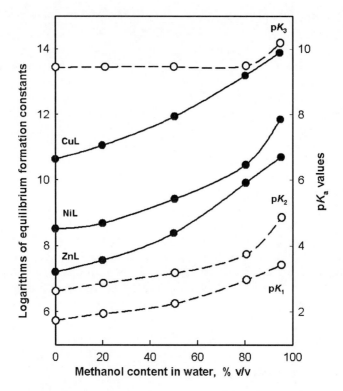

Figure 2.4 The effect of methanol content on metal complex (M:L = 1:1) formation constants and protonation constants of IDA. Adopted from Field and McBryde.[41]

CS12A column and observed a decrease in retention times for these cations. They also noted a small decrease in peak efficiencies without changes in peak asymmetry.

References

1. P. R. Haddad and P. E. Jackson, *Ion Chromatography: Principles and Applications*, Elsevier, Amsterdam, 1990.
2. J. S. Fritz and D. T. Gjerde, *Ion Chromatography*, Wiley-VCH Verlag, Weinheim, 2009.
3. H. Small, in *Ion Chromatography*, ed. D. Hercules, Plenum Press, N.Y., 1990.
4. Y. Kholin and V. Zaitsev, *Pure Appl. Chem.*, 2008, **80**, 1561–1592.
5. L. I. Budarin and K. B. Yatsimirskii, *Russ. Chem. Rev.*, 1968, **37**, 209–225.
6. A. Walcarius, M. Etienne and J. Bessière, *Chem. Mater.*, 2002, **14**, 2757–2766.
7. J. R. Jezorek and H. Freiser, *Anal. Chem.*, 1979, **51**, 366–373.

8. K. H. Faltynski and J. R. Jezorek, *Chromatographia*, 1986, **22**, 5–12.
9. M. S. Kuo and H. A. Mottola, *Anal. Chim. Acta*, 1980, **120**, 255–266.
10. H. Diebler, M. Eigen, G. Ilgenfritz, G. Maass and R. Winkler, *Pure Appl. Chem.*, 1977, **20**, 93–102.
11. E. Mentasti, E. Pelizzet and G. Saini, *J. Chem. Soc. Dalton Trans.*, 1974, 1944–1948.
12. A. K. S. Ahmed and R. G. Wilkins, *J. Chem. Soc.*, 1960, 2901–2906.
13. K. M. Saldadze and V. D. Kopylova, *Complex-forming Ion-exchangers. (Complexites)*, Khimiya, Moscow, 1980.
14. R. Raja, *Am. Lab.*, 1982, **14**, 35–37.
15. V. E. Kopylova, *Solvent Extr. Ion Exch.*, 1998, **16**, 267–343.
16. J. N. Brønsted, *J. Am. Chem. Soc.*, 1922, **44**, 877–898.
17. E. A. Guggenheim and J. C. Turgeon, *Trans. Faraday. Soc.*, 1955, **51**, 747–761.
18. G. Scatchard, *Chem. Rev.*, 1936, **19**, 309–327.
19. L. Ciavatta, *Ann. Chim. (Rome)*, 1980, **70**, 551–567.
20. K. S. Pitzer, *J. Phys. Chem.*, 1973, **77**, 268–277.
21. K. S. Pitzer, Ion Interaction Approach: Theory and Data Correlation, in *Activity Coefficients in Electrolytes Solutions*, ed. K. S. Pitzer, CRC Press, Boca Raton, 1991, Ch. 3 pp.75–153.
22. A. E. Martell and R. M. Smith, NIST Standard Reference Database 46. Version 8. 0., 2004.
23. J. Weiss, *Ion Chromatography*, Wiley-VCH, Weinheim, 2004.
24. P. Nesterenko and P. Jones, *J. Liq. Chromatogr. Relat. Technol.*, 1996, **19**, 1033–1045.
25. P. N. Nesterenko and P. Jones, *J. Chromatogr. A*, 1997, **770**, 129–135.
26. G. Bonn, S. Reiffenstuhl and P. Jandik, *J. Chromatogr.*, 1990, **499**, 669–676.
27. P. N. Nesterenko and O. A. Shpigun, *Russ. J. Coord. Chem.*, 2002, **28**, 726–735.
28. P. Jones and P. N. Nesterenko, *J. Chromatogr. A*, 1997, **789**, 413–435.
29. P. N. Nesterenko and T. A. Bol'shova, *Vestn Mosk Univ. Ser. 2. Khim.*, 1990, **31**, 167–169.
30. W. Bashir and B. Paull, *J. Chromatogr. A*, 2002, **942**, 73–82.
31. P. Jones and P. N. Nesterenko, *J. Chromatogr. A*, 2008, **1213**, 45–49.
32. D. Muraviev, A. Gonzalo and M. Valiente, *Anal. Chem.*, 1995, **67**, 3028–3035.
33. P. A. Kebets, K. A. Kuz'mina and P. N. Nesterenko, *Russ. J. Phys. Chem.*, 2002, **76**, 1481–1484.
34. C. R. Wilke and P. Chang, *AIChE J.*, 1955, **1**, 264–270.
35. W. Bashir, E. Tyrrell, O. Feeney and B. Paull, *J. Chromatogr. A*, 2002, **964**, 113–122.
36. M. J. Shaw, P. N. Nesterenko, G. W. Dicinoski and P. R. Haddad, *J. Chromatogr. A*, 2003, **997**, 3–11.
37. W. Bashir and B. Paull, *J. Chromatogr. A*, 2001, **907**, 191–200.
38. W. Bashir and B. Paull, *J. Chromatogr. A*, 2001, **910**, 301–309.

39. J. Cowan, M. J. Shaw, E. P. Achterberg, P. Jones and P. N. Nesterenko, *Analyst*, 2000, **125**, 2157–2159.
40. M. J. Shaw, P. Jones and P. N. Nesterenko, *J. Chromatogr. A*, 2002, **953**, 141–150.
41. T. B. Field and W. A. McBryde, *Can. J. Chem.*, 1981, **59**, 555–558.
42. A. I. Elefterov, M. G. Kolpachnikova, P. N. Nesterenko and O. A. Shpigun, *J. Chromatogr. A*, 1997, **769**, 179–188.
43. M. G. Kolpachnikova, N. A. Penner and P. N. Nesterenko, *J. Chromatogr. A*, 1998, **826**, 15–23.
44. P. Hatsis and C. A. Lucy, *Analyst*, 2001, **126**, 2113–2118.

CHAPTER 3
Chelating Stationary Phases

3.1 Introduction

The preparation of efficient and selective chelating stationary phases remains the most significant challenge for the various chromatographic techniques based upon surface chelation. In HPCIC the general criteria for suitable adsorbents are as follows:

1. *Mechanical stability*. This is a general requirement for all stationary phases to be used in HPLC and related techniques, where the chromatographic column is subjected to prolonged periods of high applied pressures. This criterion includes the ability to maintain *constant volume under changes of ionic strength*. As discussed later, significant increases in mobile phase ionic strength through the use of strong electrolytes is common in HPCIC for the suppression of ion exchange interactions. Therefore, minimal swelling and shrinking under such conditions is essential for the chromatographic operation of such chelating ion exchangers.
2. *Hydrolytic stability*. The hydrolysis and formation of aqua complexes have been noted for some metals (*e.g.* Fe^{3+}, Ti^{4+}, Al^{3+} *etc.*) in diluted acids. As a rule, such hydrolysis related processes decrease the separation efficiency and reproducibility of the chromatographic system. Therefore, the separation of metal ions in general should be performed under relatively acidic conditions (sometimes up to 1 M mineral acids), and stationary phases must exhibit a high degree of hydrolytic stability to withstand such mobile phase conditions. Alkaline mobile phases are not commonly seen in any chromatographic methods for the separation of non-complexed metals species, and so the high pH limitation of certain substrates, most notably silica, does not constitute a limitation here.

3. *Thermal stability*. Increased column temperature has been shown to improve separation selectivity in certain applications of HPCIC. Chelation as a retention mechanism clearly exhibits more significant temperature effects than those observed with the simple electrostatic interactions of ion exchange. Possible improved complexation kinetics achievable through elevated column temperatures can also result in an improvement in observed column efficiencies. For the above reasons, stationary phase thermal stability is important.
4. *Surface structure and efficient mass transfer in the stationary phase*. This requirement is significant for all efficient ion exchange and chelation based chromatographic phases. Ideally, uniform monolayers of either physically adsorbed or covalently bound ligands, or thin polymer layers bearing grafted chelating functional groups, result in the most efficient chelating substrates, as they facilitate the rapid two-directional diffusion of metal ions and associated improvement in separation efficiency.
5. *Homogeneous distribution of functional groups*. In the case of relatively small chelating ligands (for example, aminopropyl groups in aminopropylsilica) with a high density of surface coverage, the coordination of one metal ion by two surface bound ligands is possible. However, a chelating surface providing only 1:1 metal–ligand interactions is generally preferable for chromatographic applications, as this will result in faster chelation kinetics. It is also important to ensure that ligand distribution is homogeneous across the stationary phase surface at the particulate level, and indeed longitudinally and radially throughout the whole chromatographic bed (significant for physically adsorbed ligand based substrates).
6. *An appropriate selectivity*. Clearly, selectivity of an adsorbed or covalently bound functionality has to be suitable for the desired chromatographic application. In HPCIC a great deal of variety in selectivity exists due to the large number of available chelating ligands amenable to immobilisation. This aspect of HPCIC is given specific focus in the following sections of this Chapter.

3.2 Types and Properties of Chelating Ion Exchange Ligands

The choice of the correct functional group is crucial to the successful application of HPCIC. As the process of chelation is one involving the formation of a new complex at the surface of the adsorbent, it is important for immobilised ligands to exhibit the following properties.

3.2.1 Adsorption of Metal Ions

There should be reversible adsorption of metal ions with *suitably rapid complex association and dissociation kinetics*. In practice this means that chelating groups should form labile complexes of moderate (to relatively weak) stability. It is empirically established that lower-dentate ligands provide relatively fast kinetics of complexation, and importantly, more readily achieved and rapid dissociation of complexes. However, the conditional stability constant is a key factor in reaction/dissociation kinetics, with lower conditional constants providing faster dissociation reactions. As the vast majority of ligands are conjugate bases of weak protonic acids, the conditional stability constants will decrease rapidly with decrease in pH impacting significantly upon the complex association and dissociation kinetics. This is why protonated ligands, which most are, exhibit better efficiencies and peak shapes in low pH mobile phases.

3.2.2 Ligand Chemical Stability

Ligand chemical stability is required under the applied separation conditions. Immobilised ligands must maintain the ability to form complexes with metal ions (although the stability of such complexes will of course vary), under a variety of mobile phase conditions, such as applied extremes of temperature, pH, ionic strength, solvent polarity, pressures/flow rates, *etc*. This requirement is often more achievable with ligands exhibiting relatively simple structures, avoiding structures containing multiple chelating and non-chelating functional or potentially reactive sites.

3.2.3 Suitable Coordinating Sites

Suitable coordinating sites, with the ability to form the desired chelate, are necessary. Coordinating sites include different combinations of one or more of a small number of relatively simple groups (–OH, –COOH, –NH$_2$, *etc*.), each with two or more donor atoms (O, N, S) positioned in a way that enables sterically favourable chelates to be readily formed. The major coordinating sites used in chelating ion exchangers to date can be seen in Figure 3.1.

The total charge of the coordinating groups and suitably fast complexation kinetics are the two most important parameters for application of the resultant chelating substrates in HPCIC. As a rule, negatively charged ligands derived from the free acids (iminodiacetic acid, aminophosphonic acid, aminocarboxylic acids, salicylic acid, *etc*.) tend to better meet these requirements. Generally, it is the combination of donor atoms included within the coordinating group which can provide an approximate prediction of the eventual phase selectivity, following the basic selectivity principles established within the Pearson acid–base concept.[1]

3.2.4 Functional Selectivity

There is significant literature on the use of organic ligands in analytical chemistry detailing selectivity of different coordinating groups to a specific metal or groups of metals. In non-chromatographic applications of such

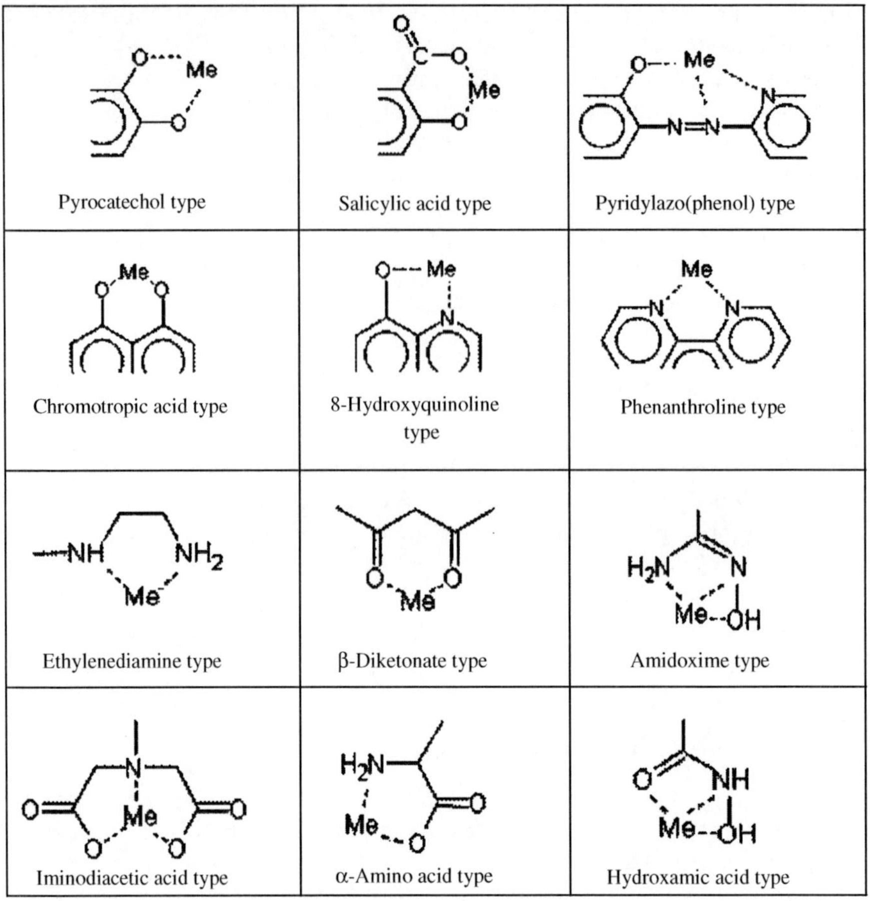

Figure 3.1 Typical coordinating groups utilised in chelating ion exchangers.

ligands, a high degree of selectivity is often the main goal. The development of chelating resins for extraction purposes also generally attempts to acheive the isolation of single, or a limited few metal species, generally avoiding non-selective ligands. However, in HPCIC group selectivity governs the suitability of the chelating stationary phase for the high-performance separation of multiple metal ions. From the practical point of view, differences in selectivity for metal ions within a target group or sample, should be suitably small to achieve the desired separation, ideally under under isocratic elution conditions, or with a reasonable concentration gradient.

One example of a chelating ion exchanger which exhibits what could be described as suitable selectivity for HPCIC is IDA functionalised silica. According to the pH-adsorption dependence of metal ions on IDA-silica, the substrate selectively interacts with a group of six metal ions within a narrow pH interval (Figure 3.2), with a reasonable chance of their successful isocratic separation. It is convenient to use pH_{50} values (at pH_{50} the amount of metal in both adsorbent phase and in aqueous solution is equal) obtained from adsorption plots published in the literature, to compare the separation selectivity of different chelating substrates. By using chromatographic terms for the volumes of the stationary phase, V_s, and of the mobile phase, V_m, the following equilibrium term can be obtained, $[\overline{MR}^{(y-n)+}]V_s = [M_{aq}^{y+}]V_m$. In practice, it equates to the retention factor, k, being equal to 1 at pH_{50}:

$$k = D_M \cdot \varphi = [\overline{MR}^{(y-n)+}]V_s/[M_{aq}^{y+}]V_m = 1 \qquad (3.1)$$

Values of pH_{50} can be used for the preliminary evaluation of the selectivity of a chelating substrate. In the case of IDA-silica, where pH values of 50% of adsorption for about five or six metals are within a narrow pH range from 1.0 and 2.0,[2] and based upon the known relationship between the logarithm of retention factors ($\log k$) and stability constants for the metal and immobilised ligand complexes (β_1), the probability of obtaining an isocratic separation of the five to six metals is high.

The pH dependences for metal ions retention on IDA, tetraethylenepentamine and amidoxime functionalised silica are presented in Figure 3.2. In the last case, the difference between the pH_{50} plots is significantly greater than for IDA and a pH range of approximately three pH units for four metals, equates to a very big difference in their retention factors, $k_{Fe}/k_{Co} = 1000$, under isocratic separation.[3] Clearly, amidoxime-silica can therefore only be used for the separation of these metals ions under gradient elution or with competitive complexing agents added to the mobile phase[3,4] (see Chapter 4). Also, the adsorption of metals on tetraethylenepentamine[5] and amidoxime-silica[3] takes place at higher pH values, so the possibility of undesirable hydrolysis of some metals should be taken into account.

Although IDA-silica shows a narrower range of stability constants and hence k values than amidoxime types, and possibly many others, it is still difficult to obtain isocratic separations with non-complexing mobile phases of more than about five metals in a reasonable analysis time. In practice, isocratic separations on IDA-silica fall into roughly three categories, the alkaline-earth metals, five or six dipositive metal ions, and the more strongly retained Pb^{2+}, Ni^{2+}, Cu^{2+} and Fe^{3+}. The more recently studied APA-silica[6-8] shows better promise for isocratic separations due an even narrower range of stability constants than IDA-silica, where 11 metals can be separated in about 20 min (see Figure 3.3).[9]

3.2.5 Suitable Acid–Base Properties and Surface Charge

The two step interaction kinetic model of adsorption of metal cations by chelating ion exchangers is described in Section 2.2. The main conclusion of this model is that negatively charged functional groups of chelating ion exchangers provide electrostatic attraction of cations and favourable association kinetics,

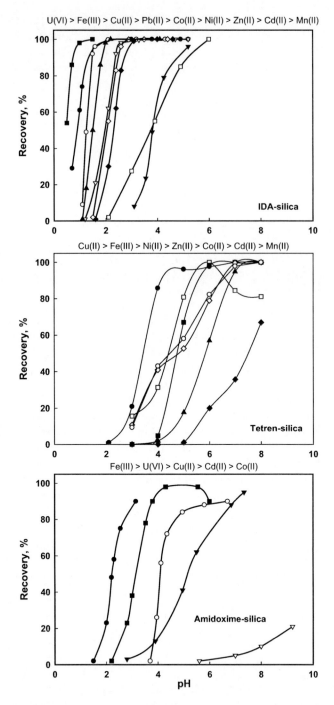

Figure 3.2 Selectivity of adsorption of transition metals on silica with bonded IDA, tetraethylenepentamine and amidoxime functional groups.[2,3,5]

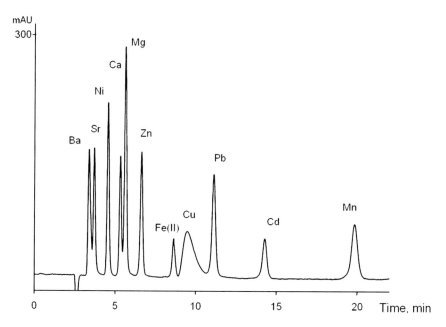

Figure 3.3 Isocratic separation of a mixture of eleven metal ions. Column: APA-silica, 150 × 4.0 mm i.d., particle size 3 μm. Mobile phase: 0.05 M nitric acid–0.8 M KNO_3. Photometric detection at 510 nm after PCR with PAR-ZnEDTA reagent. Reproduced, with permission, from Nesterenko and Jones.[9] Copyright (2007) Wiley-VCH Verlag GmbH Co, KGaA.

with resultant improved separation efficiencies in column chromatography modes. Of course, the charge of the chelating ion exchanger depends on the combination of pK_a values for the various functional groups within the immobilised ligand structure, which can also include some basic groups. For example, glutamic acid bonded silica (see adsorbent 19 in Table 3.2) exhibits cation exchange properties,[10] having tertiary amino groups present within its bound structure.

The balance between the ability of a chelating ion exchanger to retain metals and the possibility to readily elute such metal ions is also of significance. The retention of metal cations in HPCIC depends on column ion exchange capacity or amount of immobilised chelating ligand, on the pK_a values of coordinating groups within the ligand, and on stability constants of the surface complexes formed. If the ligand forms very stable complexes, there is a requirement to use stronger or more concentrated (more acidic) mobile phases to achieve elution. As the variation of acid concentration in the mobile phase will affect the dissociation/protonation of various functional groups, and thus potentially change the surface charge of the adsorbent, changes in the actual retention mechanism of metal cations and in resultant separation efficiency are possible.

To avoid hydrolysis of the majority of metal ions (particularly Al^{3+} and Fe^{3+}), the pH of the mobile phase should be kept within the range 1.0 and 3.0

(although this can be increased through the use of mobile phases containing complexing organic acids which suppress hydrolysis). Therefore, if the pK_a of certain functional groups is close to this pH range, slight changes in the acidity of the mobile phase may cause more significant changes in retention of metal ions and, possibly, in separation selectivity. For a number of dipositive metal ions showing limited tendency to hydrolyse, higher pH values can be used depending on the ligand, such as the separation of alkaline-earth metals in the alkaline pH range on ligand impregnated resins (see Chapter 7).

3.3 Chelating Ion Exchangers with Covalently Bonded Chelating Groups

3.3.1 Types of Chelating Ion Exchangers and Synthetic Methods

A three-dimensional distribution of chelating functional groups throughout adsorbent particles providing maximum ion exchange capacity was considered a priority in the preparation of new chelating resins for many years. However, such materials cannot necessarily be applied as efficient phases in HPCIC, due to poor diffusion within the relatively large particle size hydrophobic resins, resulting in slower mass transfer as discussed below. The most efficient examples of HPCIC have been demonstrated using adsorbents with chelating groups immobilised at the surface of micro-particles of organic polymers or inorganic oxides.

The most widely applied chelating ion exchangers within HPCIC are those having surface covalently bound or chemically immobilised chelating groups. In contrast to the condensation type of organo-polymer resins, or more recently developed sol–gel type adsorbents with entrapped organic reagents, this group of chelating adsorbents has functional groups only at the surface, which provides enhanced accessibility for interaction with metal ions and hence improved kinetics of mass transfer, with resultant improvements in resultant separation efficiencies.

Depending upon the immobilisation chemistry used, functional groups can be presented at the surface as a monolayer of bonded ligands, or within a covalently attached functional polymer layer. Obviously, the thickness of the polymer layer will affect diffusion of metal ions within the stationary phase, so most of the chelating ion exchangers used within HPCIC would possess a relatively thin bonded polymer layer, providing an appropriate ion exchange capacity and displaying suitably high separation efficiency for chromatographic applications.

There are several approaches used for the preparation of chelating ion exchangers.[11] The first includes the single-step grafting or 'direct immobilisation' of previously synthesised modifiers, containing both the chelating functional group/s (F) and an additional reactive functional group or 'anchor groups' (Y), which provide the stable bond(s) between the modifier molecule and the surface of the support matrix (M). The support matrix should itself exhibit active groups (X) or sites involved in the immobilisation reaction:

$$\equiv M-X + Y-R-F \rightarrow \equiv M-R-F - XY \tag{3.2}$$

For example, the covalent immobilisation of ethylenediamine functional groups onto a silica surface can be performed by reacting silica with commercially available N-2-aminoethyl-3-aminopropyltrimethoxysilane:

$$\equiv Si-OH + (CH_3O)_3Si(CH_2)_3NH(CH_2)_2NH_2 \rightarrow$$
$$\equiv Si-O-\underset{|}{Si}(CH_2)_3NH(CH_2)_2NH_2 \tag{3.3}$$

The major advantages of this direct immobilisation approach is its simplicity and resultant uniformity of adsorption sites due to the presence of only one type of target chelating group, F, together with residual reactive groups (X), at the surface of the prepared chelating ion exchanger (*e.g.* ethylenediamine – NHCH$_2$CH$_2$NH$_2$ and residual silanols $\equiv$Si–OH (eqn (3.3)). The limitation of direct immobilisation is the often complicated synthesis and purification of the modifier compound, Y–R–F, which in many instances is impractical.

For the preparation of many chelating ion exchangers, an approach involving a sequence of reactions at the surface is required. The advantage of this method, which is also termed the 'surface assembly method',[11] is the possibility to synthesise surface grafted ligands of a relatively complex structure. A typical example of this approach is the five-step surface reaction assembly of 8-hydroxyquinoline bonded silica, which was the first silica based chelating adsorbent, initially described by Weetall in 1970[12] (see Figure 3.4).

The surface assembly method makes it possible to construct very complex chelating structures at the surface, but the active concentration of bonded groups is generally lower than that achievable through a more direct immobilisation method, as the yields of each surface reaction are often not quantitative. As a result of such incomplete reactions, the presence of unwanted residual functional groups can negatively influence the chromatographic properties of the resultant chelating ion exchanger.

As mentioned above, 8-hydroxyquinoline bonded silica (Figure 3.4, **V**), provides a good illustration of this synthetic approach. Jezorek *et al.*[13–15] have investigated the preparation of this chelating ion exchanger in detail. The group found that even under optimised conditions a significant proportion of 3-aminopropyl groups (**I**) remained unreacted, which together with 8-hydroxyquinoline groups, could be responsible for additional adsorption of metals (see adsorbent 1 in Table 3.2). According to elemental analysis data reported by Jezorek *et al.*,[15] the original silica surface coverage of aminopropyl groups was approximately 0.5 µmol m^{-2}. However, the concentration of 8-hydroxyquinoline groups within the final adsorbent was only 0.15 µmol m^{-2}. Nevertheless, this adsorbent, and several similar types of chelating ion exchanger, gained considerable popularity as solid phase extraction phases for preconcentration of metal ions from complex samples, initially in flow injection analysis (FIA), and later as versatile stationary phase materials for HPCIC applications.[16,17]

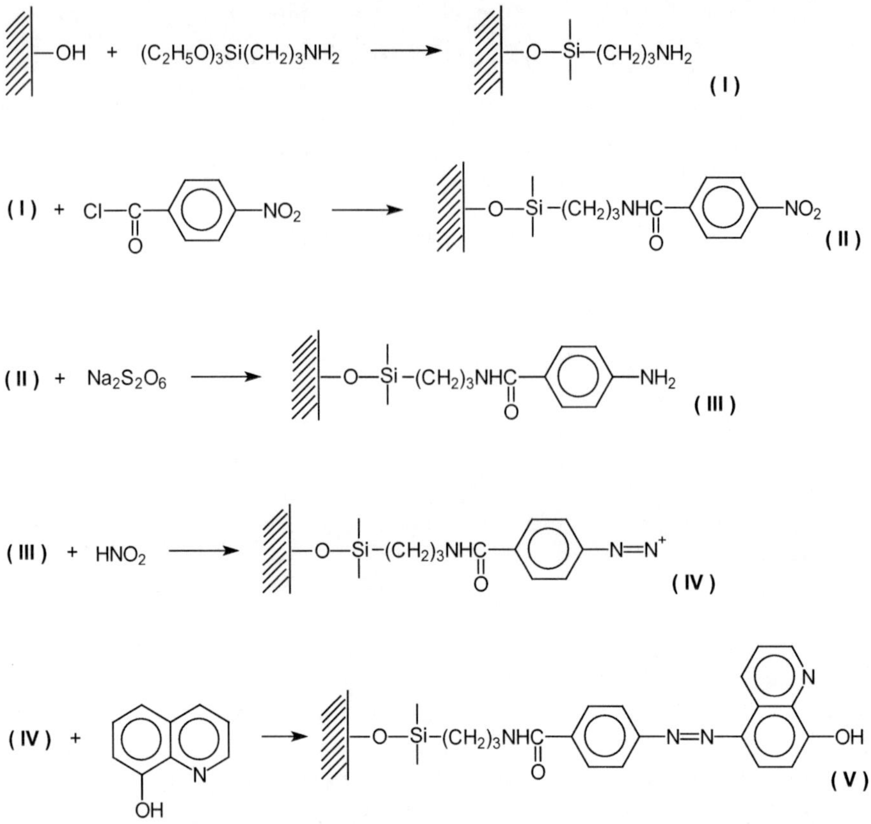

Figure 3.4 Immobilisation scheme for the preparation of a 8-hydroxyquinoline modified, silica based chelating ion exchanger (see Table 3.3).

Several attempts have been undertaken to improve the chemical homogeneity of 8-hydroxyquinoline bonded silicas. For example, Marshall and Mottola[18] simplified the immobilisation of 8-hydroxyquinoline on a silica surface by using a three-step synthesis, as shown in Figure 3.5. The key element of this work was the formation of an immobilised aromatic amine in a single step. This modified reaction series resulted in an increase of up to 2.1 µmol m^{-2} in the surface concentration of 8-hydroxyquinoline groups within the final adsorbent (**VIII**), 16 times greater than was reported previously (**V**). The yields of surface reactions in the latter scheme are also significantly greater and provided a more chemically uniform bonded layer. It should be noted that further improvement of the synthesis shown in Figure 3.5A is possible when using only one of three possible positional isomers of aminophenyltrimethoxysilane for the preparation of intermediate product (**VI**).[19] The effects of steric hindrance and differing degrees of reactivity of the aromatic amino groups in the product (**VI**) should be considered in this case.

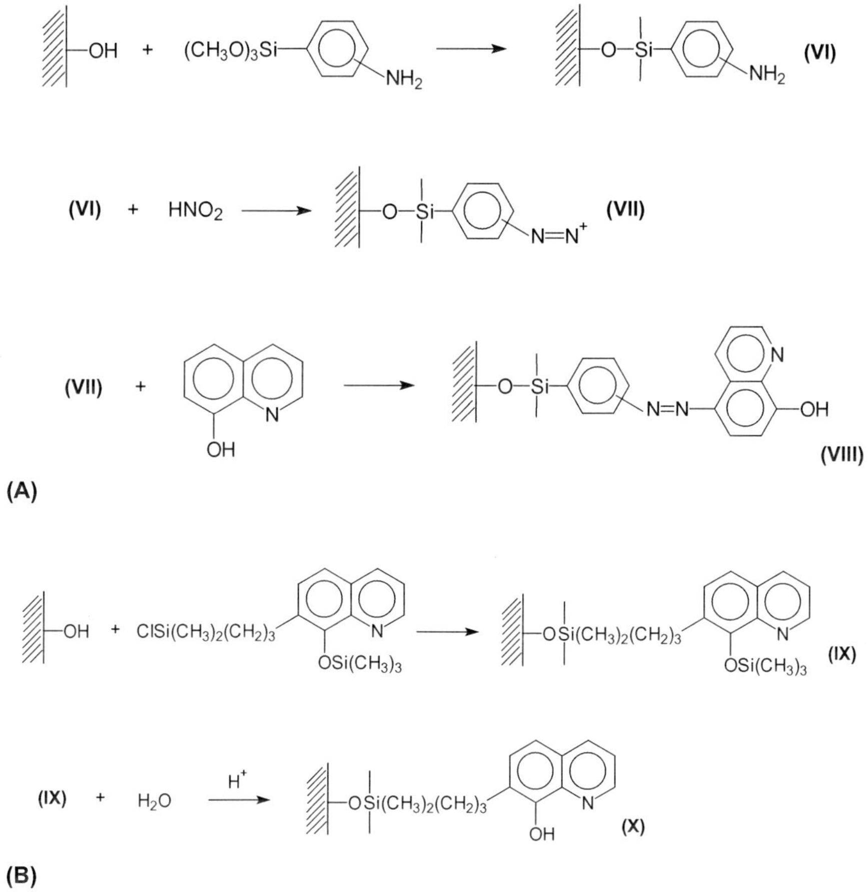

(A)

(B)

Figure 3.5 Improved scheme for the immobilisation of 8-hydroxyquinoline upon a silica surface. Adapted from Marshall and Mottola[18] and Weaver and Harris.[20]

Finally, a further simplification of the above synthetic routes was reported, resulting in a silica based adsorbent displaying only 8-hydroxyquinoline groups (**X**),[20] prepared using only a two step reaction process (see Figure 3.5B). However, in this case, the surface concentration of the final adsorbent was rather low, at only 0.33 µmol m^{-2}. Additionally, it should be mentioned that the synthesis of the 7-[3-(chlorodimethylsilyl)propyl]-8-(trimethylsilyloxy)-quinoline reagent itself, involved a four step process (a drawback mentioned above in relation to direct immobilisation approaches), and so this synthetic scheme did not gain importance in the preparation of this popular type of chelating ion exchanger.

The majority of ligands can be immobilised on silica or a polymer surface in as little as two steps through the application of oxirane chemistry. Using such

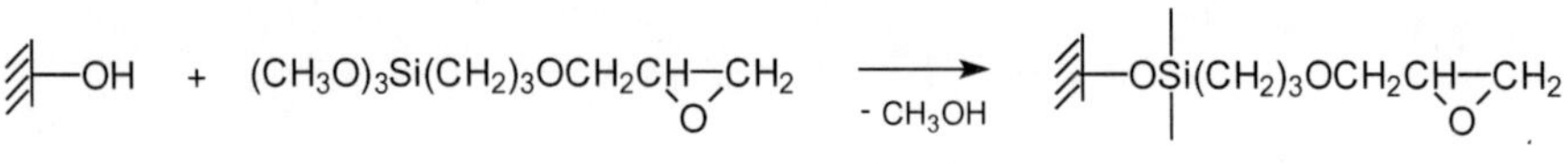

≥—OH + (CH$_3$O)$_3$Si(CH$_2$)$_3$OCH$_2$CH—CH$_2$ $\xrightarrow{-CH_3OH}$ ≥—OSi(CH$_2$)$_3$OCH$_2$CH—CH$_2$
         \O/   \O/

Scheme 3.1 Treatment of silica surface with 3-glycidoxypropyltrimethoxysilane.

an approach, the first step involves the treatment of the silica surface with 3-glycidoxypropyltrimethoxysilane, as in Scheme 3.1.

The resultant bonded layer displays the reactive oxirane ring or epoxy-groups, which can readily react with many organic molecules (see Scheme 3.5) containing primary, secondary and tertiary amino groups, and hydroxy, thiol or various other functionalities.[21] For example, silica activated with 3-glycidoxypropyl groups can be readily used for the immobilisation of IDA or various α-amino acids (see adsorbents 18–20 in Table 3.2), under mild conditions from aqueous solutions. Most importantly, unreacted epoxide groups can easily be converted into diol (–CH(OH)CH$_2$OH) groups by a simple acid wash of the final adsorbent.

3.3.2 Polymer Based Chelating Ion Exchangers

Historically, organic polymer phases covalently functionalised with various chelating groups constituted the main type group of chelating ion exchangers in both low pressure and high-performance modes of chelation based chromatography. In the former case, Chelex-100, a PS-DVB based IDA functionalised resin produced by the Dow Chemical Company, was particularly successful. The initial popularity of polymer based chelating phases could be a result of the commonly held, and somewhat misunderstood, belief of the poor hydrolytic stability of silica based adsorbents.

The large-scale application of polymer based chelating resins for trace metal extraction and concentration, particularly from environmental and industrial sample matrices, has resulted in the publication of a large number of detailed and informative monographs[22–25] and reviews[26–28] on the subject, which also describe in detail the various synthetic approaches used in their preparation and application.

There are very few published systematic investigations into the optimisation and characterisation of the phase morphology of organo-polymer adsorbents suitable for use within HPCIC. Unlike simple ion exchange chromatography, in chelation based methods the exact effects of phase pore size, the distribution and density of functional groups and other physical properties upon the resultant separation efficiency and selectivity, is still a matter for further investigation.

In practice, the above type of generic physical and chemical characterisation of organo-polymer chelating adsorbents is a challenging task, so published research in this area tends to be more specific to a particular individual, or

Chelating Stationary Phases

closely related range of adsorbents, and can usually be classified into the three following types of study:

1. Studies reporting the introduction of new chelating groups into a commercially available organo-polymer substrate and subsequent characterisation of selectivity
2. Studies reporting the optimisation of the bonding chemistry of a single type of chelating group into a variety of organo-polymer substrates
3. Studies reporting the characterisation of the selectivity of one type of chelating group bonded to the same matrix via different synthetic routes

Currently, there is a considerable research effort continuing to synthesis new and ever more selective chelating groups for their incorporation into new chelating ion exchangers, together with optimisation of surface attachment strategies and conformational design. Taking into consideration the fact that any new, or indeed well known, organic chelating ligand can be attached to one of a large number of supporting polymer substrates, using one of a variety of immobilisation chemistries, each producing effectively a new chelating phase with unique selectivity, it is clear to see how the possible number of such new phases could rapidly equal, or become a multiple of the current number of known organic chelating ligands.

In the early stages of the development of HPCIC, a significant number of fundamental investigations into the selectivity of new medium to high performance chelating organo-polymer phases were carried out. For example, Fritz et al. were pioneers in this area, using macro-porous PS-DVB resins as the matrix, preparing chelating adsorbents with functional groups of o-hydroxyoxime,[29] amide,[30] hydroxamic acid,[31,32] thioglycolic acids,[33,34] propylenediaminetetraacetic acid,[35] and investigating their suitability for the high-performance separation of metals.

However, there exist a few disadvantages associated with the use of neutral hydrophobic PS-DVB as the supporting substrate. Firstly, PS-DVB must be first activated through the introduction of reactive functional groups, which may involve multi-step and complex synthetic procedures. For example, chloroformyl, which can be obtained by the three-step reaction shown as Scheme 3.2.

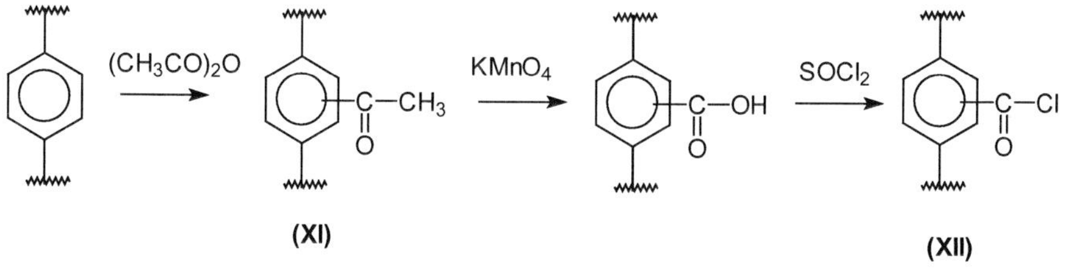

Scheme 3.2 Activation of PS-DVB to provide reactive chloroformyl groups.

The incorporation of an acetyl group (product **XI**) can occur at a variety of carbon atoms within aromatic rings, which imparts some differentiation in the reactivity of these groups. Additionally, the yields of some surface reactions are not quantitative and some by-products may appear, for example during the oxidation step. Nevertheless, the resultant activated substrate (**XII**) has been successfully used by Fritz et al.[31,35] for the attachment of various chelating ligands.

Secondly, the hydrophobicity of PS-DVB can also result in reduced separation efficiency due to adsorption of neutral organic species originating from within the sample matrix. Such unwanted contamination of the surface can, over time, reduce stationary phase capacity and efficiency, and chromatographic selectivity.

For this reason, more polar and hydrophilic organo-polymers such as poly(acrylonitrile-DVB), and polymethacrylates grew in popularity as substrate materials for the preparation of chelating ion exchangers. Poly(acrylonitrile-DVB) co-polymer contains reactive nitrile groups, which can be readily converted into carboxylic acid functional groups by alkaline hydrolysis and further converted to hydroxamic acid groups via reaction with hydroxylamine in methanol, as in Scheme 3.3.[36]

γ-Aminohydroxamate resin (named **P-13**), with a longer linker group between the bonded ligand and the polymer substrate was prepared using the following additional steps to increase chain length (Scheme 3.4).

Resin **P-13** and its *N*-methylated analogue (*N*-methyl **P-13**) were both used for the preconcentration of metals and their subsequent chromatographic separations[37,38] (see adsorbents 8 and 9 in Table 3.1).

Poly(glycidylmethacrylate) gel (PGMA) practically satisfies a number of specific requirements for use in the preparation of chelating ion exchangers. The hydrophilic PGMA has a macro-porous structure and terminal glycidoxy groups, which are available for covalent attachment of chelating ligands having reactive hydroxyl, amino or thiol groups, under mild conditions, similar to

$$R-C\equiv N \xrightarrow{OH^-} R-COOH \xrightarrow{NH_2OH} R-CONHOH$$

Scheme 3.3 Two step conversion of reactive nitrile groups to hydroxamic acid groups.

$$R-C\equiv N \xrightarrow{OH^-} R-COOH \xrightarrow{SO_2Cl} R-COCl \xrightarrow{H_2N(CH_2)_3COOH}$$

$$\longrightarrow R-CONH(CH_2)_3COOH \xrightarrow{NH_2OH} R-CONH(CH_2)_3CONHOH \quad (\textbf{P-13})$$

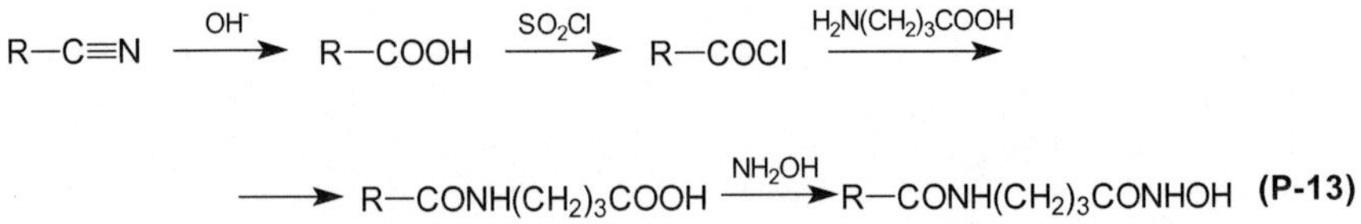

Scheme 3.4 Four step conversion of reactive nitrile groups to give γ-aminohydroxamate sites.

Chelating Stationary Phases

Table 3.1 Polymer based chelating ion exchangers applied to the chromatographic separation of metal ions.

No	Functional groups	Matrix, type of resin[a]	Column, particle size	Separated ions	Reference
1	Ph–COCH[–CH$_2$N(CH$_2$COOH)$_2$][–CH$_2$N(CH$_2$COOH)$_2$] with C=O	PS-DVB, 5 nm, 0.45 mmol groups g^{-1}	28 × 6 mm, 44–56 μm	Mg^{2+}, Mn^{2+}, Ni^{2+}, Zn^{2+}, Cd^{2+}, Co^{2+}, Pb^{2+}, Cu^{2+}, UO$_2^{2+}$, Th^{4+}, Zr^{4+}	35
2	Ph–CH$_2$N[–CH$_2$CH$_2$N(CH$_2$COOH)$_2$][–CH$_2$CH$_2$N(CH$_2$COOH)$_2$]	PS-DVB, 2%DVB, 3.0 mmol Cu^{2+} g^{-1}	95 × 10 mm, 37–74 μm	Co^{2+}, Ni^{2+}	43
3	Ph–CH$_2$OCCH$_2$SH with C=O	PS-DVB, 5 nm, 1.5 mmol Ag$^+$ g^{-1}	100 × 2 mm, <75 μm	Au^{3+}, Ag$^+$, Hg^{2+}, Bi^{3+}, Cd^{2+}, Cu^{2+}, Pb$^+$, Sn^{2+}, Sb^{3+}, As(v)	33
4	Ph–CO(CH$_2$)$_6$OCCH$_2$SH with both C=O	PS-DVB, 5 nm, 0.17 mmol Au^{3+} g^{-1}	33 × 6 mm, 44–56 μm	Sn^{2+}, Cu^{2+}, Fe^{3+}, Pb^{2+}, Bi^{3+}, Ag$^+$, Hg^{2+}, Au^{3+}, Ru(III), Cd^{2+}, Sb^{3+}	34
5	Ph–C(OH)=CHC(=S)SH	PS-DVB, 0.05 mmol Cu^{2+} g^{-1}	50 × 6 mm, 75–150 μm	Ag$^+$, Au^{3+}, Cd^{2+}, Cu^{2+}, Fe^{3+}, Hg^{2+}, Ni$^+$, Pb^{2+}, Pt(IV), Zn^{2+}	44
6	Ph–N=N–(8-hydroxyquinolin-5-yl)	PS-DVB, non-porous, 0.615 mmol Cu^{2+} g^{-1}	250 × 3 mm, 7–10 μm	Mn^{2+}, Fe^{2+}, Co^{2+}, Zn^{2+}, Ni^{2+}, Cu^{2+}	45

Table 3.1 (Continued).

No	Functional groups	Matrix, type of resin[a]	Column, particle size	Separated ions	Reference
7	(phenol with OH, NOH, CH–CH3 groups on two rings)	PS-DVB, 5 nm, 1.26 mmol groups g^{-1}	70 × 2 mm, 44–56 μm	Zn^{2+}, Ni^{2+}, Cu^{2+}, $Mo(VI)$	29
8	–CNH(CH₂)₃C(=O)N(OH)CH₃ *N*-methyl P-13 resin	Macro-porous poly(acrylonitrile-DVB), 8% cross-linking	250 × 3 mm, 44–56 μm	Fe^{3+}, Zn^{2+}, Mn^{2+}, Cd^{2+}, Er^{3+}, Eu^{3+}, Sm^{3+}, La^{3+}	38, 37
9	–CNH(CH₂)₃C(=O)N(OH)H P-13 resin	Macro-porous poly(acrylonitrile-DVB), 8% cross-linking	50 × 4 mm, 44–63 μm	$V(IV)$, $Mo(VI)$, $W(VI)$	46
10	CH₃–CH(CH₂NH–C(=O)CH₃)– (phenyl)	PS-DVB, 20 nm, 1.33 mmol groups g^{-1}	25 × 5 mm, 74–95 μm	Au^{3+}, Pr^{3+}, UO_2^{2+}, $Ru(IV)$, Th^{4+}, Zr^{4+}, $Rh(III)$, $Te(VI)$, Pd^{2+}, $Ir(IV)$, Cu^{2+}, Zn^{2+}, Pb^{2+}, Mn^{2+}, Co^{2+}, Ni^{2+}, $Pt(II)$	30
11	–C(=O)–N(CH₃)–OH (phenyl)	PS-DVB, 5 nm, 0.69 mmol Cu^{2+} g^{-1}	100 × 2 mm, 44–74 μm	Mg^{2+}, Ca^{2+}, Mn^{2+}, Co^{2+}, Zn^{2+}, Ni^{2+}, Cu^{2+}, Cr^{3+}, Al^{3+}, Fe^{3+}, La^{3+}, Eu^{3+}, Lu^{3+}, Sc^{3+}, Ti^{4+}, Th^{4+}, $V(IV)$, $V(V)$, UO_2^{2+}	31
12	–C(=O)–OCH₂CH(OH)CH₂NH(CH₂)₄CH[N(CH₂COOH)₂](CH₂COOH)	Cross-linked PGMA, 27 μmol Cu^{2+} g^{-1}	150 × 4.6 mm, 10 μm	REE	40

Chelating Stationary Phases

	Structure	Support	Dimensions	Analytes	Ref
13	Ar-CH₂N(CH₂COOH)(CH₂)₄CH(CH₂COOH)(COOH)	Macro-porous chloro-methylated PS-DVB, 0.06 mmol Cu^{2+} g^{-1}	50 × 4.6 mm, 10 µm	Mn^{2+}, Cd^{2+}, Co^{2+}, Zn^{2+}, Ni^{2+}, Cu^{2+}, UO_2^{2+}, Pb^{2+}, Sb^{3+}, $Mo(vi)$	47
14	—COCH₂CHCH₂NH(CH₂)₄CH(NHCH₂COOH)(COOH) with OH	Cross-linked PGMA, 0.22 mmol Cu^{2+} g^{-1}	125 × 4.6 mm, 10 µm	Lanthanides	41
15	—COCH₂CHCH₂N(CH₂COOH)₂ with OH	Crosslinked PGMA, 0.41 mmol Cu^{2+} g^{-1}	40 × 4.6 mm, 10 µm	Lanthanides	41
16	Ar-CH₂N(CH₂COOH)₂	Macro-porous chloro-methylated PS-DVB, 0.40–0.45 mmol Cu^{2+} g^{-1}	125 × 4.6 mm, 10 µm; 50 × 4.6 mm, 10 µm	Lanthanides; Mn^{2+}, Cd^{2+}, Co^{2+}, Zn^{2+}, Ni^{2+}, Cu^{2+}, UO_2^{2+}, Pb^{2+}, Sb^{3+}, $Mo(vi)$	41; 47
17	—COCH₂CHCH₂NH-Ar-CH₂-CH(N(CH₂COOH)₂)(N(CH₂COOH)₂) with OH	Macro-porous cross-linked PGMA, 0.41 mmol Cu^{2+} g^{-1}	125 × 4.6 mm, 10 µm	Lanthanides	39
18	—COCH₂CHCH₂NH-Ar-CH₂N(CH₂CH₂NH₂)₂	PS-DVB, 2% DVB, 2.9 mmol Cu^{2+} g^{-1}	95 × 10 mm, 37–74 µm	Zn^{2+}, Ni^{2+}, Cu^{2+}	43

[a] Including pore size, ion exchange capacity or concentration of bonded groups.

reactions of epoxy-activated silica. Suzuki et al. reported upon the preparation of a number of chelating ion exchangers, based upon 10 μm PGMA particles with attached 1-aminobenzyl-1,2-diaminopropane-N,N,N',N'-tetraacetic acid[39] and lysine-N,N-diacetic acid.[40] The resins were applied to the separation of REE and the separation selectivity compared to that of a PS-DVB resin with bonded IDA functional groups.[41] A PGMA resin having a longer spacer group between the chelating ligands and the polymer surface exhibited the most promising results. The separation of REE using this PGMA based chelating ion exchanger is shown as Figure 3.6, obtained using a nitric acid gradient in 22 min on a 150 × 4.6 mm i.d. column. It should be noted that similar improvements in performance using resins with longer spacer groups have been also demonstrated in works by Liu et al. for poly(acrylonitrile-DVB) resins functionalised with hydroxamic acids.[36,38]

Although the above studies have demonstrated how such organo-polymer chelating phases can indeed produce efficient separations of metal ions, it should be noted that the limited hydrolytic stability of PGMA based resins, under highly acidic conditions (pH < 1), restricts some possible applications of these types of chelating ion exchangers to the separation of strongly retained metals, as their elution requires the use of strongly acidic conditions.

Recently, evaluation of a new poly-IDA functionalised polymer 10 μm resin (Dionex ProPac IMAC-10), for potential application within HPCIC has been

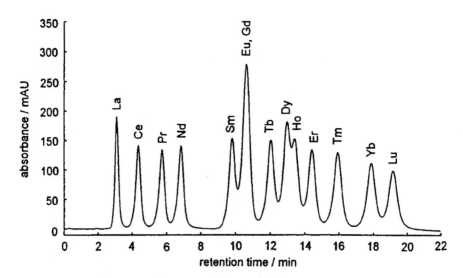

Figure 3.6 Chelation ion chromatogram of lanthanides on a 150 × 4.6 mm i.d. column packed with 10 μm PGMA gel with covalently bonded lysine-N,N-diacetic acid. Gradient elution with 0.02 M nitric acid at 0 min to 0.08 M nitric acid at 25 min under a 1.0 mL min^{-1} flow rate. Column temperature 40 °C. Photometric detection at 660 nm after PCR with chlorophosphonazo III. Reproduced, with permission, from Inoue et al.[40] Copyright (1996) American Chemical Society.

reported by Barron et al.[42] This stationary phase, typically used for the separation of proteins, is composed of long chain poly-IDA groups grafted to a cross-linked poly(acrylate) hydrophilic layer surrounding 10 μm non-porous PS-DVB beads (55% DVB). Following a column conditioning procedure to ensure the availability of free poly-IDA chains at the surface, the separation selectivity exhibited for a range of alkaline earth and transition metals was found to be comparable to that shown by IDA chelating resins of a more traditionally singly surface bound structure. Figure 3.7 shows overlaid separations of seven transition and heavy metal cations in only 2.5 min obtained using gradient elution on a short column (50 × 2 mm i.d.) packed with the 10 μm poly-IDA functionalised polymer resin.

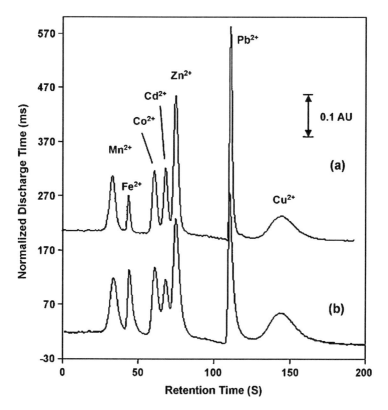

Figure 3.7 Overlaid chelation ion chromatograms of seven transition and heavy metal cations on a short column (50 × 2 mm i.d.) packed with 10 μm poly-IDA functionalised polymer resin. Gradient elution conditions: Mobile phase A: 0.01 M KCl–0.25 mM HNO_3; Mobile phase B: 0.01 M KCl–15 mM HNO_3–0.4 mM picolinic acid. Gradient conditions: 0–10% B for 60 s; 10–12% B for 27 s; 100% B at 87 s and hold for 153 s. Re-equilibration time: 2.5 min. Flow rate 2.5 mL min^{-1}. Photometric detection at 510 nm after PCR with 0.4 mM PAR, 0.5 M NH_3, pH 10.5. Reproduced, with permission, from Barron et al.[42] Copyright (2008) Elsevier.

3.3.3 Silica Based Chelating Ion Exchangers

A considerable body of work detailing the immobilisation of different complexing functional groups upon silica surfaces has emerged since the pioneering work of Weetall on the preparation of 8-hydroxyquinoline modified controlled porous glass substrates in 1970.[12] In 1973, Hill[48] suggested the use of the same ligand, but instead immobilised on silica gel for the chromatographic separation of metal ions. A series of extensive investigations into silica based chelating ion exchangers were carried by Leyden,[49,50] who applied the resultant adsorbents to the preconcentration of trace metals. In many of these early studies, the advantageous kinetics of adsorption and desorption from the various modified silica based chelating phases was noted.

In the preparation of silica based chelating ion exchangers, the presence of unreacted functional groups on the surface of the modified silica following the series of surface reactions has to be considered. For example, as mentioned above, a variety of ligands can be immobilised onto silica in a two step process using oxirane chemistry. The reactive oxirane ring or epoxy groups can react with primary, secondary and tertiary amino- groups, or hydroxy, thiol or other functional groups.[21] Silica activated with glycidoxypropyl groups has been used for the preparation of chelating ion exchangers, including the immobilisation of IDA,[51] various α-amino acids, including glutamic acid and lysine[52] and oligoethyleneamines[5] (see Table 3.2). The reaction proceeds under mild conditions from aqueous solutions according to Scheme 3.5.

Importantly, unreacted epoxide groups within the bonded layer can be readily converted into diole groups ($-CH(OH)CH_2OH$), with a simple acid wash step, which show no affinity towards metal ions.

A second class of functionalised silica commonly utilised for further chemical modification is aminated silica. The commercial availability of 3-aminopropyltrialcoxysilanes and 3-(2-aminoethylamino)propyltrialcoxysilanes makes this synthetic route an attractive alternative to the above oxirane chemistry. Chelating adsorbents with β-diketonate (adsorbent 6 in Table 3.2) and dithiocarbamate (adsorbents 27 and 28 in Table 3.2) functionalities can be prepared by reactions with alkylacetoacetates[53] or carbon disulfide,[50,53] respectively. Aminophosphonic acid functionalised silica (adsorbent 22 in Table 3.2) has been prepared by the Mannich reaction of 3-aminopropylsilica with formaldehyde and hypophosphoric acid,[6,7] as shown in Scheme 3.6.

3-Aminopropylsilica has also been used for the immobilisation of crown ether type molecules,[54,55] and as the supporting substrate for multi-step surface immobilisation reactions[16] with various aromatic chelating reagents, similar to the four-step attachment of 8-hydroxyquinoline shown earlier (Figure 3.4).

A very simple approach achieving the conversion of commercially available nitrile or cyanoalkyl bonded silica to an alkylamidoxime modified phase has also been reported (see adsorbent 11 in Table 3.2), through heating with an alcohol solution of hydroxylamine, according to the reaction shown as Scheme 3.7.[56]

Chelating Stationary Phases

Table 3.2 Separation of metal ions on silica based chelating ion exchangers with covalently bonded functional groups.

No	Structure of bonded groups	Matrix (type, pore size, surface area, capacity)	Column and particle size	Separation	Reference
1	≡–O–Si–(CH$_2$)$_3$NH$_2$	Separon SGX-NH$_2$, 8 nm, 500 m^2 g^{-1}	150 × 3.0 mm, 5 µm	MeHg$^+$, PhHg$^+$, Hg^{2+}, Cu^{2+}	65
		Baker silica gel 7, 6 nm	TLC plates, 5–40 µm	Fe^{3+}, Cu^{2+}, Ni^{2+}, Zn^{2+}	53, 66
2	≡–O–Si–(CH$_2$)$_3$NHCH$_2$CH$_2$NH$_2$	Baker silica gel 7, 6 nm	TLC plates, 5–40 µm	Fe^{3+}, Cu^{2+}, Ni^{2+}, Zn^{2+}	53, 66
3	≡–O–Si–(CH$_2$)$_3$OCH$_2$CHCH$_2$N(CH$_2$CH$_2$NH)$_n$, OH, n = 0-4	Silochrom S-120, 40 nm, 120 m^2 g^{-1}	300 × 10.0 mm, 100–160 µm	Mn^{2+}, Cd^{2+}, Zn^{2+}, Ni^{2+}, Cu^{2+}	67
		Silasorb Si300, 10 nm, 300 m^2 g^{-1}, 0.28 mmol H$^+$ g^{-1}	250 × 4.6 mm, 10 µm	Mg^{2+}, Cd^{2+}, Zn^{2+}, Co^{2+}	68
4	≡–O–Si–(CH$_2$)$_3$NHC(=O)–C$_6$H$_4$–N=N–naphthyl–NHCH$_2$CH$_2$NH$_2$	Adsorbosil-LC, 7 nm, 480 m^2 g^{-1}, 0.02 mmol Cu^{2+} g^{-1}	250 × 4.0 mm, 10 µm	Mn^{2+}, Cd^{2+}, Zn^{2+}, Cu^{2+}, Fe^{2+}, Pb^{2+}	16

Table 3.2 (Continued).

No	Structure of bonded groups	Matrix (type, pore size, surface area, capacity)	Column and particle size	Separation	Reference
5	$(CH_3)_3SiO-[Si(CH_3)-O]_n[Si(CH_3)-O]_m Si(CH_3)_3$ with $(CH_2)_2 OCH_2 CH(OH)CH_2-$ linked to a tetraazacyclotetradecane (cyclam) ring	Silica gel 60, 6 nm, 550 m^2 g^{-1}, 0.08 mmol H$^+$ g^{-1}	100 × 4.6 mm, 25–40 μm	Mn^{2+}, Cd^{2+}, Zn^{2+}, Cu^{2+}, Pb^{2+}, Co^{2+}, Ni^{2+}, Hg^{2+}	69
6	$-O-Si(CH_2)_3 NHCH_2 CH_2 NHCCH_2 C-R$ with C=O groups, where R = –CF$_3$, –CH$_3$, –phenyl	Baker silica gel 7 6 nm	150 × 2.0 mm, 5–40 μm	Cu^{2+}, Zn^{2+}, Ni^{2+}	70
			TLC plates, 40 μm	Fe^{3+}, Cu^{2+}, Ni^{2+}, Zn^{2+}	53, 66
7	$-O-Si-(CH_2)_3 CH(COCH_3)(COCH_3)$ (acetylacetonate-type bonded ligand)	Nucleosil Si 100-5, 10 nm, 350 m^2 g^{-1} 0.885 mmol groups g^{-1}	135 × 4.6 mm 5 μm	Co^{2+}, Cd^{2+}, Cu^{2+}	71
8	$-O-Si-(CH_2)_3 NHC(=O)-$C$_6$H$_3$(OH)-N=N-C$_6$H$_4$-C(=O)CH$_2$CCH$_3$(=O)	Adsor bosil-LC, 7 nm, 480 m^2 g^{-1}, 0.047 mmol Cu^{2+} g^{-1}	250 × 4.0 mm, 10 μm	Mn^{2+}, Cd^{2+}, Zn^{2+}, Co^{2+}, Ni^{2+}, Pb^{2+}	16

Chelating Stationary Phases

#	Structure	Support	Column	Metal ions	Ref.
9	−O−Si−(CH₂)₃NHC(=O)−C₆H₄−N=N−(thiophene)−C(=O)−O−CCH₂CCF₃ (=O)	Adsorbosil-LC, 7 nm, 480 m² g⁻¹	250 × 4.0 mm, 10 μm	Mn^{2+}, Cd^{2+}, Zn^{2+}, Co^{2+}, Ni^{2+}, Pb^{2+}	16
10	−O−Si−(CH₂)₃NH−CH₂−C(=O)−O−CH₂CH₂−O−CH₂−C(=O)−N(CH₃)(C₄H₉)	LiChrosorb NH₂, 0.314 mmol groups g⁻¹	77 × 5.0 mm, 10 μm	Ca^{2+}, Ba^{2+}, Sr^{2+}, Mg^{2+}	54
11	−O−Si−(CH₂)₃CH(=NOH)(NH₂)	Silasorb Si600, 6 nm, 570 m² g⁻¹, 0.56 mmol H⁺ g⁻¹	250 × 4.0 mm, 5 μm	Cr^{3+}, Cu^{2+}, Fe^{3+}, Ca^{2+}, UO_2^{2+}, Cd^{2+}, Mn^{2+}, Pb^{2+}	4
			62 × 2.0 mm, 5 μm	Ni^{2+}, Cu^{2+}, Fe^{3+}, $Mo(VI)$, Cr^{3+}, UO_2^{2+}, $W(VI)$	3
			100 × 3.0 mm, 5 μm	Zn^{2+}, Co^{2+}, Ni^{2+}, Cu^{2+}, Hg^{2+}, Fe^{3+}	72, 73
12	−O−Si−(CH₂)₂C(=O)−NHOH	Analytichem International	100 × 4.0 mm, 40 μm	Cu^{2+}	74
13	−O−Si−(CH₂)₃NHC(=O)−C₆H₄−N=N−(8-hydroxy-2-methylquinolin-5-yl)	Polygosil 60, 6 nm, 500 m² g⁻¹	250 × 4.0 mm, 10 μm	Mn^{2+}, Cd^{2+}, Zn^{2+}, Cu^{2+}, Fe^{2+}, Pb^{2+}	16

Table 3.2 (Continued).

No	Structure of bonded groups	Matrix (type, pore size, surface area, capacity)	Column and particle size	Separation	Reference
14	–O–Si–(CH$_2$)$_3$NHC(O)–C$_6$H$_4$–N=N–(8-hydroxyquinoline)	Porasil B, 15 nm, 180 m^2 g^{-1}, 0.054 mmol Cu^{2+} g^{-1}	250 × 4.0 mm, 37–74 μm	Mn^{2+}, Cd^{2+}, Pb^{2+}, Zn^{2+}, Co^{2+}, Ni^{2+}, La^{3+}, Gd^{3+}, Yb^{3+}	17
		Polygosil 60, 6 nm, 500 m^2 g^{-1}, 0.010–0.156 mmol Cu^{2+} g^{-1}	250 × 4.0 mm, 10 μm	Mn^{2+}, Tl$^+$, Zn^{2+}, Ni^{2+}, Co^{2+}, Cd^{2+}, Pb^{2+}	16,75,76
		Adsorbosil-LC, 7 nm, 480 m^2 g^{-1}, 0.04 mmol Cu^{2+} g^{-1}	250 × 4.0 mm, 10 μm	Mn^{2+}, Cd^{2+}, Pb^{2+}, Zn^{2+}	77
15	–O–Si(CH$_2$)$_3$NHCH$_2$CH$_2$NHC(O)–C$_6$H$_4$–NH$_2$	Baker silica gel 7, 6 nm	TLC plates, 5–40 μm	Fe^{3+}, Cu^{2+}, Ni^{2+}, Zn^{2+}	53
16	–O–Si–C$_6$H$_4$–NH–N=CH–(pyridyl)	Pierce CPG, 24 nm, 130 m^2 g^{-1}, 0.025 mmol Cu^{2+} g^{-1}	250 × 4.6 mm, 37–74 μm	Fe^{2+}, Co^{2+}, Ni^{2+}, Cu^{2+}	78
		Chromosorb LC-6, 12 nm, 400 m^2 g^{-1}, 0.034 mmol Cu^{2+}/column	250 × 4.6 mm, 5 μm	Mn^{2+}, Fe^{2+}, Cd^{2+}, Zn^{2+}, Co^{2+}, Pb^{2+}, Cu^{2+}	79
17	–O–Si–C$_6$H$_4$–NH–N=CH–(1,10-phenanthrolinyl)	Chromosorb LC-6, 12 nm, 400 m^2 g^{-1}, 0.039 mmol Cu^{2+} g^{-1}	250 × 4.6 mm, 5 μm	Mn^{2+}, Pb^{2+}, Fe^{2+}, Cd^{2+}, Zn^{2+}, Co^{2+}	80

Chelating Stationary Phases

18	≡Si−(CH$_2$)$_3$OCH$_2$CH(OH)CH$_2$N(CH$_2$COOH)$_2$	Silasorb Si300, 10 nm, 300 m^2 g^{-1}; 0.13 mmol H$^+$ g^{-1}	250 × 4.0 mm, 6 μm	Mg^{2+}, Ca^{2+}, Sr^{2+}, Ba^{2+}	81
				REE	64,82
				Mg^{2+}, Ca^{2+}, Mn^{2+}, Co^{2+}, Cd^{2+}, Zn^{2+}, Ni^{2+}, UO$_2^{2+}$, Pb^{2+}	83,84
			150 × 3.0 mm, 6 μm	Na$^+$, K$^+$, Mg^{2+}, Ca^{2+}, Ba^{2+}, Sr^{2+}	81,85
				Fe^{2+}, Co^{2+}, Zn^{2+}, Cd^{2+}, Pb^{2+}	73
			250 × 3.0 mm, 6 μm	Co^{2+}, Zn^{2+}, Cd^{2+}, Pb^{2+}, Cu^{2+}	86
				Mg^{2+}, Ca^{2+}, Sr^{2+}, Mn^{2+}, Ba^{2+}, Be^{2+}, Co^{2+}, Cd^{2+}, Zn^{2+}	81,87-89
		Nucleosil 100-5, 10 nm, 360 m^2 g^{-1}	100 × 4.6 mm, 5 μm	Na$^+$, K$^+$, Mg^{2+}, Ca^{2+}, Sr^{2+}, Ba^{2+}, Mn^{2+}, Fe^{2+}, Co^{2+}, Cd^{2+}, Zn^{2+}, Cu^{2+}	90
		Nucleosil 50-5, 5 nm	150 × 4.6 mm 250 × 4.6 mm, 7 μm	Al^{3+} Co^{2+}, Cd^{2+}, Fe^{2+}, alkali metal cations	63,91 92
		Chromolith, 12 nm, 300 m^2 g^{-1}, 0.075 mmol Cu^{2+}/column	100 × 4.6 mm, monolith	Li$^+$, Na$^+$, K$^+$, Cs$^+$, Mg^{2+}, Ca^{2+}, Sr^{2+}, Ba^{2+}, Mn^{2+}, Co^{2+}, Cd^{2+}, Zn^{2+}, Pb^{2+}, Cu^{2+}	93,94

Table 3.2 (Continued).

No	Structure of bonded groups	Matrix (type, pore size, surface area, capacity)	Column and particle size	Separation	Reference
19	≡−O−Si−(CH$_2$)$_3$OCH$_2$CHCH$_2$NHCH(COOH)(CH$_2$COOH) with OH	Silasorb Si300, 10 nm, 300 m^2 g^{-1}, 0.038 mmol Cu^{2+} g^{-1}	100 × 4.6 mm, 5 μm	Mg^{2+}, Ca^{2+}, Mn^{2+}, Co^{2+}, Cd^{2+}, Zn^{2+}, Pb^{2+}, Cu^{2+}	10
20	≡−O−Si−(CH$_2$)$_3$OCH$_2$CHCH$_2$NHCH(CH$_2$)$_3$NH$_2$ with COOH and OH	Chromolith, 12 nm, 300 m^2 g^{-1}	100 × 4.6 mm, monolith	Mn^{2+}, Co^{2+}, Cd^{2+}, Zn^{2+}	95
21	≡−O−Si−(CH$_2$)$_3$−P(=O)(OH)(O$^-$)	Grade 710 silica, 6 nm, 500 m^2 g^{-1}	150 × 3.0 mm, 9.5–11.0 μm	Lanthanides	96
22	≡−O−Si−(CH$_2$)$_3$NHCH$_2$−P(=O)(OH)(O$^-$)	Silasorb, 6 nm, 500 m^2 g^{-1}; 0.1 mmol H$^+$ g^{-1}	50 × 4.6 mm, 5 μm	Ni^{2+}, Zn^{2+}, Cd^{2+}, Mn^{2+}, Al^{3+}, Be^{2+}, La^{3+}, Lu^{3+}	8
				Ni^{2+}, Co^{2+}, Zn^{2+}, Cu^{2+}, Fe^{3+}, Cd^{2+}, Pb^{2+}	6
			250 × 4.6 mm, 5 μm	Ba^{2+}, Sr^{2+}, Ca^{2+}, Mg^{2+}, Ni^{2+}, Co^{2+}, Zn^{2+}, Cu^{2+}, Pb^{2+}, Cd^{2+}, Mn^{2+}	7

Chelating Stationary Phases

No.	Structure	Support	Dimensions	Ions	Ref.
23	—O—Si—(CH$_2$)$_3$NH$^+$ with two CH$_2$-P(=O)(OH)(O$^-$) groups	SG60, 300 m^2 g^{-1}, 0.56 mmol H$^+$ g^{-1}	64 × 2.0 mm, 40–100 μm	Cu^{2+}	97
24	—O—Si—(phenyl)—N=N—(thiazole)—N=N—(phenyl-OH, OH)	Silasorb Si600, 6 nm, 570 m^2 g^{-1}, 0.03 mmol Cu^{2+} g^{-1}	250 × 4.6 mm, 10 μm	Mn^{2+}, Cd^{2+}, Pb^{2+}, Zn^{2+}, Co^{2+}, Cu^{2+}	98, 99
25	—O—Si—(phenyl)—N=N—(thiazole)—N=N—(phenyl-OH)—N(C$_2$H$_5$)$_2$	Silasorb Si600, 6 nm, 570 m^2 g^{-1}	250 × 4.6 mm, 10 μm	Mn^{2+}, Cd^{2+}, Pb^{2+}, Zn^{2+}, Co^{2+}, Cu^{2+}	98, 99
26	—O—Si—(CH$_2$)$_3$NHC(=O)—(phenyl)—N=N—(phenyl)—N=N—C(=S)—NH—NH—(phenyl)	Adsorbosil-LC, 7 nm, 480 m^2 g^{-1}, 0.035 mmol Cu^{2+} g^{-1}	250 × 4.0 mm, 10 μm	Mn^{2+}, Cd^{2+}, Zn^{2+}, Cu^{2+}, Fe^{2+}, Pb^{2+}	16
27	—O—Si(CH$_2$)$_3$NHCH$_2$CH$_2$NHC(=S)—S$^-$	Baker silica gel 7, 6 nm	TLC plates, 5–40 μm	Fe^{3+}, Cu^{2+}, Ni^{2+}, Zn^{2+}	53
28	—O—Si(CH$_2$)$_3$NCH$_2$CH$_2$NHC(=S)—S$^-$ with C(=S)(S$^-$) branch	Baker silica gel 7, 6 nm	TLC plates, 5–40 μm	Fe^{3+}, Cu^{2+}, Ni^{2+}, Zn^{2+}	53

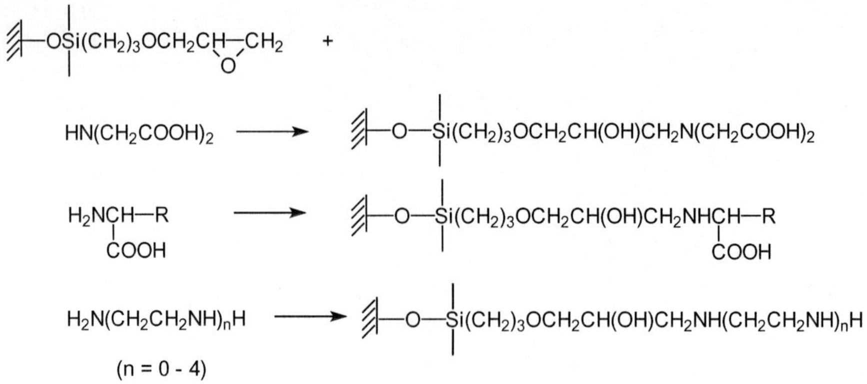

≡−OSi(CH₂)₃OCH₂CH—CH₂ +
            _O_/

HN(CH₂COOH)₂ ⟶ ≡−O−Si(CH₂)₃OCH₂CH(OH)CH₂N(CH₂COOH)₂

H₂NCH—R ⟶ ≡−O−Si(CH₂)₃OCH₂CH(OH)CH₂NHCH—R
   | |
  COOH   COOH

H₂N(CH₂CH₂NH)ₙH ⟶ ≡−O−Si(CH₂)₃OCH₂CH(OH)CH₂NH(CH₂CH₂NH)ₙH
(n = 0 - 4)

Scheme 3.5 Immobilisation of ligands onto silica using oxirane chemistry.

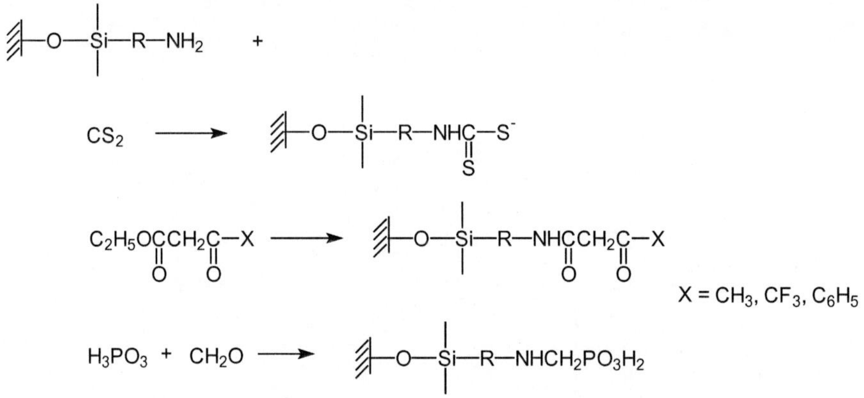

≡−O−Si—R—NH₂ +

CS₂ ⟶ ≡−O−Si—R—NHC—S⁻
                        ‖
                        S

C₂H₅OCCH₂C—X ⟶ ≡−O−Si—R—NHCCH₂C—X
     ‖   ‖   ‖   ‖
     O   O   O   O
                                                    X = CH₃, CF₃, C₆H₅

H₃PO₃ + CH₂O ⟶ ≡−O−Si—R—NHCH₂PO₃H₂

Scheme 3.6 Reactions of silica bound amino groups with carbon disulfide, acetoacetates, formaldehyde and hypophosphoric acid resulting in dithiocarbamate, β-diketone and aminophosphonic acid functional groups.

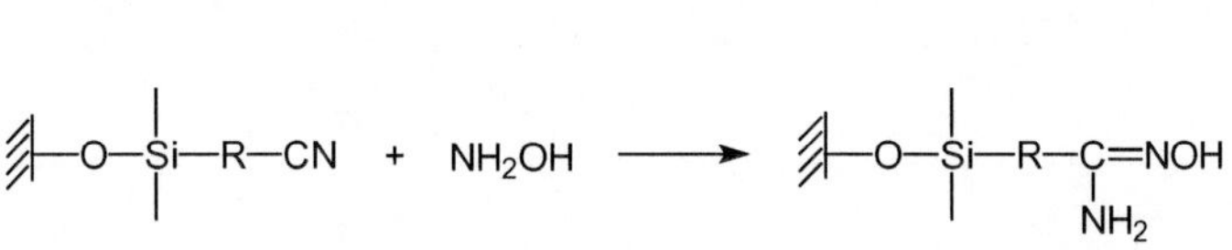

≡−O−Si—R—CN + NH₂OH ⟶ ≡−O−Si—R—C=NOH
                                                     |
                                                    NH₂

Scheme 3.7 Conversion of cyanoalkyl bonded silica to an alkylamidoxime modified phase.

The preparation of a large number of silica based adsorbents modified with chelating groups has been reported[57,58] and utilised for the preconcentration of trace metals. The hydrolytic stability of these chemically modified silicas has been often mistakenly considered as very poor and their use restricted to a very narrow pH range from 1 to 7. However, this is incorrect for the majority of silica based stationary phases used in liquid chromatography and in cation exchange chromatography. The hydrolytic stability of the covalently bonded layer is defined by the stability of the siloxane ($\equiv$Si–O–SiR) bonds, formed between the silanol groups at the surface and the silane anchor group. In turn, the hydrolytic stability of this bond is connected with the dissociation of residual silanols, and with the type and conformational availability of the functional group from the bonded molecule/layer and its interaction with the siloxane bond. Hydrolytic stability has been carefully studied for 3-aminopropylsilica,[59] where hydrolysis has been observed even at pH 5.7. In this case, the protonated amino group was electrostatically attracted to dissociated residual silanols at the silica surface, with the formation of hydrogen bonds, which catalysed the cleavage of the siloxane bond, with the subsequent release of the bonded ligand (see Figure 3.8). This effect is even more dramatic when the chemically modified silica contains primary or secondary amino groups in the γ-position, or positioned within a flexible organic linker group, between the functional group and the surface.

Bare silica exhibits weak cation exchange properties, due to the presence of approximately 4.6–4.8 silanol groups per nm^2, with pK_a values of ~ 6.8.[60] Due to steric hindrance at the surface, only approximately 50% of these surface groups can react with silanes, so the majority of silica based bonded phases contain at least 2.0–2.2 residual silanol groups per nm^2.[61] According to studies using capillary electrophoresis, the appearance of electroosmotic flow within fused silica capillaries, which is associated with the dissociation of silanols, begins at pH 2.0–2.2 and increases, obtaining a stable value at pH $>$ 9.0.[62] For 3-aminopropylsilica, this means that below pH 2, both silanols at the silica surface and amino groups within the bonded layer are protonated, such that both electrostatic interactions and hydrogen bonding (Figure 3.8) are minimal. Therefore, the siloxane bond is actually reasonably stable at pH$<$2. However, some small decreases in column capacity due to partial hydrolysis of bonded layer has been reported for some chelating ion exchangers prepared from 3-aminopropylsilica,[8] when using 0.5–1.0 M nitric acid containing mobile phases. Alternatively, very good stability of IDA bonded silica phases (adsorbent 18 in Table 3.2), under very acidic conditions, has been noted in a number of publications.[63,64] Therefore, taking into consideration the necessity of using acidic mobile phase conditions (pH$<$3) in HPCIC, to avoid hydrolysis of the actual

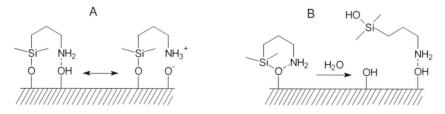

Figure 3.8 Interactions between amino groups from within the bonded layer and silica surface silanols. (A) Hydrogen bond coordination. (B) Cleavage of the siloxane bond in aminopropylsilica.

metal ions during their separation, it can be concluded that silica based chelating ion exchangers do meet the requirements of hydrolytic stability, and their limited stability at higher pH values, >7, does not constitute a significant problem. Perhaps the only disadvantage of silica gel based adsorbents worth specifically mentioning here is a high affinity for fluoride, which interacts strongly with silica, and so special care should be taken when analysing fluoride-containing samples.

The variety of silica based chelating ion exchangers developed for, and used within HPCIC, are detailed within Table 3.2. Adsorbents with negatively charged or ampholitic functional groups, such as IDA or aminophosphonic acid (APA) have tended to produce better selectivity and efficiency from the studies reported. Some silica based chelating or complexing phases have now been released commercially, and these are considered separately within Section 3.1.3.

3.3.4 Commercially Available Chelating Ion Exchange Phases

The development of cation exchangers for the simultaneous high-performance ion exchange chromatography of alkali and alkaline-earth metal cations has resulted in the widespread application of carboxylic acid type cation exchangers, which, in most cases exhibit some degree of complexing capability (Table 3.3). These complexing properties are even more developed in cation exchangers with a grafted polymer chain of carboxylic acid groups. Examples include poly(butadiene-maleic acid) (PBDMA) or poly(itaconic acid) functionalised resins, as shown in Figure 3.9.

In the example of grafted poly(itaconic acid), structural analogues of butanedioic or succinic acid can be identified if the carboxyl linking via carbon atoms 1 and 2 is considered, or pentanedioic or glutaric acid, with the linkage through carbon atoms 1, 3 and 4. In the case of bonded PBDMA, structural fragments of succinic acid can be easily identified.

The number of chromatographic columns specifically designed for separation of metals by HPCIC is very limited. BioChemMack (Russia) and JPP Chromatography (UK) produce silica based phases with IDA and amidoxime functional groups. Interestingly, various complexing groups, including crown ethers or phosphonic acid groups have also been used to improve the ion exchange selectivity of some commercial ion exchangers for IC of alkali and alkaline-earth metal cations (Table 3.4). Examples include Dionex IonPac CS12A and IonPac CS15 columns. Pohl et al.[100] used the introduction of phosphonic acid functional groups into a carboxylic type cation exchanger to improve the separation selectivity of Mn^{2+} and alkaline-earth metal cations. The introduction of crown-ether groups was used by the same authors to increase the retention of the K^+ and thus improve the separation selectivity for the K^+/NH_4^+ peak pair.

However, many commercially available chelating columns have been developed more for the separation of organic molecules by other liquid chromatographic techniques, including ligand-exchange liquid chromatography,

Table 3.3 Chemical and physical properties of commercially available weak acid cation exchangers for the IC of metal ions.

Cation exchanger	Functional groups	D_p (μm)	S ($m^2\,g^{-1}$)	d_{pore} (nm)	Ion exchange capacity ($\mu eq\,g^{-1}$)
Silica based					
Shodex IC YF-421/YK-421	PBDMA	5		12.5/2	—/1800
LiChrosil IC CA	PBDMA	5			20–200
Universal Cation	PBDMA	7			20–200
Universal Cation HR	PBDMA	3			20–200
Waters IC-Pak C M/D	PBDMA	5			60^a (3.9×150)
Super Sep C 1-2	PBDMA	5	350	10	
Metrosep Cation 1-2 IC	PBDMA	7	350	10	122^a (4.0×125)
Metrosep C 1	-COOH	5			123^a (4.6×125)
Metrosep C 2	-COOH	7			117^a (4.0×150)
Metrosep C 4	-COOH	5			
Deltabond UCX	PBDMA	5	200	12	
Waters IC Pak C M/D	PBDMA	5			60
Nucleosil-5- 100-PBDMA	PBDMA	5	350	10	—
IonPac SCS 1	PBDMA	4.5	300	12	318^a (4.0×250)
Methacrylate based					
TSKgel Super IC-Cation	-COOH	5			1.0^b
TSKgel IC-Cation I/II HR	-COOH	5			
TSKgel Super IC-A/C	-COOH	3–4			0.2^b
TSKgel OApak	-COOH	5			0.1^b

Table 3.3 (*Continued*).

Cation exchanger	Functional groups	D_p (μm)	S ($m^2\,g^{-1}$)	d_{pore} (nm)	Ion exchange capacity ($\mu eq\,g^{-1}$)
PS-DVB or PS-EVB based					
IonPaC CS12	acrylic –COOH	8.5	450	6	2800^a (4.0 × 250)
IonPaC CS14	acrylic –COOH	8	450		1300^a (4.0 × 250)
IonPaC CS16	acrylic –COOH	5	450		8400^a (5.0 × 250)
IonPaC CS17	Maleic –COOH, EVB-DVB, 55%	7	450	15	1450^a (4.0 × 250)
IonPaC CS18	Maleic –COOH, EVB-DVB, 55%	6	450		1450^a (4.0 × 250)
Poly(vinyl alcohol) based					
Shodex IC YS-50 6D	–COOH	5			3000
Metrosep C3	–COOH	5			

[a] Ion exchange capacity per column.
[b] Ion exchange capacity in meq mL^{-1}.

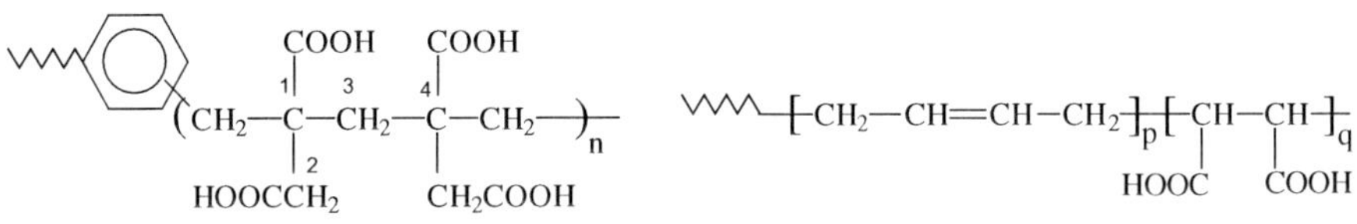

Figure 3.9 Structure of bonded polymer layers of some weak acid cation exchangers used in IC, which exhibit complexing selectivity.

immobilised metal affinity chromatography (IMAC), chromatofocusing of peptides and proteins, hydrophilic interaction liquid chromatography (HILIC) and chiral phase separations. In these cases, the preparation of stationary phases include immobilisation of the same chelating reagents, such as IDA, 8-hydroxyquinoline, oligoethyleneimines and poly(aspartic acid) (Table 3.4) as in HPCIC. In fact some of them have indeed been successfully applied to the separation of metal ions using a developed HPCIC method as discussed elsewhere.

3.4 Stationary Phase Matrix Effects

The immobilisation of chelating ligands upon the surface of a suitable supporting substrate or matrix material often results in a heterogeneous system, the chromatographic performance and selectivity of which is sensitive to the physico-chemical properties of the supporting matrix and resultant modified phase, including its porous structure (pore size, pore volume, surface area), the homogeneity of the distribution of the attached ligands at the surface (including the type of coverage, and the density of bonded groups), the matrix hydrophobicity or hydrophilicity (including residual charged sites, such as unreacted silanols in chemically modified silica) and the mobility of the attached ligand, particularly in surface grafted polymer phases.

The immobilisation of chelating ligands can have a significant effect upon their complexing ability and selectivity. These effects may result from one or more of the following reasons:

1. A change in the acid–base properties of ionogenic groups within the immobilised chelating ligand resulting from the close proximity of polar groups within the supporting matrix, or from the bonded charged ligands themselves, interacting with each other, in the case of phases displaying a dense surface coverage
2. A change in the complexing ability of monodentate ligands due to the pseudo-chelation effect resulting from the possible formation of cyclic structures, involving several monodentate ligands, immobilised in close proximity and displaying some degree of flexibility (see Figure 3.10)

Table 3.4 Commercially available chelating ion exchangers suitable for the high-performance separation of metal ions.

No	Structure of bonded groups	Matrix (type, pore size, surface area, capacity)	Column and particle size	Metal separation	Reference
1	Iminodiacetic acid Si100 Polyol, Serva, Germany	Silica, 10 nm, 300 m² g⁻¹, 0.8 mmol Cu²⁺ g⁻¹	250 × 4.6 mm, 5 μm	Mg^{2+}, Fe^{2+}, Co^{2+}, Cd^{2+}, Zn^{2+}, Pb^{2+}, UO_2^{2+}, Cu^{2+}	101
2	Diasorb IDA-130, BioChemMak, Russia	Silica, 10 nm, 300 m² g⁻¹, 0.13 mmol H⁺ g⁻¹	250 × 4.0 mm, 6 μm	Na^+, K^+, Mg^{2+}, Ca^{2+}, Ba^{2+}, Sr^{2+}, Be^{2+}, Mn^{2+}, Fe^{2+}, Co^{2+}, Cd^{2+}, Zn^{2+}, Ni^{2+}, UO_2^{2+}, Pb^{2+}, Cu^{2+}, REE, In^{3+}, Th^{4+}	64,73,81–89,102–104
3	IDA-silica, JPP Chromatography, UK	Silica, 10 nm, 360 m² g⁻¹	50, 100, 150 × 4.0 mm, 5 μm	Na^+, K^+, Mg^{2+}, Ca^{2+}, Sr^{2+}, Ba^{2+}, Mn^{2+}, Fe^{2+}, Fe^{3+}, Co^{2+}, Cd^{2+}, Zn^{2+}, Pb^{2+}, Cu^{2+}, Al^{3+}, REE	9,63,91,105
4	ProPac IMAC-10, Dionex, USA	Nonporous PS-DVB beads with grafted poly(IDA) layer, 0.04 mmol Cu²⁺ g⁻¹	250 × 4.0 mm, 10 μm	Fe^{2+}, Mn^{2+}, Zn^{2+}, Co^{2+}, Cd^{2+}, Pb^{2+}, Al^{3+}	42
5	IonPac CS12A/CG12A mixed phosphonate–carboxylic acid groups, Dionex, USA	PEVB-DVB, 15 nm, 450 m² g⁻¹, 2.8 meq per column	250 × 4.0 mm, 8.5 μm	Fe^{2+}, Mn^{2+}, Zn^{2+}, Co^{2+}, Ni^{2+}, Cd^{2+}, Cu^{2+}, Pb^{2+}, REE	106,107
6	IonPaC CS15 mixed phosphonate–carboxylic acid–crown ether groups, Dionex, USA	PEVB-DVB, 15 nm, 450 m² g⁻¹, 2.8 meq per column	250 × 4.0 mm, 8.5 μm	Li, Na, K, Mg^{2+}, Ca^{2+}, Sr^{2+}, Ba^{2+}	108
7	PolyCat A, poly(aspartic)acid functionalised silica, PolyLC, USA	Silica grafted polymer, 6 or 30 nm	200 × 4.6 mm, 5 μm	Mg^{2+}, Ca^{2+}, Sr^{2+}, Ba^{2+}	109-110
8	PRP-X800, itaconic acid functionalised macro-porous polymer, Hamilton, USA	PS-DVB, 3.7 meq g⁻¹	150 × 4.0 mm, 7 μm	Mg^{2+}, Ca^{2+}, Fe^{3+}, Mn^{2+}, Cd^{2+}, Zn^{2+}, Co^{2+}, Pb^{2+}, Cu^{2+}	111,112

Chelating Stationary Phases

#	Column	Support	Dimensions	Analytes	Ref.
9	HRLC MA7C, BioRad Lab., USA www(CH$_2$CH$_2$NH)$_n$-(CH$_2$CH$_2$N)$_m$ $\mid$ CH$_2$CH$_2$COOH	Non-porous polymethacrylate	50 × 7.8 mm, 7 µm	Mg^{2+}, Zn^{2+}, Cd^{2+}, Cr^{3+}, Al^{3+}, Pb^{2+}	113
10	Universal Cation column, poly(butadiene–maleic acid) coated silica, Alltech, USA	Silica	100 × 4.6 mm, 5 µm	Mg^{2+}, Ca^{2+}, Mn^{2+}, Cd^{2+}, Pb^{2+}, Zn^{2+}, Co^{2+}, Ni^{2+}, Cu^{2+}	114
11	TSK-Gel Chelate 5 PW, IDA functionalised macro-porous hydrophilic polymer, Tosoh, Japan	Macro-porous hydrophilic polymer, 100 nm, 24 µmol Cu^{2+} mL^{-1}	75 × 7.5 mm, 10 µm	Mg^{2+}, Ca^{2+}, Sr^{2+}, Ba^{2+}, Mn^{2+}, Co^{2+}, Cd^{2+}, Zn^{2+}, Ni^{2+}, UO$_2^{2+}$, Pb^{2+}	83, 115
12	PBE-94, polysaccharide with bonded olygoethylenimines, Pharmacia, Sweden	Polysaccharide, 32 µmol mL^{-1} Ph unit	300 × 10.0 mm, 65–132 µm	Cd^{2+}, Fe^{2+}, Ni^{2+}, Cu^{2+}, Co^{2+}	116
13	8-Hydroxyquinoline Sil100 Polyol, Serva, Germany	Silica, 10 nm, 300 m^2 g^{-1}	250 × 4.6 mm, 5 µm	No separations reported	
14	Diasorb Amidoxim, BioChemMak, Russia	Silica, 6 nm, 570 m^2 g^{-1}, 0.56 mmol H$^+$ g^{-1}	250 × 4.0 mm, 7 µm	Cr^{3+}, W(vi), Fe^{3+}, UO$_2^{2+}$, Ca^{2+}, Cd^{2+}, Mn^{2+}, Pb^{2+}, Ni^{2+}, Mo(vi), Zn^{2+}, Co^{2+}, Ni^{2+}, Hg^{2+}, Cu^{2+}	3,4,72,73
15	Spheron Oxine 1000, Lachema, Czech Republic	Spheron Oxine 1000, HEMA, 37–50 nm, 0.53 nmol Pd^{2+} g^{-1}	150 × 5.0 mm, 40–63 µm	Pd^{2+}, Cu^{2+}	117
16	CIM IDA Disk, Bia Separations, Slovenija	Poly(glycidyl methacrylate-co-ethylene dimethacrylate-co-ethyl methacrylate), 23±3 µmol Cu^{2+} mL^{-1}	3 × 12.0 mm, monolithic	No separations reported	

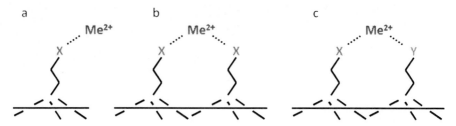

Figure 3.10 Coverage density effects upon complex stoichiometry and surface complex formation for immobilised monodentate ligands. (a) Isolated attached ligand X, or (b) two similar type X ligands in close proximity, or (c) two different ligands X and Y attached at the surface in close proximity.

3. Environments resulting in reduced stoichiometry of the surface complexes, as compared with complexes formed within homogeneous systems under similar conditions

3.4.1 Surface Distribution of Covalently Bound Ligands

The chromatographic properties of all chelating stationary phases strongly depend on the bonding density and distribution of the immobilised functional groups. A three-dimensional distribution of functional groups is generally obtained following condensation type synthesis of ion exchange resins, or within co-polymers, where one of the monomers has a target functional group within the molecule. Early chelating resins were produced containing a three dimensional distribution of the chelating ligands throughout the whole volume of the substrate particle. Although such materials possess high loading and exchange capacities, for reasons such as long diffusion pathways, restricted and varying conformational mobility, leading to the formation of metal complexes of differing stochiometry, such resins are for the most part unacceptable for high-performance chromatographic separations, as excessive band broadening is observed.

Surface coated or chemically modified porous and non-porous adsorbents, displaying a thin bonded layer, present a two-dimensional distribution of the available functional groups. The probability of the simultaneous interaction of a metal cation with two bonded ligands is significantly reduced in such a structure, although still possible in the case of high capacity adsorbents, or within some area of the surface with a particularly high density of bonded groups. According to a proposed description,[118] three models of surface coverage with a functional modifier can be envisaged (see Figure 3.11): (i) phases with statistically random coverage, where attachment of the modifier molecules is independent of each other; (ii) phases with 'island like' or cooperative coverage, where the first attached molecule promotes the attachment of the next molecule in close proximity, so-called ligand-to-ligand attraction mode; and

(iii) phases displaying a uniform coverage, where the attachment of any new molecule takes place at some uniform and independent distance from the previously attached molecule, or the so-called ligand-to-ligand repulsion mode. Clearly, the latter type of coverage, which provides a more homogeneous distribution of chelating ligands on the surface also results in a more uniform stoichiometry of resultant metal complexes at the surface, theoretically producing better column efficiency when applied within HPCIC.

However, the characterisation of the type of distribution of immobilised ligands upon the stationary phase surface is not a trivial task. Theoretically, a more homogeneous distribution should result from the immobilisation of charged chelating ligands, due to electrostatic repulsion during immobilisation. Random or island-like coverage is more likely for chelating ligands exhibiting possible hydrogen bonding or hydrophobic based interactions.[61] However, in all cases, it should be noted that differences in chromatographic performance between chelating phases displaying these three types of ligand distribution, would be more significant, and more readily identified, for phases of relatively low overall capacity (Figure 3.11).

The density and type of the distribution of attached functional groups upon the stationary phase surface can play an important role in the effective chelating capacity and observed selectivity of the prepared adsorbents. Clearly, in the case of chelation ion exchange, the evaluation of phase capacity is not simply a case of determining exchange equivalents, as the effects of surface complex stoichiometry will be reflected in the complex stability and therefore in metal ion retention. Effective capacity, measured for a specific chelating phase and metal ion pair, under controlled pH conditions, is a more useful measurement.

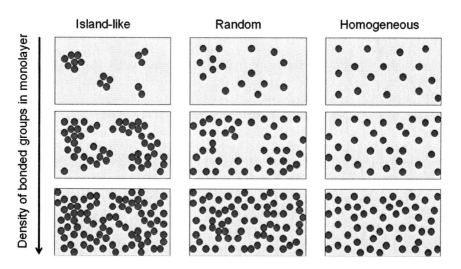

Figure 3.11 Distribution models for ligands immobilised at the phase surface, displaying differing bonding density and surface coverage. Adapted from Lisichkin et al.[61]

A good illustration is primary amino or carboxylic functionalities immobilised upon the surface via short methylene chains of limited flexibility (Figure 3.10). The 'chelate effect' is the gain in stability of metal complexes formed from one polydentate ligand, in relation to complexes formed with two monodentate ligands, coordinated to the same central metal ion. On a stationary phase surface, depending upon the distance between attached groups, a similar stabilising effect can result, with the stationary phase substrate effectively acting as a rigid spacer between the partially mobile chelating functional groups. In the case of two similar ligands, (**X–X**, Figure 3.10) attached in close proximity, it could form the equivalent structure to a diamino- or dicarboxylic acid functionality. In the case of a mixed amino and carboxylic acid functionalised surface, the formation of mixed (**X–Y**) ligand complexes, could be envisaged as directly analogous to metal binding to free amino acids.

Jezorek et al.[13,14] compared the coverage of various silica substrates possessing different porous structures, with 3-aminopropyl ligands, for their ion exchange capacity for Cu^{2+}. The researchers found that the actual bound ligand concentration upon the silica, calculated as $\mu mol\,m^{-2}$, was approximately twice the value measured for the adsorption of Cu^{2+} (effective capacity), which clearly suggested coordination of each metal ion with two aminopropyl moieties at the surface. These observations would suggest a random type of coverage of the silica surface with the 3-aminopropyltriethoxysilane modifier. In this example, even for the relatively small molecular weight of the bonded ligands, adjacent amino- groups were still capable of interacting with metal ions on a 2:1 basis. However, in general terms, the capacity for surface complexation displaying 2:1 or greater complex stoichiometry, should increase with an increase in the conformational mobility of any linker groups between the functional group and the surface.

Busche et al.[119] covalently attached an IDA terminated organosilane to a surfactant templated nano-porous silica, forming a self-assembled monolayer of the chelating groups. When comparing the selectivity of the new chelating ion exchanger (named IDAA SAMMS) to Chelex-100, the retention of selected transition metal ions on the modified silica phase was notably higher. The authors concluded this effect may result from the formation of 2:1 ligand–metal complexes at the surface. The authors suggest that at a loading density of 0.91 silanes nm^{-2}, the average distance between the silane anchors (and hence IDA groups) is approximately 1 nm, meaning the attached IDA chains have only to achieve a 0.5 nm 'bend' to achieve such a complex, which in this case can easily be facilitated by the flexibility of the alkyl chain.

It should also be noted that the density of charged bonded ligands can also effect the acid–base dissociation–protonation constants of immobilised functional groups (in a similar way such changes can be seen in polyelectrolyte molecules), which directly impacts upon resultant conditional stability constants of the surface complexes.[120] Undoubtedly, the protonation of acidic functional groups from within the bonded layer with a decrease in pH, acts to decrease the charge density of the surface as a whole, as the attached ligands are obviously immobile, which will therefore influence the pK_a values of the remaining dissociated functional groups. This effect can be illustrated by the changes in the protonation constants of oligoethylenediamines (Table 3.5).

3.4.2 Effects of Phase and Particle Porosity

The surface of porous micro-particles displays a rather complex morphology, exhibiting both concave or convex regions, and rarely the planer surface shown in many schematic representations (Figure 3.12). This complex surface structure may also have an effect upon the properties of immobilised chelating ligands, particularly where the phase is formed from a homogenously dispersed monolayer of the bound (or indeed adsorbed) chelating ligand. These matrix curvature effects have been proposed as one potentially significant reason for observed inconsistencies in the adsorption properties of different alkylsilicas.[122] However, the situation can be even more complex in the case of chelating ion exchangers, as here, complex stoichiometry may be affected.

For micro-porous materials, limitations exist in obtaining complete surface coverage, as it is difficult to obtain the complete modification of the surface within the micro-pores themselves ($d_{pore} < 2$ nm), especially with large linker groups and bulky chelating ligands. In this case, a great deal of the surface could remain unmodified and still available for secondary interactions with

Table 3.5 Protonation constants of oligoethylenediamines.[121]

Amine	pK_1	pK_2	pK_3	pK_4	pK_5
Aminopropyl	10.7	—	—	—	—
Ethylendiamine	10.1	7.0	—	—	—
Diethylenetriamine	9.9	9.1	4.3	—	—
Triethylenetetramine	9.9	9.2	6.7	3.3	—
Tetraethylenepentamine	9.9	9.1	7.9	4.3	2.7

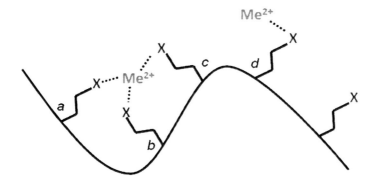

Figure 3.12 The effect of surface and pore curvature on stoichiometry of complexes formed at the surface of chelating ion exchangers with bonded monodentate ligands (X).

Table 3.6 Dependence of effective capacity of various IDA functionalised silicas of varying porous structure. Adapted from Nesterenko and Jones[102] and Bonn et al.[124]

				Effective capacity	
Silica	D_{pore} (nm)	V_{pore} (mL g^{-1})	S (m^2 g^{-1})	μeq Zn^{2+} m^{-2}	meq Zn^{2+} g^{-1}
Nucleosil 300	30	0.8	100	0.60	0.06
KSK-G	15	0.9	230	0.56	0.13
Develosil Silica 100	12	1.0	350	0.22^a	0.08^a
Nucleosil 100	10	1.0	350	0.26	0.09
Develosil Silica 60	6	0.75	500	0.19^a	0.09^a
Nucleosil 50	5	0.8	420	0.20	0.09
Develosil 30	2.8	0.5	639	0.14^a	0.10^a

^{a}Concentration of bonded groups as calculated from elemental analysis data.

metal cations. For inorganic oxides such as silica, these interactions could be quite significant, in this case resulting from reactive silanol groups electrostatically interacting with cations.[123]

In the case of phases displaying a complete homogeneously distributed layer of chelating groups, the coordination of metal ions with more than one ligand is possible, indeed probable, even within meso-pores of diameter of 2 nm and greater, due to their non-planar and curved surfaces. The length and conformational mobility of the linker between the functional group and the surface will also impact upon this complex structure.

In practice, the relationships between particle and phase porous structure, functional group density and the measured effective capacity, have yet to be clearly established, and a great deal of conflicting data has been reported. According to a comparative study of the adsorption of Zn^{2+} on various types of silica chemically modified with IDA functional groups using the same type of bonding chemistry (Table 3.6) a maximum effective capacity, expressed as μequiv Zn^{2+} m^{-2}, was achieved with a macro-porous silica, the pores within which would obviously display the smallest degree of surface curvature. At the same time, all of the prepared adsorbents had similar effective capacities, ranging between 0.06 and 0.13, when normalised to meq Zn^{2+} g^{-1}.

3.4.3 Effect of Stationary Phase Structure upon Separation Selectivity

For chelating ion exchangers possessing the same functional groups, but immobilised following various means on differing substrates, the separation selectivity exhibited for the same group of metal ions may be markedly different.[83] Figure 3.13 shows the dependence of the logarithm of retention factor (log k) of transition metal ions on the logarithm of the concentration of maleic acid (log C), on a chromatographic column packed with PS-DVB particles

Chelating Stationary Phases

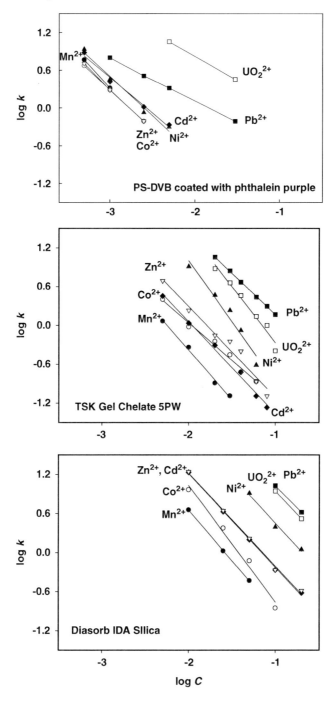

Figure 3.13 Dependence of the retention of transition metals on different IDA functionalised stationary phases on the concentration of mobile phase maleic acid. Adapted and modified from Nesterenko and Jones.[83]

impregnated with the IDA based metallochromic ligand, phthalein purple (top), a commercial TSK gel chelate 5PW column, packed with IDA functionalised poly(vinylalcohol) particles (middle), and an IDA-silica packed column (bottom). Evidently, metal ion retention, expressed in terms of log k values, depends on column effective capacity, so all of chelating ion exchangers used for this comparison had similar capacities. Maleic acid has a relatively weak complexing ability and reasonable buffering capacity in the studied pH range. As shown in Figure 3.13, the overall retention of metal ions decreased according to the following order, IDA-silica > TSK Gel Chelate 5PW > phthalein purple column, with the latter column showing a differing elution order. A poorer separation selectivity for the Zn^{2+}/Cd^{2+} pair, but improved selectivity for the Cd^{2+}/Co^{2+} pair was shown by the silica based column, as compared to the TSK Gel Chelate 5PW column. The phthalein purple column exhibited a considerably poorer selectivity overall, which can be attributed to its weaker complexing ability, as the surface adsorbed ligand will have a low conformational mobility in its adsorbed form, in comparison with other bonded phases (see Section 3.4).

A similar investigation to the above was recently conducted by Kumagai et al.,[41] who also investigated the chromatographic properties of a variety of stationary phases with different bonding chemistry of IDA functional groups. More recently still, Barron et al.[42] compared selectivity, retention and efficiency on a variety of IDA functionalised phases, both silica and polymer based. Efficiency values for transition metals on a short (2 mm × 50 mm) column packed with a poly-IDA acid functionalised polymer 10 μm resin (Dionex ProPac IMAC-10®) was compared to those typically possible with a 5 μm IDA-silica phase and an IDA-silica monolithic column. The data, although acquired using a variety of mobile phase conditions, shows that in general the silica phases were significantly more efficient, although a 5 μm version of the Dionex resin could well result in comparable efficiencies. Table 3.7 shows the comparative column efficiencies, expressed in plates per metre ($N\,m^{-1}$).

Table 3.7 Comparison of column efficiency (theoretical plates per metre) of different IDA-functionalised columns.[42]

Metal	ProPac 10-IMAC, 10 μm, 50 × 2 mm i.d.[a]	IDA-silica, 5 μm, 100 × 4.0 mm i.d.[b]	Merck Chromolith IDA-silica, 100 × 4.6 mm i.d.[93c]
Mn^{2+}	3470	48 480	16 300
Fe^{2+}	7780	—	—
Co^{2+}	4500	22 510	27 530
Cd^{2+}	10 060	43 280	9400
Zn^{2+}	10 420	37 720	16 450
Pb^{2+}	2240[d]	16 200	18 930
Al^{3+}	1200[e]	6870[f] [63]	—

Mobile phases: [a] 0.25 mM HNO_3–0.01 M KCl;
[b] 0.01 M nitric acid;
[c] 0.01 M HNO_3–0.2 M KCl;
[d] 15 mM HNO_3;
[e] 7.25 mM HNO_3–0.01 M KCl;
[f] 0.03 M HNO_3–0.5 M KCl.

3.4.4 Monolithic Chelating Ion Exchangers

Monolithic materials have gained considerable popularity in various modes of liquid chromatography, including ion exchange.[125,126] A recent review by Ščančar and Milačič gives a very good overview of the application of monolithic phases in elemental analysis and speciation.[127] A few studies have been carried out detailing the modification and application of monolithic chelating ion exchangers, most for the separation of proteins using IDA modified polymer monoliths, applying immobilised metal affinity chromatography (IMAC).[128,129] Additionally, some interesting monolithic phases with attached ligands clearly capable of metal ion coordination have also been used for other modes of liquid chromatography, for example, chiral chromatography, as reported by Preinerstorfer et al.[130]

Not as much has been reported in relation to the direct separation of metal ions on monolithic chelating ion exchangers by HPCIC. Sugrue et al.[94] covalently attached IDA groups to a PEEK lined bare monolithic silica column (Performance SI, Merck KGaA) of 10 cm length and 4.6 mm i.d., with a surface area of $300\,m^2\,g^{-1}$, and a 2 µm macro-porous and 13 nm meso-porous structure. Modification involved activation of the surface via flushing the column with distilled water (DW) under 60 °C for 4 h, followed by modification with IDA groups through recycling 80 ml of an aqueous solution of 3-glycidoxypropyltrimethoxysilane and IDA through the column at 70 °C. Finally, the column was washed with dilute nitric acid for approximately 1 h and equilibrated with the mobile phase before use. The resultant column was investigated for selectivity toward alkaline-earth metal and transition metal cations, and compared directly to an IDA-silica particle packed column. Figure 3.14 illustrates the very similar selectivity exhibited by the two IDA modified columns for alkaline-earth metals, which in this case appears almost independent of the structural and porous nature of the supporting silica substrate. The monolithic IDA was also applied to the separation of common transition and heavy metal ions, and the separation of Mn^{2+}, Cd^{2+}, Co^{2+}, Zn^{2+} and Pb^{2+} (in the order stated) was possible using a 0.4 M KCl mobile phase with pH 2.5.

However, under the above conditions, the peak shape for Zn^{2+} was extremely poor, being considerably broader than the later eluting peak for Pb^{2+}, indicating some form of specific secondary interaction with the supporting silica monolith. Following an elemental analysis investigation along the length of the column, a longitudinally heterogeneous distribution of attached IDA groups was identified.[93] Clearly this resulted in a considerable concentration of free silica surface and silanol groups along the length of the second half of the column, which could contribute to cation retention and band broadening for certain cations, through unwanted secondary interactions. This effect of the 'on-column' modification technique used also meant a reduced overall effective chelating ion exchange capacity for the column, and clearly highlights a need for improved modification procedures for the modification of pre-housed monolithic phases.

Sugrue et al. also modified 100 × 4.6 mm i.d. silica monolith column through in situ covalent attachment of lysine (2,6-diaminohexanoic acid) groups.[95] The

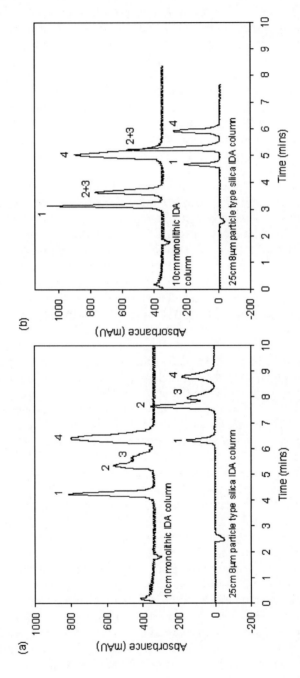

Figure 3.14 Chelation ion chromatograms showing the separation of 1, Mg^{2+}; 2, Sr^{2+}; 3, Ba^{2+}; and 4, Ca^{2+}; using a mobile phase of (a) 0.3 M KNO_3, adjusted to pH 4.2 on an IDA silica gel column and pH 4.85 on an IDA monolithic column, and (b) 0.5 M KNO_3, adjusted to pH 4.2 for the silica gel column and pH 4.85 on an IDA monolithic column. Flow rate: 1 mL min^{-1}. Photometric detection at 570 nm after PCR with phthalein purple. Reproduced, with permission, from Sugrue et al.[94] Copyright (2003) The Royal Society of Chemistry.

prepared column exhibited zwitterionic properties and could be applied to both anion and cation separations. The metal ions investigated included Mn^{2+}, Co^{2+}, Cd^{2+}, Zn^{2+} and Pb^{2+}. The separation selectivity indicated complexation between metal ions and lysine groups as the dominant retention mechanism.

3.5 Dynamically Modified and Impregnated Stationary Phases

Although a number of highly successful chemically bonded chelating stationary phases have been produced and utilised in HPCIC (see Section 3.3), the process of chemically bonding functional groups to supporting substrates can often prove difficult and rather unpredictable in terms of resultant capacity and selectivity. Often the chemistries utilised to bond the ligand to the stationary phase support can alter the selectivity of the bound ligand or provide additional exchange sites or unwanted secondary interactions. An alternative approach is to exploit impregnation of the surface and throughout the internal pore spaces of the stationary phase substrate with suitable organic chelating ligands, which can present a simpler alternative to covalent attachment and result in a coating which can remain stable in aqueous solutions over a wide range of ionic strength and pH.

Chelating ligands when immobilised onto stationary phase supports, either through semi-permanent impregnation (pre-coating) of the stationary phase particle, or through dynamic modification resulting from the ligands inclusion within the mobile phase, can maintain the ability to selectively form chelates with metal ions and thus selectively extract them or separate them chromatographically. The selectivity of these chelating substrates is based upon the conditional stability constants of the chelates being formed, and as with covalently bonded chelating phases, their extraction and elution depends predominantly on control of the pH of the mobile phase.

As with covalently bonded chelating phases, the most important parameter in obtaining efficient chromatographic separations in HPCIC when using dynamically modified or impregnated ligand phases is the rate at which the metal ligand chelates are formed and dissociated. This not only depends upon the nature of the functional group, the concentration and nature of competing ligands within the mobile phase, the mobile phase pH and column temperature, but also upon the confirmation and surface structure of the adsorbed ligand layer or layers, and the stability of these layer(s). In the case of dynamically modified phases, the chelating ligand is present in both the stationary and mobile phase, which results in a complex system of competing equilibria, precise control over which is needed to obtain desired selectivity and separations.

One of the earliest publications on the use of an impregnated chelating stationary phase, was that produced by Yamazaki et al.,[131] who modified a reversed-phase column with N-n-dodecyliminodiacetic acid and applied the resultant column to the separation of alkaline-earth metal ions. The approach provided an efficient separation of Ca^{2+}, Mg^{2+}, Sr^{2+} and Ba^{2+} within just 8 min, using a dilute tartaric acid containing mobile phase with PCR detection. The publication also provided one of the first demonstrations of the potential of HPCIC for the analysis of high ionic strength samples, with the developed method being applied to the direct determination of Ca^{2+} and Mg^{2+} in seawater (see Chapter 7).

A much more recent example using a similar approach, is that produced by Yasui et al.,[132] who reported the use of bidentate and tridentate heterocyclic azo compounds, having long alkyl chains, as coating reagents for the chelation ion exchange chromatography of Mn^{2+}, Zn^{2+} and Cd^{2+} ions. The prepared azo compounds having a long alkyl chain, were used to modify a reversed-phase stationary phase, orientating themselves in such a way that the coordinating sites of the reagents were available for metal ion complexation. Bidentate ligands showed sharp peaks, but almost no resolution of Mn^{2+} and Cd^{2+}; however, a tridentate ligand strongly retained all three metal ions, which could be separated within 10 min when using a mobile phase containing a competing complexing agent and optimised pH.

3.5.1 Impregnated (Pre-coated) Phases Using Metallochromic Ligands

Early investigations into impregnated phases based upon chelating metallochromic ligands (dyestuffs) were carried out for solid phase extraction applications, often using the impregnated substrates within mini-chelating columns in flow injection analysis systems. This was initially carried out using silica based substrates[133,134] or large particle size macro-porous PS-DVB resins,[135,136] with the resulting application generally being metal ion preconcentration or selective matrix elimination.[137] Following on from these earlier studies, the commercial availability of high-performance grade PS-DVB resins led to a number of papers being published detailing the impregnation of these phases with chelating metallochromic ligands (see Table 3.8). These high-performance chelating exchangers were then applied to both, in fact simultaneous (or perhaps sequential operation on a single column), metal ion preconcentration and high-performance separations.

Jones and Schwedt[138] were among the first to investigate this approach with the use of a column coated with chromazurol S (CAS) for the single column preconcentration and separation of divalent and trivalent metal ions. Their study showed how using a pH step gradient elution programme, a series of relatively high efficiency separations of alkaline earth and common transition metal ions was possible based upon a chelation ion exchange retention mechanism. Retention was predominantly pH dependent with ionic strength having only minor influence upon observed selectivity. Some typical early chromatograms from this early work on CAS impregnated 10 μm neutral PS-DVB resin (Benson BPI-10) are shown below within Figure 3.15. The resins

Chelating Stationary Phases

Table 3.8 Chemical structures and physical data for a range of metallochromic ligands used in HPCIC.

Name: Xylenol orange (XO)
Synonyms: m-Cresolphthalexon S, o-Cresolphthalexon S

Chemical name (IUPAC): 2-[[5-[3-[3-[[bis(carboxymethyl)amino]methyl]-4-hydroxy-5-methylphenyl]-1,1-dioxobenzo [c]oxathiol-3-yl]-2-hydroxy-3-methylphenyl]methyl-(carboxymethyl)amino]acetic acid

Molecular weight	Molecular formula				Ligating atoms		CAS registration number	
672.66	$C_{31}H_{32}N_2O_{13}S$				O,N		1611-35-4	

Ligand dissociation constants[a]

pK_{a1}	pK_{a2}	pK_{a3}	pK_{a4}	pK_{a5}	pK_{a6}	pK_{a7}	pK_{a8}	pK_{a9}
-SO$_3$H	$\equiv$OH$^+$	-COOH	-COOH	-COOH	-COOH	-OH	$\equiv$NH$^+$	$\equiv$NH$^+$
-1.74	-1.09	0.76	1.15	2.58	3.23	6.40	10.46	12.58

Name: Methylthymol blue (MTB)
Synonyms: Methylthymol blue complexone

Chemical name (IUPAC): 2-[[5-[3-[[bis(2-oxido-2-oxoethyl)amino]methyl]-4-hydroxy-2-methyl-5-propan-2-ylphenyl]-1,1-dioxobenzo[c]oxathiol-3-yl]-2-hydroxy-6-methyl-3-propan-2-ylphenyl] methyl-(2-oxido-2-oxoethyl)amino]acetate

Molecular weight	Molecular formula				Ligating atoms		CAS registration number	
844.74	$C_{37}H_{44}N_2O_{13}S$				O,N		1945-77-3	

Ligand dissociation constants[a]

pK_{a1}	pK_{a2}	pKa_3	pK_{a4}	pK_{a5}	pK_{a6}	pK_{a7}	pK_{a8}	pK_{a9}
-SO$_3$H	$\equiv$OH$^+$	-COOH	-COOH	-COOH	-COOH	-OH	$\equiv$NH$^+$	$\equiv$NH$^+$
-1.76	-1.11	-0.78	1.13	2.60	3.24	7.20	10.96	12.93

Table 3.8 (Continued).

Name:	Synonyms:	Chemical name (IUPAC):
Phthalein purple (PP)	o-Cresolphthalein complexone	2-[[5-[1-[3-[[bis(carboxymethyl)amino]methyl]-4-hydroxy-5-methylphenyl]-3-oxo-2-benzofuran-1-yl]-2-hydroxy-3-methylphenyl]methyl-(carboxymethyl)amino]acetic acid

Molecular weight	Molecular formula			Ligating atoms		CAS registration number
636.61	$C_{32}H_{32}N_2O_{12}$			O,N		2411-89-4

Ligand dissociation constants[a]

pK_{a1}	pK_{a2}	pK_{a3}	pK_{a4}	pK_{a5}	pK_{a6}	pK_{a7}	pK_{a8}	pK_{a9}
=OH$^+$	-COOH$^+$	-COOH	-COOH	-COOH	-COOH	-OH	≡NH$^+$	≡NH$^+$
			2.2	2.9	7.0	7.8	11.4	12.0

Name:	Synonyms:	Chemical name (IUPAC):
Chromazurol S (CAS)	Solochrome brilliant blue B, Alberon	5-[(E)-(3-carboxylato-5-methyl-4-oxo-1-cyclohexa-2,5-dienylidene)-(2,6-dichloro-3-sulfonatophenyl)methyl]-2-hydroxy-3-methylbenzoate

Molecular weight	Molecular formula			Ligating atoms	CAS registration number
605.29	$C_{23}H_{16}Cl_2O_9S$			O,O	1667-99-8

Ligand dissociation constants[a]

pK_{a1}	pK_{a2}	pK_{a3}	pK_{a4}	pK_{a5}
-SO$_3$H	=OH$^+$	-COOH	-COOH	-OH
−3.0	−1.2	2.25	4.88	11.75

Name: Glycine cresol red
Synonyms:

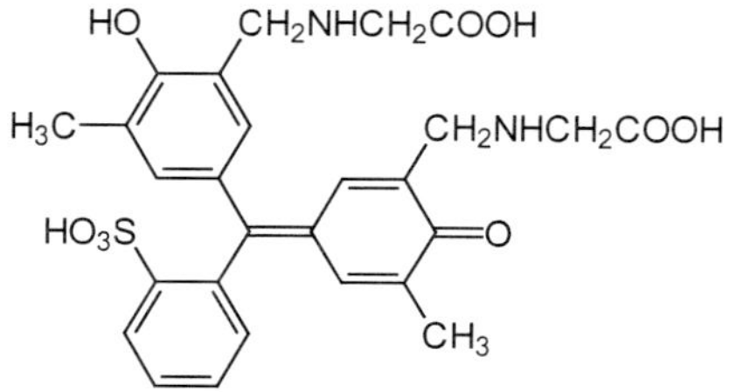

Chemical name (IUPAC):
2-[(Z)-[3-[(carboxymethylamino)methyl]-4-hydroxy-5-methylphenyl]-[3-[(carboxymethylamino)methyl]-5-methyl-4-oxocyclohexa-2,5-dien-1-ylidene]methyl]benzenesulfonate

Molecular weight	Molecular formula	Ligating atoms	CAS registration number
578.57	$C_{27}H_{28}N_2O_9S$	O,N	77031-64-2

Ligand dissociation constants[a]

Name: Aurin tricarboxylic acid (ATA)
Synonyms: Aluminon, Lysofon

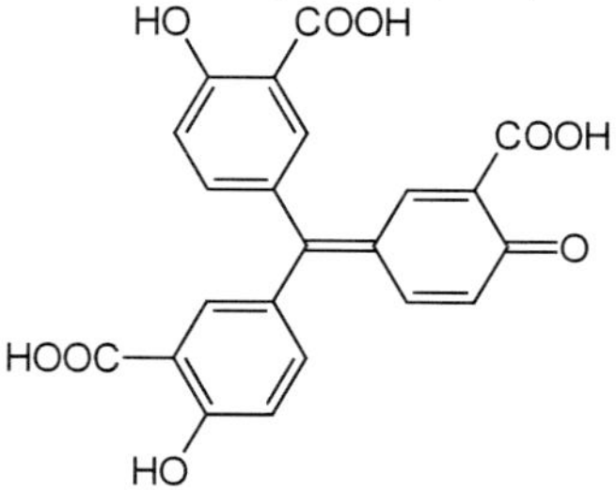

Chemical name (IUPAC):
5-[(3-carboxylato-4-hydroxyphenyl)-(3-carboxylato-4-oxocyclohexa-2,5-dien-1-yl)methyl]-2-hydroxybenzoate

Molecular weight	Molecular formula	Ligating atoms	CAS registration number
475.45	$C_{22}H_{25}N_3O_9$	O,O	569-58-4

Ligand dissociation constants[a]

Name: Calmagite (CAL)
Synonyms:

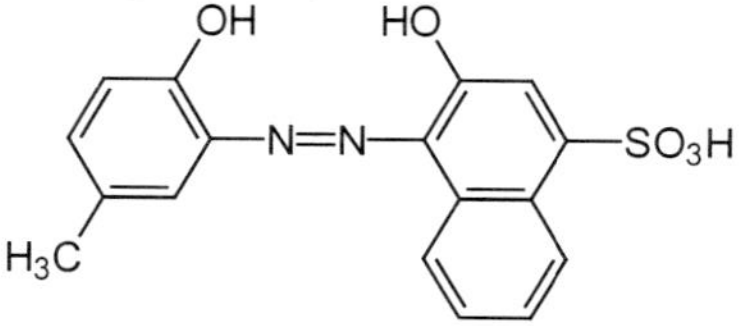

Chemical name (IUPAC):
(4Z)-4-[(2-hydroxy-5-methylphenyl)hydrazinylidene]-3-oxonaphthalene-1-sulfonic acid

Molecular weight	Molecular formula	Ligating atoms	CAS registration number
358.37	$C_{17}H_{14}N_2O_5S$	O,N	3147-14-6

Ligand dissociation constants[a]

pK_{a1}	pK_{a2}	pK_{a3}
-SO$_3$H	-OH(phenol)	-OH(naphthol)
—	7.93	12.35

Table 3.8 (*Continued*).

Name:	Synonyms:	Chemical name (IUPAC):
4-(2-pyridylazo) resorcinol (PAR)		(4E)-3-hydroxy-4-(pyridin-2-ylhydrazinylidene)cyclohexa-2,5-dien-1-one

Molecular weight	Molecular formula	Ligating atoms	CAS registration number
215.21	$C_{11}H_9N_3O_2$	O,N	1141-59-9

Ligand dissociation constants[a]

pK_{a1}	pK_{a2}	pK_{a3}
≡NH⁺	p-OH	o-OH
2.85	5.71	12.02

Name:	Synonyms:	Chemical name (IUPAC):
SPPH		2-(3-Sulfobenzoyl)-pyridine 2-pyridyl-hydrazone (SPPH)

Molecular weight	Molecular formula	Ligating atoms	CAS registration number
390.41	$C_{17}H_{14}N_4O_3S$	N,N	123333-58-4

Ligand dissociation constants[a]

[a] Data averaged from Ueno et al.[171] and Hulanicki et al.[179]. $I = 0.1$ to 1.0.

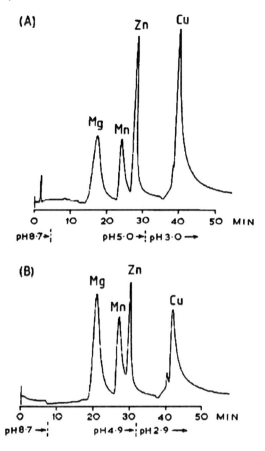

Figure 3.15 Chelation ion chromatogram of divalent metal ions on chromazurol S impregnated neutral 10 µm PS-DVB resin using pH step gradient elution. Mobile phase: 1 M KNO$_3$ containing 5 mM lactic acid adjusted to required pH. (a) = 100 µL injection of 5–20 mg L^{-1} mixed cation standard, (b) 7 mL injection of 10–20 µg L^{-1} mixed cation standard. Flow rate: 1 mL min^{-1}. Photometric detection at 610 nm after PCR with Calmagite. Reproduced, with permission, from Jones and Schwedt.[138] Copyright (1989) Elsevier.

were later used for to demonstrate how the preconcentration and separation of Al^{3+} in tap water on a single chelating column was possible.

Jones and co-workers were also responsible for much of the later work which followed the above pioneering study, going on to successfully apply HPCIC using impregnated polymer resins to the determination of trace metals in a number of sample types.[139–143] Several methods were developed by the above research group using 100 × 4.6 mm i.d. columns packed with 8 µm PS-DVB resin and pre-impregnated with either CAS or more often, xylenol orange (XO).

3.5.1.1 Xylenol Orange

Xylenol orange (XO) is a metallochromic ligand which is well known for its use as a metal ion indicator in compleximetric titrations (see Table 3.8). In addition to this, it has also been applied in a number of ways to the determination of metal ions using various modes of liquid chromatography. For example, one application of XO has been as a highly sensitive post-column reagent for the visible detection of the lanthanide metal ions, following their separation using standard IC.[144,145] In the work on HPCIC reported by Jones and co-workers, it was found, somewhat surprisingly, that the separation of metal ions, in particular the alkaline-earth metal ions, Ba^{2+}, Sr^{2+}, Mg^{2+} and Ca^{2+}, obtained using a XO-impregnated resin phase, exhibited similar efficiency to that obtained when carrying out the same separation on a covalently bonded commercially available chelating column (Tosoh TSK-GEL Chelate-5PW). Later, the same XO column was also used to demonstrate the separation of a number of transition and heavy metal ions, with Cd^{2+} and Pb^{2+} being separated at a pH of 2.5.[146,147] Further work,[142,143] as with the early CAS work above,[138] concentrated on the development of pH step gradient elution systems for the separation of a larger range of metal ions. The injection of the sample solution at high pH, under which conditions the metal ions would be totally retained on the front of the column, were followed by subsequent steps down in mobile phase pH in order to elute the metal ions, two or three at a time as sharp individual peaks, until the most strongly retained metal ions were eluted under the most acidic conditions. Using this approach, nine metal ions, namely Ba^{2+}, Sr^{2+}, Mg^{2+}, Ca^{2+}, Mn^{2+}, Cd^{2+}, Zn^{2+}, Ni^{2+} and Cu^{2+}, could be separated in under 40 min. This impressive early separation is shown in Figure 3.16.

The system was developed further again to allow the simultaneous preconcentration and separation of metal ions on a single column, through the injection of large sample volumes. The potential of such an approach was then demonstrated through the application of the XO impregnated column to the determination of trace levels of alkaline earth and transition metal ions in concentrated industrial KCl and NaCl brines, and to the separation of a number of transition metal ions spiked into a seawater matrix (see Chapter 7). This preliminary investigation into seawater analysis using HPCIC with the XO column prompted a further and more detailed study into this particular application.[140] It was found that through the preconcentration of between 10 and 30 cm^3 of sample onto the front of the XO column, followed by step gradient elution, it was possible to detect sub-µg L^{-1} concentrations of Mn^{2+}, Zn^{2+}, Pb^{2+}, Ni^{2+} and Cu^{2+} in coastal seawater samples.

3.5.1.2 Methylthymol Blue

As with the structurally similar XO, methylthymol blue (MTB) is a triphenylmethane based ligand, capable of complexing metal ions via either of two IDA groups. As might be expected, MTB pre-impregnated resins exhibit very similar selectivity to their XO counterparts, and comparable separations can be

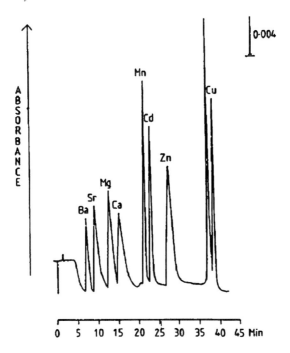

Figure 3.16 Chelation ion chromatogram of nine metals on a xylenol orange impregnated PLRP-S resin column, using pH step gradient elution. Mobile phase: 1 M KNO_3, pH adjusted using 0.05 M lactic acid. Sample: 5 mg L^{-1} Mg^{2+}, Ca^{2+}, Mn^{2+}, Ni^{2+}, Cu^{2+}, 10 mg L^{-1} Cd^{2+}, Zn^{2+}, 20 mg L^{-1} Ba^{2+} and Sr^{2+}. Flow rate: 1 mL min^{-1}. Photometric detection at 490 nm after PCR with PAR-ZnEDTA. Reproduced, with permission, from Challenger et al.[142] Copyright (1993) Elsevier.

achieved. For example, Jones et al.,[141] applied a MTB pre-impregnated PS-DVB based column (100 × 4.6 mm i.d., length, 9 μm particle size) to the separation of alkaline-earth metal ions, which could be separated in the order Ba^{2+}, Sr^{2+}, Mg^{2+} and Ca^{2+}, in just under 20 min, although the peak shape for Ca^{2+} was unusually broad. The selectivity shown for the alkaline-earth metal ions by the MTB column (the reverse of that exhibited by simple cation exchange resins) meant the column was suited to the determination of Ba^{2+} and Sr^{2+} in samples containing excess Mg^{2+} and Ca^{2+}. Two applications exploited this advantage, namely the determination of low μg L^{-1} concentrations of Ba^{2+} and Sr^{2+} in bottled mineral waters,[141] and the determination of Ba^{2+} and Sr^{2+} in offshore oil-well brines[139] (see details in Chapter 7).

Further work using a MTB pre-impregnated polymeric column, similar to the one used above, has been applied to the determination of uranium(VI) (as the uranyl ion UO_2^{2+}).[148] Uranyl was eluted from the MTB column at a retention time of $t_R = 4.5$ min with a mobile phase comprising 0.5 M KNO_3 (pH 1.4). Reducing the pH to 1.25 also allowed Cu^{2+} to be eluted from the column ($t_R = 10$ min) (see Chapter 7 for full details of applications of MTB impregnated phases).

3.5.1.3 Production and Stability of Impregnated Columns

For the majority of work to date using impregnated or coated phases for HPCIC, the approach used to produce the chelating phase has been remarkably simple. The basic experimental approach is to pump through a pre-packed column an aqueous solution (often containing a small amount (5–15%) of an appropriate organic modifier to wet the stationary phase surface) containing the chelating ligand at relatively high concentration (10–50 mM), together with a buffer to maintain the desired coating pH and some concentration of an inorganic salt to maintain a desired ionic strength. Variations include columns being impregnated in a single direction, in both directions consecutively, at high or low ionic strength and higher or lower pH. In each case exact conditions are developed to suit the particular chemistry of the ligand itself and/or stationary phase. However, it is reasonable to conclude form the available literature that no truly optimised approach has been reported, or if indeed possible given the complexity and variety of ligands available. Impregnated phases are subsequently washed with various aqueous wash solutions of varying pH to remove unstable portions of adsorbed ligand, often requiring extended wash cycles to fully eliminate ligand bleed. The resultant adsorbed chelating layers have not – so far – been investigated fully physically, with limited understanding of the nature of such a layer's structure, including whether the ligand is adsorbed as a monolayer or multilayer structure, whether such coatings are longitudinally homogeneous, or the degree to which the ligand penetration extends fully and uniformly into the pore structure of the porous resins used.

As discussed previously, XO as a chelating ligand has been extensively utilised in HPCIC. Jones and co-workers first began producing XO impregnated phases in 1992, and some experimental effort was later directed towards greater understanding of the impregnation procedures used. In early studies by Jones et al.,[143] using XO impregnated neutral PS-DVB resins (Polymer Laboratories PLRP-S resin), it was reported that stable ligand loadings of ~ 48 mg per 0.9 g of resin (10 μm, 10 nm pore size resin) could be achieved. These 100×4.6 mm i.d. XO columns were applied to the high-performance separations of transition, heavy and alkaline-earth metals, in various high ionic strength sample matrices. However, later attempts to reproduce this degree of loading on the same resin, using impregnation procedures outlined in the above work proved unsuccessful, resulting in loadings of between 17 and 28 mg, and a subsequent low capacity insufficient to produce efficient metal separations. An investigative study into impregnation procedures was carried out, focussing on pH and concentration of the ligand solution.[149] In earlier work XO was impregnated using solutions of pH 6.0 to 8.5. Reducing this to pH 5.0 caused an increase in the loadings achieved due to a reduction in the net charge on the weak acid ligand. Reducing the pH further also increased the loading, although the additional loading was unstable and bled from the column during the subsequent wash cycle. The ligand solution concentration was varied from the 0.2% used in previous work, to 0.5%. This increase in the concentration of the chelating ligand was aimed to help force increased numbers of molecules onto the surface and into the pores of the resin. The combination of adjustments to pH and concentration led to increased loadings (up to 39 mg) and enabled the production of a XO column which exhibited the same retention characteristics and separation efficiencies as the column produced by Jones et al.[138]

An interesting approach was taken by Shaw et al. to improve the ligand impregnation of polymer resins.[150] In this work modification of a 7 μm neutral PS-DVB resin (Hamilton PRP-1) with the chelating dye aurin tricarboxylic acid (ATA) was carried out in batch experiments, using ultra-sonic agitation of loose resin in a buffered solution of the ligand. Clearly the availability of the surface of the resin for adsorption of the ATA molecules is greater in such batch experiments compared with the above packed column method. Shaw et al. varied ligand concentration (0.1–0.4%), methanol concentration (0–30%) and pH (pH 3–7), finding a 2% ATA solution in 10% methanol at pH 5 provided the highest loadings. The study concluded that the use of ultra-sonication during batch impregnation experiments resulted in a four fold increase in adsorbed ATA and that loading took considerably less time (40%) using this approach. Compared to in-column modification, the batch approach also was reported to decrease washing times, although obviously subsequent column packing is required for chromatographic evaluation.

Using the above optimised batch impregnation approach Shaw et al.,[150] went on to pack a 250 × 4.6 mm i.d. column, which when used with a pH step gradient with a 0.1 M KNO_3 mobile phase, provided the separation of six transition and heavy metals, the chromatogram of which is shown in Figure 3.17.

The above short studies, although specific for individual chelating metallochromic ligands, were significant for the fact they clearly show that differing impregnation procedures and conditions, will inevitably lead to variations in column capacities and overall loadings, exhibiting variations in observed separation efficiencies between columns. It is therefore critical in order to reproduce chelating resins of similar loadings and selectivity that detailed impregnation procedures are both reported and followed in subsequent studies.

3.5.1.4 Effects of Resin Structure upon Ligand Impregnation

As with the above brief investigations into impregnation procedures, some work has been reported comparing high-performance resins for structural effects, most notably surface area and porosity, upon obtained ligand loadings. For example, XO was used in a brief study to investigate the degree of loading achieved on PS-DVB substrates of varying particle and pore size.[149] The results show a significant increase in ligand loading with the PLRP-S resins containing the smaller pore sizes of 10 nm and 30 nm, compared to the larger pore size resins (up to 400 nm). This was likely to be purely a surface area effect, although

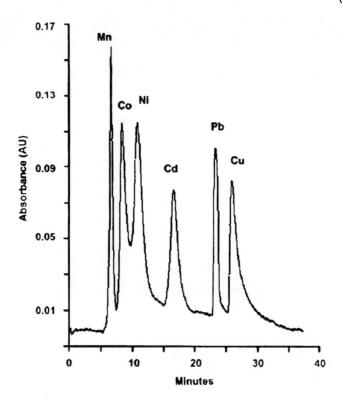

Figure 3.17 Chelation ion chromatogram showing the pH gradient separation of Mn^{2+}, Co^{2+}, Ni^{2+}, Cd^{2+}, Pb^{2+} and Cu^{2+} on a 250×4.6 mm i.d. aurin tricarboxylic acid impregnated column. Elution program: 0.1 M KNO_3, pH 4.5 for 15 min; sample injection (100 µL); pH 3.5 for 10 min; pH 1.0 final. Flow rate: 1 mL min^{-1}. Photometric detection at 520 nm after PCR with PAR. Reproduced, with permission, from Shaw et al.[150] Copyright (2008) Taylor and Francis.

it was also considered possible that the smaller pore spaces contributed to phase loading through the physical trapping of the ligand molecule within them. The comparative results obtained within this study can be seen in Table 3.9.

The study also reported how the impregnation of the Dionex neutral resins resulted in lower loadings than the above PLRP-S resins of similar pore size, but again illustrated the trend of higher loadings being achieved with smaller pore size resins.

The above study also attempted to evaluate the effective capacity of the ligand impregnated resins, based upon the sorption/extraction of Zn^{2+} pumped through the column within a aqueous buffer solution (pH 8.0). By calculating the number of Zn^{2+} moles required to fill all the theoretical chelation sites, and using the number of moles of ligand known to be impregnated within the resin, it was possible to calculate the effective capacity, or the relative amount of active sites.

Chelating Stationary Phases

Table 3.9 Loadings of xylenol orange achieved on neutral PS-DVB resins (Polymer laboratories PLRP-S and Dionex macro-porous resin) of varying particle and pore size.[149]

Pore size (nm)	Surface area ($m^2 g^{-1}$)	Particle size (μm)	Amount impregnated (mg ligand g^{-1} resin)
Polymer laboratories PLRP-S resin			
400	139	10	21
100	267	10	21
30	384	10	31
10	414	10	31
Dionex macro-porous neutral resin			
30	60	7.0	22
12	470	8.8	25

Table 3.10 Metallochromic ligand total loading and effective capacity on hydrophobic neutral 8.8 μm PS-DVB resin (Dionex).

Metallochromic ligand	Amount impregnated (μmol ligand g^{-1} resin)	Capacity (μmol Zn^{2+} g^{-1} impregnated resin)	Effective capacity (%)
XO	33	10	30
SXO	29	—	—
MTB	48	13	27
PP	157	22	14
GCR	99	7	7
CAS	35	3	9
CAL	28	5	18
PAR	396	19	5
OII	111	2	2
SPPH	36	1	3

Due to steric effects within the column a large percentage of the impregnated dye is unable to complex with the metal ions present. For similar resin substrates, this inactive proportion is thus dependent upon the type and structure of the impregnated dye. This was proven within the above study[149] as a variety of metallochromic ligands, of both triphenylmethane and azo type, were used to impregnate the hydrophobic neutral 8.8 μm PS-DVB, 12 nm pore size Dionex resin. The results of this study, together with data from similar studies are shown in Table 3.10. As can be seen a large variation in loading capacity is evident, although this should be expected with a range of structurally different chelating ligands. However, it is noticeable that with the exception of calmagite (CAL), most ligands with the highest per cent effective capacities were triphenylmethane type, namely xylenol orange (XO), methylthymol blue (MTB), phthalein purple (PP), glycine cresol red (GCR) and chromazurol S (CAS). This observation was clearly related to the presence on each ligand of two chelating sites and a relatively flexible structure, compared to the azo type ligands investigated. Of the azo based dyes, 4-(2-pyridylazo)-resorcinol (PAR) produced a particularly high loading, resulting in the second highest capacity of the columns investigated, but only 4.8% of the impregnated ligand was actively chelating.

Interestingly, when the ligand impregnated phases of such diverse capacity detailed in Table 3.10 were characterised for selectivity for a range of alkaline earth, transition, heavy divalent and trivalent metal ions, very similar trends emerged, with the general selectivity order as follows, $Ba^{2+} < Sr^{2+} < Ca^{2+} < Mg^{2+} < Mn^{2+} < Cd^{2+} < Zn^{2+} < Co^{2+} < Pb^{2+} < Ni^{2+} < Cu^{2+} < Al^{3+} < Ga^{3+} < In^{3+} < Fe^{3+} < Bi^{3+}$. The chelating phases were characterised according to the pH at which each metal ion was fully retained. Figure 3.18A shows the resultant data sets graphically for the XO, SXO, MTB, PP, GCR and CAS impregnated phases, with the data for the CAL, PAR, orange II (OII) and 2-(3-sulfobenzoyl) pyridine-2-pyridylhydrazone (SPPH) phases shown in Figure 3.18B. Clearly, the selectivity of the chelating ligands for the divalent and trivalent transition and heavy metal cations (light and dark grey circles) over the alkaline-earth metal ions (black circle) is very apparent, this of course being a fundamental advantage of chelation ion exchange over simple ion exchange, particularly in the analysis of these metals in samples containing high levels of alkaline-earth metals such as seawater. This selectivity gap is more apparent for the first class of triphenylmethane ligands (A) than for the second group of mostly azo type ligands (B), although with each phase investigated, bar PAR, it can be seen that full retention of all the divalent and trivalent transition and heavy metal ions can be obtained at least 1 pH unit before that which sees strong retention of the alkaline earth (and alkali) metals. Figure 3.18 also illustrates the large variation in pH values required for retention of the above metals, which therefore limits the number of metals which can be separated isocratically (without a pH gradient) using HPCIC.

From the selectivity data collected in this study[149] it was apparent that the metal selectivities of the various resins corresponded relatively closely to that predicted from known conditional stability constants for several of the chelating ligands investigated, indicating that the dominant retention mechanism taking place was indeed chelation with very little contribution from simple ion exchange. This was expected as any ion exchange sites within the ligand molecule, such as the sulfonic acid groups present on several of the triphenylmethane type ligands, would be 'swamped' through use of high ionic strength mobile phase, in this case, 1 M KNO_3.

Sutton et al. looked at the impregnation of a variety of macro-reticular polystyrene based resins with PAR alone, determining overall ligand loading and effective capacity, the latter here determined using mmoles of Cu^{2+} bound per gram of PAR impregnated resin.[151] Four resins were compared, namely Amberlite IRA 904, a strong basic anion exchange PS-DVB resin with ion exchange capacity 0.7 meq g^{-1}, Amberlite XAD-2, a neutral macro-porous PS-DVB resin with average pore diameter 9 nm and specific surface area $330 \text{ m}^2 \text{ g}^{-1}$, Purolite MN100, a weak basic anionic exchange hyper-crosslinked

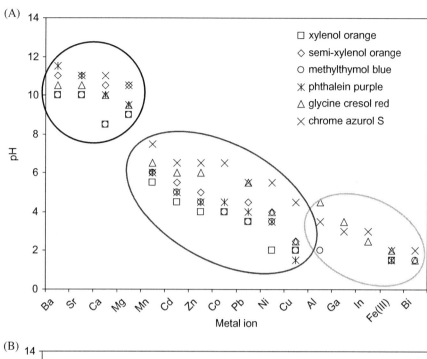
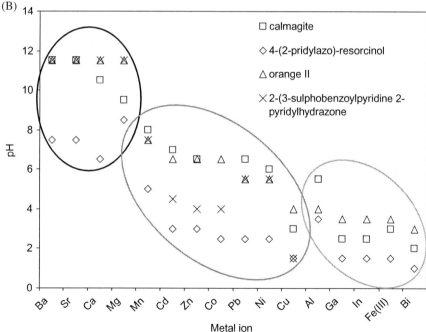

Figure 3.18 pH required for the complete retention of metal ions using hydrophobic neutral 8.8 μm PS-DVB resin impregnated with various metallochromic ligands. Data sourced from Paull and Jones.[149]

PS-DVB resin and Purolite MN200, a neutral hyper-cross-linked polystyrene resin. Sutton et al.[151] reported the lowest total loadings for the two anion exchange resins, with the strong Amberlite IRA 904 resin also exhibiting a very low (4.4%) effective capacity. The neutral XAD-2 resin exhibited the highest ligand loading, although a very poor effective capacity (7.9%). However, it was the MN200 neutral hyper-crosslinked polystyrene resin which demonstrated both a high overall ligand loading (0.2 mmol of PAR per gram of resin) and an almost quantitative effective capacity (95.3%), retaining 0.147 mmoles of Cu^{2+} per gram of resin. The high loading and effective capacity shown for the MN200 resin, results from the combination of well developed micro-porous structure and large surface area (1000–1500 $m^2 g^{-1}$) of the substrate.[152]

The selectivity exhibited by each of the above PAR impregnated polymer phases produced by Sutton et al.[151] are shown graphically for the Zn^{2+} and Mg^{2+} ions in Figure 3.19. The graph again shows the clear selectivity advantage of such chelating phases for transition metal ions over the alkaline-earth metal ions, with the black circle showing the similar selectivity for Mg^{2+} exhibited for the MN100, MN200 and XAD-2 PAR impregnated resins, and the light grey circle showing the relatively greater selectivity for Zn^{2+}. However, the graph

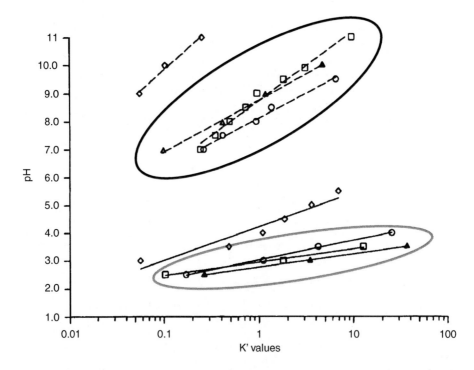

Figure 3.19 Increase in capacity (retention) factor with pH for PS-DVB resins impregnated with PAR. □, MN100; ○, MN200; ◇, IRA 904; △, XAD-2; ——, Zn^{2+}; ----, Mg^{2+}. Adapted from Sutton et al.[151]

also shows clearly the marked difference in selectivity apparent for the IRA 904 phase in comparison to these three alternative phases investigated.

Most of the above work was carried out on relatively large diameter particles compared with those available today. It would be interesting in the near future to evaluate effective capacities and efficiencies on impregnated resins of 5 μm diameter or less.

3.5.2 Dynamically Modified Phases

3.5.2.1 Theory

The dynamic modification of the stationary phase involves inclusion of the chelating ligand itself within the mobile phase during the chromatographic process, resulting in a system analogous to the well established ion-interaction chromatography, only in this case a bulky chelating ligand replaces the bulky ionic surfactant used in ion-interaction mode. However, in such a dynamic chelation ion exchange system, understanding the exact retention mechanisms and controlling parameters upon selectivity and retention are obviously quite complex, and less well documented than related chromatographic techniques.

In any chromatographic system consisting of a hydrophobic adsorbent and a buffered mobile phase, which also contains a complexing ligand, together with an organic solvent, and possibly an inorganic salt as an ionic strength regulator, a number of complex equilibria can exist,[153–155] when applied to the separation of transition metals (see Scheme 3.8). Within this series of interdependent interactions it is possible to outline the following individual equilibria:

1. *Surface adsorption of chelating ligand* in protonated (HL) and deprotonated (L) forms, as described by:

$$\theta_0 + L_m \leftrightarrow L_s \quad (3.4)$$

$$\theta_0 + (HL)_m \leftrightarrow (HL)_s \quad (3.5)$$

where θ_0 is the equilibrium free surface area of an adsorbent.

2. *Chelating ligand dissociation*, as described by:

$$HL \leftrightarrow H^+ + L^- \quad (3.6)$$

(For simplicity here only monoprotic ligands are considered.) These equilibria, dependent upon the dissociation/protonation constant for the chelating ligand, E_a^{HL}, which changes both the hydrophobicity and ionic state/charge of the ligand, can be further described using the distribution coefficients of the corresponding forms of the ligand between mobile and stationary phases, K_C^L and K_C^{HL}.

3. *Surface adsorption of organic solvent* (S), as described by:

$$\theta_0 + S_m \leftrightarrow S_s \quad (3.7)$$

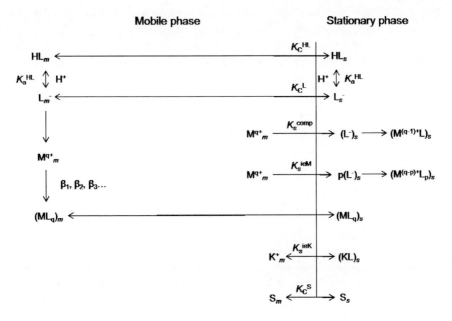

Scheme 3.8 Possible equilibria occurring in HPCIC using a dynamically modified chelating ion exchange column, with a mobile phase containing the complexing ligand and organic solvent, and a hydrophobic stationary phase.

Introduction of metal ions into the above system results in a further series of equilibria based upon the metal–ligand interactions and complex formation within both the mobile and stationary phase. The observed retention of metal ions thus results from the following contributing equilibrium processes, each to various extents.

4. *Formation of various complex forms*, exhibiting different composition and charge, formed within the mobile phase, as described by:

$$M_m^{q+} + qL_m^- \leftrightarrow (ML_q)_m \qquad (3.8)$$

and by the equilibrium constant K_m, and associated stability constants β_1, β_2, ... β_{q+j}.

5. *Adsorption of the complex* $(ML_q)_m$ from mobile phase to the stationary phase due to hydrophobic interactions, as described by:

$$\theta_0 + (ML_q)_m \leftrightarrow (ML_q)_s \qquad (3.9)$$

and by the equilibrium constant E_s^{hydroph}.

6. *Complexation upon the sorbent surface* between the uncomplexed metal ions, M_m^{q+}, and adsorbed ligand molecules in either dissociated, L_s, or non-dissociated, $(HL)_s$, forms, as described by:

$$M_m^{q+} + (HL)_s \leftrightarrow (M^{(q-1)+}L)_s + H^+ \quad (3.10a)$$

$$M_m^{q+} + (L^-)_s \leftrightarrow (M^{(q-1)+}L)_s \quad (3.10b)$$

Complexation on the surface differs from that within a mobile phase because of sterical hindrance, meaning the coordination of a single metal ion with multiple ligands adsorbed on a surface is less likely, therefore the balanced equilibria can be written down in a simplified form and described by the equilibrium constant E_s^{comp}.

7. *Possible ion exchange interactions* exist between free metal ions and adsorbed ligand molecules displaying ion exchange functional groups:

$$M_m^{q+} + p \cdot (L^-)_s \leftrightarrow (M^{(q-p)+}L_p)_s + p \cdot H^+ \quad (3.11)$$

There is possible similarity between equilibria (6) and (7) but the chemistry of the formed surface associates $(M^{(q-p)+}L_p)_s$ is very different, resulting in the formation of either a complex or salt, respectively. Similar interactions between the ligand and free metal ions, M_m^{q+} may also occur within the mobile phase. Obviously, it is impossible to develop a comprehensive metal retention model in this system taking into consideration all possible interactions. However, it is necessary to note that in the presence of a high concentration of electrolyte, for example, KNO_3, the contribution of ion exchange to the retention of specified transition metal ions can be minimised.

3.5.2.2 Dynamically Modified Phases Using Metallochromic Ligands

In each of the examples of pre-impregnated chelating ion exchangers discussed in Section 3.5.1, the mobile phase used during chromatographic application of the stationary phase, often consisted of a 0.5–1 M solution of an inorganic salt, commonly KNO_3, together with some additional acid or buffer to maintain and control pH. Provided the chelating ligand itself remained permanently or semi-permanently adsorbed onto the surface of the resin, this simple approach has been relatively successful.

However, although achieving some promising results, the pre-impregnation of the stationary phase approach does suffer from certain drawbacks. Certainly in comparison to covalently bonded high performance chelating stationary phases, it could be argued that to date the efficiencies of chelating columns produced through pre-impregnation have been relatively poor.[64,88,90,102] This translates to a maximum of five to six peaks being separating isocratically, since peak shapes tend to broaden rapidly if the metal is retained for over 5–10 min.

In addition to the efficiency limitation, the constant issue with all surface coated/impregnated stationary phases in liquid, and indeed gas chromatography, is ligand bleed over time. Although not investigated quantitatively in the studies described in Section 3.5.1.3, some loss in capacity over longer periods of use is likely due to instability of the sorbed layer(s), which may or may not be readily remediated through recoating/re-impregnation procedures.

As discussed previously, the dynamic modification of the stationary phase, via the inclusion of the chelating ligand itself within the mobile phase during the chromatographic process, results in a large number of complex and competing equilibria, originating from the presence of the chelating ligand in both the mobile and stationary phase simultaneously. Despite this complexity, stationary phases dynamically modified with a range of metallochromic ligands have been shown to exhibit similar retention characteristics to those of the above pre-impregnated counterparts, achieving similar, and in some cases superior high-performance separations of metal ions. Some identified advantages to this approach include: (i) a stabilised and increased column capacity, leading in some instances to an improvement in separation efficiency; (ii) an alternative selectivity exhibited for metal ions (through the existence of competitive equilibria in the mobile phase); and (iii) under certain circumstances direct visible detection of eluting metal–ligand complexes, negating the need for post-column reagent addition.

This latter advantage is perhaps the most significant aspect of using dynamically modified phases in HPCIC. A number of workers have previously investigated the use of mobile phases containing metallochromic ligands in standard IC, using either simple anion or cation exchangers within the separation column. For example, Zenki was the first to investigate such an approach in 1981 using a chlorosulfonazo III containing mobile phase for the determination of alkaline-earth metal ions.[156] Later, a number of colour-forming metallochromic ligands were investigated as mobile phase additives by Toei and co-workers, namely PP,[115,157,158] arsenazo III[159,160] and XO.[161] Of the above work, the majority focused on the detection of alkaline-earth metals,[115,156–160] with only XO being investigated as a colour-forming mobile phase for the detection of transition and heavy metals.[161] In this study Toei used an anion exchange separation column with a XO containing mobile phase. Injection of metal ions resulted in the on-column formation of anionic metal–XO complexes and anion exchange was said to be predominantly responsible for retention. The partial separation of Ni^{2+}, Co^{2+}, Zn^{2+} and Cu^{2+} was shown using a mobile phase of 0.1 mM XO with 0.05 M phosphate buffer (pH 5.2) at 50 °C. Further work by Dasgupta et al.[162] investigated the use of 8-HQS as a component of the mobile phase with cation exchange chromatography for the separation and fluorescent detection of metal ions. The study focussed on the alkaline-earth metals, although retention data was determined for several other divalent and trivalent metal ions. The exact retention mechanism in Dasgupta's system was unclear, with a combination of ion exchange and adsorption of the 8-HQS onto the stationary phase suggested. More recently, Wada et al.[163] used a 0.2 mM chlorophosphonazo III containing mobile phase with a sulfonated cation exchange column, once again for the determination of alkaline-earth metal ions, with visible detection at 665 nm. Whilst Bashir et al.[164] used acetate buffered mobile phases containing either 0.2 mM XO or MTB with a strong cation exchange column for the separation and visible detection at 572 nm of Zn^{2+} and Pb^{2+}.

3.5.2.3 Phthalein Purple

Phthalein purple (PP), (also known as o-cresolphthalein complexone) has been historically used as a photometric reagent for the determination of Ca^{2+}, and to a lesser extent Mg(II), in a wide variety of samples.[165-167] Jones et al. used PP to pre-impregnate neutral polymeric reversed-phase substrates for the separation of alkaline-earth metal ions[141] and selected transition metal ions.[149] An alkaline mobile phase pH of 8–10 was required for the separation of the alkaline-earth metal ions, which eluted in the order of their increasing stability constants with the ligand, retention = $Ba^{2+} < Sr^{2+} < Ca^{2+} < Mg^{2+}$. A separation of the transition metal ions Mn^{2+}, Cd^{2+}, Zn^{2+} and Pb^{2+} was possible at an acidic pH of 3.7.

The first dynamically modified HPCIC method using PP was based on the use of a porous graphitic carbon (PGC) column and a water/methanol mobile phase containing 0.4 mM PP, buffered to pH 10.[168,169] The strongly hydrophobic nature of the PGC resulted in a high dynamic capacity causing excessive retention for Mg^{2+} and Ca^{2+}, requiring up to 60% methanol to reduce the dynamic loading of PP and thus reduce retention to under 8 min., with direct detection of the eluting PP–metal complexes at 575 nm. Detection limits of 50 µg L^{-1} for Mg^{2+} and 100 µg L^{-1} for Ca^{2+} in samples containing over 2.0 g L^{-1} Na$^+$ were reported (see Chapter 7).

When the PGC column was replaced with a polymeric PS-DVB column in a later study[170] the column exhibited a much lower loading, so that less methanol (25%) was required in the mobile phase for the timely elution of Mg^{2+} and Ca^{2+}. However, although the elution order obtained for Mg^{2+} and Ca^{2+} using a dynamic system was the same on both PGC and polymer columns, this order was the reverse of that obtained when using a pre-coated polymeric PP column. Figure 3.20 shows this effect, where Ca^{2+} is seen to elute before Mg^{2+} with a pre-impregnated PP column and after Mg^{2+} with the dynamic system. This effect is a characteristic effect of dynamic HPCIC as the ligand present in the mobile phase of the dynamic system is also acting as a complexing agent, thereby reducing the retention of those metal ions with which it formed the strongest complexes, here Mg^{2+}, (log K_{MHL} = 8.9), relative to Ca^{2+} (log K_{MHL} = 7.8).[171]

The retention of Ba^{2+} and Sr^{2+} was also considerably less on the polymeric column than had been observed with the PGC column, with Ba^{2+} and Sr^{2+} eluting before Mg^{2+} and Ca^{2+} without the addition of methanol. Under the conditions required for the separation Ba^{2+} and Sr^{2+}, two other alkaline-earth metals were retained completely. Application of a pH gradient could be used for the determination of Ba^{2+} and Sr^{2+} in samples containing (excess) Ca^{2+} and Mg^{2+}. This was illustrated by the determination of less than 1 mM Sr^{2+} in a saline lake sample containing over 3 M NaCl and 0.5 M Mg^{2+} (see Chapter 7).

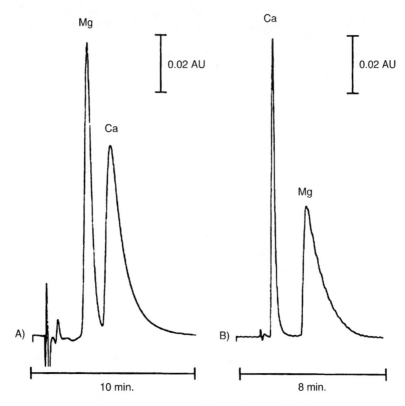

Figure 3.20 Chelation ion chromatograms of Mg^{2+} and Ca^{2+} standards obtained using phthalein purple (A) dynamically modified and (B) pre-impregnated PS-DVB column. Flow rate: $1\,mL\,min^{-1}$. Photometric detection at 575 nm. Reproduced, with permission, from Paull et al.[170] Copyright (1998) Elsevier.

Some of the advantages of the dynamic HPCIC approach can be seen from Figure 3.21, which shows the separation of Ba^{2+} and Sr^{2+} obtained using the above dynamic system (A), compared to the separation of the two metal ions obtained using a similar pre-impregnated PP column (B). The addition of the ligand to the mobile phase resulted in a higher stationary phase loading and thus increased retention and greater resolution. Additionally, as there was no need for PCR detection with the dynamic approach, the removal of unwanted post-column reagent pump pulsation can be seen in the comparison of the two baselines shown.

3.5.2.4 Methylthymol Blue

The most successful demonstration of dynamic HPCIC using metallochromic ligands has been achieved using MTB.[148,172] Using a similar organopolymeric column to that used above with PP, and a mobile phase containing 0.5 M

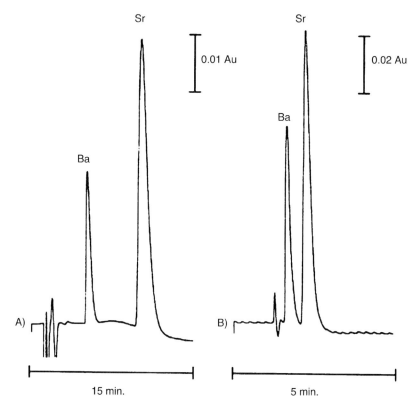

Figure 3.21 Chelation ion chromatograms of Ba^{2+} and Sr^{2+} standards obtained using phthalein purple (A) dynamically modified and (B) pre-impregnated PS-DVB column. Flow rates $1\,mL\,min^{-1}$. Photometric detection at 575 nm. Reproduced, with permission, from Paull et al.[170] Copyright (1998) Elsevier.

KNO_3 and 0.2 mM MTB (pH 2.1), the separation of five metal ions, namely Mg^{2+} (unretained), Mn^{2+}, Zn^{2+}, Cd^{2+} and Pb^{2+} was possible in under 15 min. The eluted metal ions could be detected as their respective MTB complexes at 600 nm through the post-column addition of a pH 5.9 buffer solution. In the study it was shown that increasing the concentration of MTB in the mobile phase, and/or the mobile phase pH, lead to an increase in retention for each of the above metal ions except Mg^{2+}. These observations confirmed the retention mechanism to be predominantly stationary phase chelation, with the mobile phase MTB in the concentration range explored, acting to stabilise a relatively high stationary phase concentration of the ligand, more than behaving as a strong complexing mobile phase.

The developed method was shown to be relatively sensitive for Zn^{2+} and Pb^{2+}, giving linear responses over the range 0.01–0.2 mM, and detection limits of 1 μM (100 μL inj. vol.). These detection limits were later improved by using an ODS column with a similar mobile phase, see Figure 3.22, as it was found

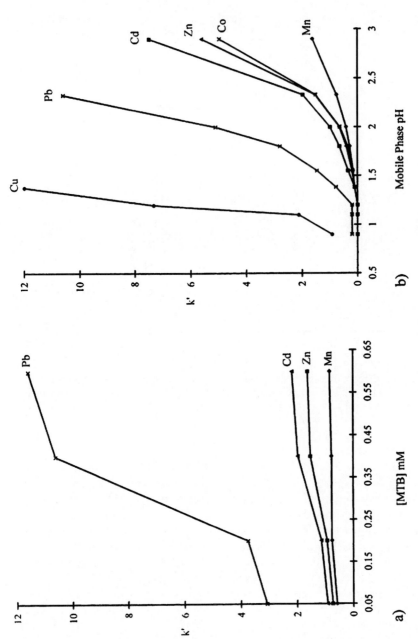

Figure 3.22 Effect of (a) concentration of MTB added to the mobile phase on retention of Mn^{2+}, Zn^{2+}, Cd^{2+} and Pb^{2+} on a 250 × 4.6 mm i.d. 5 μm ODS column (other conditions: 0.5 M KNO_3, pH 2.3); and (b) mobile phase pH on retention (other conditions: 0.5 M KNO_3, 0.4 mM MTB). Flow rate: 1 mL min^{-1}. Photometric detection at 600 nm after post-column change of pH. Reproduced, with permission, from Paull et al.[172] Copyright (1998) Elsevier.

Chelating Stationary Phases

that the dynamic loading achieved on the ODS column was greater than that obtained previously, achieving a similar separation at a more acidic mobile phase pH of 1.9. The resultant peaks, particularly Pb^{2+}, were noticeably sharper despite having longer retention times.[172] Figure 3.23 shows a metal ion separation, Mg^{2+}, Mn^{2+}, Zn^{2+}, Cd^{2+} and Pb^{2+}, achieved using the MTB modified ODS column produced using the dynamic modification approach. The chromatogram shown illustrates the improvements in peak efficiency possible with the dynamic approach very clearly, particularly for the later eluting peaks, here Pb^{2+}. In addition, a difference in selectivity for Zn^{2+} and Cd^{2+} between that exhibited by a pre-impregnated column and shown here with the dynamic system was noted. Zn^{2+} forms the stronger complex with MTB and, as expected, is eluted after Cd^{2+} on the pre-impregnated column. However, with MTB also in the mobile phase, Zn^{2+} is eluted before Cd^{2+}, due to the ligand again acting as a complexing mobile phase, in the same way as discussed previously using PP.

Using a mobile phase containing 0.5 M KNO_3, 0.2 mM MTB (pH 2.1) and an elevated column temperature, the above dynamic system was applied to the separation of a number of the lanthanide ions.[172] Seven of the lanthanides could be separated isocratically, namely La^{3+}, Ce^{3+}, Nd^{3+}, Sm^{3+}, Tb^{3+}, Tm^{3+} and Lu^{3+}, which were eluted in the above order at retention times of 22–42 min. Figure 3.24 shows the isocratic separation of a mixture of alkaline earth ions, transition metal ions and several lanthanide ions using the dynamically modified MTB column. To date this is the largest number of metal ions which have been separated isocratically using metallochromic ligand modified chelating columns.

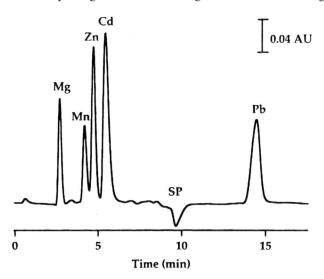

Figure 3.23 Chelation ion chromatogram showing the separation of Mg^{2+}, Mn^{2+}, Zn^{2+}, Cd^{2+} and Pb^{2+} on a 250×4.6 mm i.d. 5 μm ODS column with a mobile phase of 0.5 M KNO_3, 0.4 mM MTB, pH 1.9. Flow rate: 1 mL min^{-1}. Photometric detection at 600 nm after post column change of pH. Reproduced, with permission, from Paull et al.[172] Copyright (1998) Elsevier.

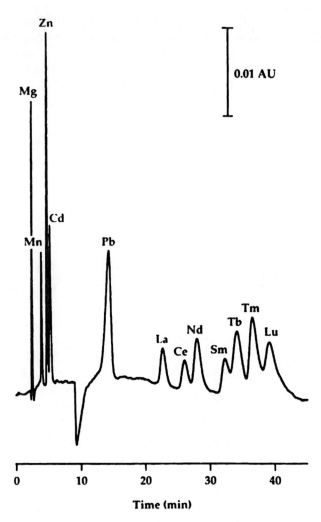

Figure 3.24 Chelation ion chromatogram of 12 metals on a MTB dynamically modified ODS column. Mobile phase conditions: 0.5 M KNO$_3$, 0.2 mM MTB, pH 2.1. Column temperature: 65 °C. Flow rate: 1 mL min^{-1}. Photometric detection at 600 nm after post column change of pH. Metal concentrations: 2 mM Mg^{2+}, 0.15 mM Mn^{2+}, 0.1 mM Cd^{2+}, 0.02 mM Zn^{2+}, Pb^{2+}, La^{3+}, Nd^{3+}, Tb^{3+}, Tm^{3+}, Lu^{3+}, 0.01 mM Ce^{3+}, Sm^{3+}. Reproduced, with permission, from Paull et al.[172] Copyright (1998) Elsevier.

3.5.2.5 Aromatic and Heterocyclic Acids

The strong retention and stability of bulky hydrophobic ligands on neutral PS-DVB based substrates, discussed above in relation to triphenylmethane based metallochromic ligands, is based upon a combination of hydrophobic and strong π–π interactions between the aromatic rings with the ligands and the benzene rings within the resin. These strong π–π interactions are such that relatively hydrophilic small substituted aromatic and heterocyclic acids can also be used to dynamically modify PS-DVB resins and produce stable ion exchange and chelation ion exchange systems. Such acids have also been used to dynamically modify ODS phases, although in this case, lower capacity chelating phases result, as the degree of stationary phase modification is based only upon hydrophobic adsorption of the ligand, without additional π–π interactions.

Elchuck et al.[173] used a mobile phase containing mandelic acid (oxygen–oxygen chelating ligand, pK_a 3.85) for the determination of inorganic cations, via what the authors termed 'dynamic ion exchange and hydrophobic interaction chromatography'. However, from the results reported, the chromatography clearly displays all the expected characteristics of stationary phase complexation, making this an early example of dynamic HPCIC. Mandelic acid displays similar complexation chemistry to the much explored α-hydroxyisobutyric acid (α-HIBA), which was used extensively for inclusion within mobile phases for the reversed-phase separation of transition metals and lanthanide α-HIBA complexes. However, mandelic acid, being a more hydrophobic ligand, results in the significant adsorption of the ligand itself onto the stationary phase (dynamic modification). The authors studied the retention behaviour of transition metals, lanthanides and actinides. Retention of the actinides was dependent on the concentration and pH of the mandelic acid mobile phase, the column temperature and the concentration of organic modifier. Near-baseline separation was achieved for Np(V), Am(III), U(VI), Th(IV), Pu(IV) and Np(IV) with a mandelic acid mobile phase using isocratic conditions.

Later work by Hao et al.[174] also utilised mandelic acid mobile phases with an ODS column for the separation of Th^{4+} and UO_2^{2+}. The chromatographic characteristics of the dynamic system were investigated in detail to elucidate the exact retention mechanism. The study concluded the system displayed typical reversed-phase characteristics, with a linear relationship displayed between the logarithm of the retention factor and the percentage of organic modifier. However, the effect of mobile phase pH was also significant, showing a considerable increase in retention of both species, with modest increases in mobile phase pH. This combination of the above two observations would suggest the observed retention in this system is predominantly based upon mobile phase complexation and the subsequent hydrophobic (reversed-phase type) adsorption of the formed complexes.

Shaw et al. and Nesterenko et al. studied a number of heterocyclic carboxylic acids as ligands for the dynamic modification of both PS-DVB resins and ODS phases.[175–177] These included picolinic acid (2-pyridinecarboxylic acid), dipicolinic acid (2,6-pyridinecarboxylic acid) and quinaldic acid (2-quinolinecarboxylic acid), each of which was capable of coordination with metal ions through a phenolic nitrogen and one or two carboxylic acid groups. The structures of these three chelating acids and their corresponding pK_a data are shown in Table 3.11. The results of these studies showed the potential of using relatively small acids, in place of the relatively large metallochromic ligands. A number of separations of Pb^{2+}, Cd^{2+}, Co^{2+}, Mn^{2+}, Ni^{2+} and Zn^{2+} were

Table 3.11 Structures and properties of some aromatic acids used as mobile phase chelating ligands in dynamic HPCIC.

Picolinic (2-pyridinecarboxylic) acid, pK_a 1.07; log P 0.72	4-Chlorodipicolinic (4-chloro-2,6-pyridine-carboxylic) acid, log P 1.21	Dipicolinic (2,6-pyridinecar-boxylic) acid, pK_{a1} 2.16; pK_{a2} 4.76; log P 0.57
Mandelic (2-hydroxy-2-phenylacetic) acid pK_a 3.85; log P 0.57		Quinaldic (2-quinoline-carbo-xylic acid, pK_a 1.82; log P 2.02

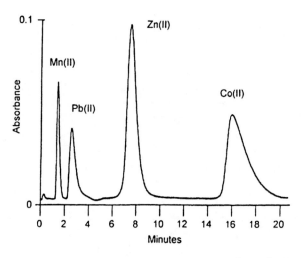

Figure 3.25 Chelation ion chromatogram of 0.5 mg L^{-1} Mn^{2+}, 5 mg L^{-1} Pb^{2+}, 15 mg L^{-1} Zn^{2+} and 5 mg L^{-1} Co^{2+}, on a 100 × 4.6 mm i.d. column containing 8 µm PLRP-S polystyrene divinylbenzene resin. Mobile phase: 1 M KNO$_3$, with 0.1 mM quinaldic acid at pH 2.6. Flow rate: 1 mL min^{-1}. Photometric detection at 490 nm after PCR with PAR. Reproduced, with permission, from Shaw et al.[175] Copyright (1999) Elsevier.

achieved, using high ionic strength mobile phases, containing 1 M KNO$_3$ and between 1 and 0.1 mM of the chelating ligand. In general, the studies reported that the retention order of metal ions on the modified polymer and silica phases were similar, although interestingly the larger 8 µm polymer substrate provided

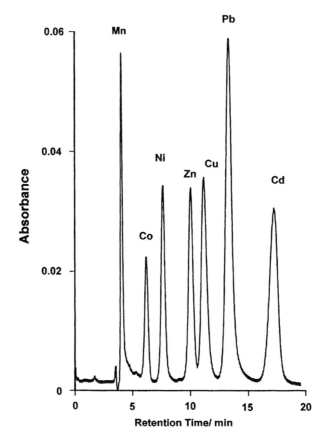

Figure 3.26 Separation of 0.5 mg L^{-1} Mn^{2+} 0.5 mg L^{-1} Co^{2+}, 0.5 mg L^{-1} Ni^{2+}, 2 mg L^{-1} Zn^{2+}, 1 mg L^{-1} Cu^{2+}, 10 mg L^{-1} Pb^{2+}, and 20 mg L^{-1} Cd^{2+} on a 300 × 4.6 mm i.d column containing 7 μm Hamilton PRP-1 PS–DVB resin. Mobile phase: 1 M KNO$_3$, 0.25 mM 4-chlorodipicolinic and 6.25 mM nitric acid, pH 2.2. Flow rate: 1 mL min^{-1}. Photometric detection at 520 nm after PCR with PAR. Reproduced, with permission, from Shaw et al.[178] Copyright (2002) Elsevier.

more efficient separations under the dynamic HPCIC conditions, than the 5 μm Hypersil ODS phase. Obviously, the dynamic coating of PS-DVB resin with aromatic heterocyclic acids is more stable than with ODS phases, due to π–π interactions. Figure 3.25 shows a chelation ion chromatogram of Mn^{2+}, Pb^{2+}, Zn^{2+} and Co^{2+} on a 100 × 4.6 mm i.d. column containing a 8 μm PS-DVB phase, with a 1 M KNO$_3$ mobile phase containing 1 × 10^{-4} quinaldic acid, at pH 2.6.

Shaw et al.[178] took the approach of dynamic HPCIC using small aromatic acids further, with the use of a 4-chlorodipicolinic acid containing mobile phase with a 7 μm PS-DVB packed column. The chromatographic behaviour of common transition metal ions, together with Al^{3+}, UO_2^{2+} and the lanthanides was investigated, and a complicated retention factor dependence upon mobile phase pH was determined. 4-Chlorodipicolinic acid resulted in a significantly greater dynamic surface loading than the acids investigated within the earlier study,[175] although as with any such weak acid, its dynamic loading upon the stationary phase was shown to be somewhat dependent upon mobile phase pH. However, for such amphoteric acids, maximum dynamic loading will occur at the isoelectric point, which in this case Shaw et al. showed to be close to between pH 1 and 2. Within this pH range the dynamic loading of the reagent upon the polymer phase appeared relatively stable, however, above this pH the dynamic loading of the column was shown to reduce drastically. Countering this effect, the formation of more stable metal complexes both on the stationary phase, and to a lesser extent within the mobile phase, caused retention to increase for most common transition and heavy metal ions up to pH 2. After which point retention times began to stabilise and show slight reductions. The complex and varied pH dependences shown for these cations allowed manipulation of selectivity to achieve some impressive separations. Selectivity for Pb^{2+} and Cd^{2+} on the 4-chlorodipicolinic acid dynamically modified phase was unusually high, when compared to most chelation ion exchangers, eluting after Cu^{2+}. Figure 3.26 shows the isocratic chelation ion chromatogram obtained for a group of seven common transition and heavy metal ions using the developed method. The excellent peak shapes and efficiencies shown in this study, demonstrate how dynamic HPCIC really is an attractive practical alternative to covalently bound chelating ion exchangers.

References

1. R. G. Pearson, *Chemical Hardness: Applications from Molecules to Solid*, Wiley-VCH, Weinheim, 1998, p. 210.
2. G. V. Kudryavtsev, G. V. Lisichkin and V. M. Ivanov, *J. Anal. Chem. USSR (Engl. Transl.)*, 1983, **38**, 16–25.
3. P. N. Nesterenko, T. I. Tikhomirova, V. I. Fadeeva, I. B. Yuferova and G. V. Kudryavtsev, *J. Anal. Chem. USSR (Engl. Transl.)*, 1991, **46**, 800–806.
4. P. N. Nesterenko, I. B. Yuferova and G. V. Kudryavtsev, *Zh. Vses. Khim. O va im D. I. Mendeleeva.*, 1989, **34**, 426–427.
5. P. N. Nesterenko, A. V. Ivanov, N. A. Galeva and G. B. C. Seneveratne, *J. Anal. Chem. (Engl. Transl.)*, 1997, **52**, 736–742.
6. P. N. Nesterenko, O. S. Zhukova, O. A. Shpigun and P. Jones, *J. Chromatogr. A*, 1998, **813**, 47–53.
7. P. N. Nesterenko, M. J. Shaw, S. J. Hill and P. Jones, *Microchem. J.*, 1999, **62**, 58–69.
8. M. J. Shaw, S. J. Hill, P. Jones and P. N. Nesterenko, *J. Chromatogr. A*, 2000, **876**, 127–133.
9. P. N. Nesterenko and P. Jones, *J. Sep. Sci.*, 2007, **30**, 1773–1793.

10. A. I. Elefterov, M. G. Kolpachnikova, P. N. Nesterenko and O. A. Shpigun, *J. Chromatogr. A*, 1997, **769**, 179–188.
11. G. V. Lisichkin, G. V. Kudryavtsev and P. N. Nesterenko, *J. Anal. Chem. USSR (Engl. Transl.)*, 1983, **38**, 1288–1307.
12. H. H. Weetall, *Biochim. Biophys. Acta*, 1970, **212**, 1–7.
13. C. Fulcher, M. A. Crowell, R. Bayliss, K. B. Holland and J. R. Jezorek, *Anal. Chim. Acta*, 1981, **129**, 29–47.
14. J. R. Jezorek, J. W. Tang, W. L. Cook, R. Obie, D. J. Ji and J. M. Rowe, *Anal. Chim. Acta*, 1994, **290**, 303–315.
15. J. R. Jezorek, C. Fulcher, M. A. Crowell, R. Bayliss, B. Greenwood and J. Lyon, *Anal. Chim. Acta*, 1981, **131**, 223–231.
16. K. H. Faltynski and J. R. Jezorek, *Chromatographia*, 1986, **22**, 5–12.
17. J. R. Jezorek and H. Freiser, *Anal. Chem.*, 1979, **51**, 366–373.
18. M. A. Marshall and H. A. Mottola, *Anal. Chem.*, 1983, **55**, 2089–2093.
19. M. A. Marshall and H. A. Mottola, *Anal. Chim. Acta*, 1984, **158**, 369–373.
20. M. R. Weaver and J. M. Harris, *Anal. Chem.*, 1989, **61**, 1001–1010.
21. S. H. Chang, K. M. Gooding and F. E. Regnier, *J Chromatogr.*, 1976, **120**, 321–333.
22. M. Marhol, *Ion Exchangers in Analytical Chemistry: Their Properties and Use in Inorganic Chemistry*, Elsevier, Amsterdam, 1982, p. 585.
23. G. V. Myasoedova and S. B. Savvin, *Chelate Forming Sorbents*, Nauka, 1984, p. 171.
24. K. M. Saldadze and V. D. Kopylova, *Complex-forming Ion-Exchangers. (Complexites)*, Khimiya, Moscow, 1980, p. 336.
25. R. Hering, *Chelatbildende Ionenaustausherharze.*, Akademie Verlag, Berlin (East), 1967, p. 267.
26. G. V. Myasoedova and S. B. Savvin, *CRC Crit. Rev. Anal. Chem.*, 1986, **17**, 1–63.
27. V. E. Kopylova, *Solvent Extr. Ion Exch.*, 1998, **16**, 267–343.
28. S. K. Sahni and J. Reedijk, *Coord. Chem. Rev.*, 1984, **59**, 1–139.
29. J. N. King and J. S. Fritz, *J Chromatogr*, 1978, **153**, 507–516.
30. C. Pohlandt and J. S. Fritz, *J. Chromatogr.*, 1979, **176**, 189–197.
31. R. J. Phillips and J. S. Fritz, *Anal. Chim. Acta*, 1982, **139**, 237–246.
32. R. J. Phillips and J. S. Fritz, *Anal. Chim. Acta*, 1980, **121**, 225–232.
33. J. S. Fritz and R. J. Phillips, *Anal. Chem.*, 1978, **50**, 1504–1508.
34. E. M. Moyers and J. S. Fritz, *Anal. Chem.*, 1976, **48**, 1117–1120.
35. E. M. Moyers and J. S. Fritz, *Anal Chem.*, 1977, **49**, 418–423.
36. C. Y. Liu, M. J. Chen, N. M. Lee, H. C. Hwang, S. T. Jou and J. C. Hsu, *Polyhedron*, 1992, **11**, 551–558.
37. C. Y. Liu, N. M. Lee and T. H. Wang, *Anal. Chim. Acta*, 1997, **337**, 173–182.
38. C. Y. Liu, N. M. Lee and J. L. Chen, *Anal. Chim. Acta*, 1998, **369**, 225–233.
39. H. Kumagai, T. Yokoyama, T. M. Suzuki and T. Suzuki, *Analyst*, 1999, **124**, 1595–1597.

40. Y. Inoue, H. Kumagai, Y. Shimomura, T. Yokoyama and T. M. Suzuki, *Anal. Chem.*, 1996, **68**, 1517–1520.
41. H. Kumagai, Y. Inoue, T. Yokoyama, T. M. Suzuki and T. Suzuki, *Anal. Chem.*, 1998, **70**, 4070–4073.
42. L. Barron, M. O'Toole, D. Diamond, P. N. Nesterenko and B. Paull, *J. Chromatogr. A*, 2008, **1213**, 31–36.
43. T. M. Suzuki and T. Yokoyama, *Polyhedron*, 1984, **3**, 939–945.
44. J. J. Lih, K. Y. Yeh and C. Y. Liu, *Fresenius J Anal Chem.*, 1990, **336**, 12–15.
45. P. M. Jonas, D. J. Eve and J. R. Parrish, *Talanta*, 1989, **36**, 1021–1026.
46. C. Y. Huang, N. M. Lee, S. Y. Lin and C. Y. Liu, *Anal. Chim. Acta*, 2002, **466**, 161–174.
47. H. Kumagai, M. Yamanaka, T. Sakai, T. Yokoyama, T. M. Suzuki and T. Suzuki, *J. Anal. At. Spectrom.*, 1998, **13**, 579–582.
48. J. M. Hill, *J. Chromatogr*, 1973, **76**, 455–458.
49. D. E. Leyden, G. H. Luttrell and T. A. Patterson, *Anal. Lett.*, 1975, **8**, 51–56.
50. D. E. Leyden and G. H. Luttrell, *Anal Chem.*, 1975, **47**, 1612–1617.
51. F. B. Anspach, *J. Chromatogr. A*, 1994, **676**, 249–266.
52. G. Gubitz and S. Mihellyes, *Chromatographia*, 1984, **19**, 257–259.
53. K. T. DenBleyker and T. R. Sweet, *Chromatographia*, 1980, **13**, 114–118.
54. J. Chmielowiec and W. Simon, *Chromatographia*, 1978, **11**, 99–101.
55. P. Grossmann and W. Simon, *J. Chromatogr.*, 1982, **235**, 351–363.
56. I. B. Yuferova, G. V. Kudryavtsev, T. I. Tikhomirova and V. I. Fadeeva, *J. Anal. Chem. USSR (Engl. Transl.)*, 1988, **43**, 1329–1333.
57. J. F. Biernat, P. Konieczka and B. J. Tarbet, *Sep. Purif. Meth.*, 1994, **23**, 77–348.
58. P. K. Jal, S. Patel and B. K. Mishra, *Talanta*, 2004, **62**, 1005–1028.
59. K. S. Abou-El-Sherbini, *J. Sol–Gel Sci. Technol.*, 2009, **51**, 228–237.
60. R. K. Iler, *The Chemistry of Silica*, Wiley New York, 1979, p. 866.
61. G. V. Lisichkin, A. Y. Fadeev, A. A. Serdan, P. N. Nesterenko and D. B. Furman, *Chemistry of Surface Grafted Compounds*, Fizmatlit, Moscow, 2003, p. 592.
62. J. Kohr and H. Engelhardt, *J. Chromatogr. A*, 1993, **652**, 309–316.
63. J. Tria, P. R. Haddad and P. N. Nesterenko, *J. Sep. Sci.*, 2008, **31**, 2231–2238.
64. P. N. Nesterenko and P. Jones, *J. Chromatogr. A*, 1998, **804**, 223–231.
65. M. Foltin, S. Megová, T. Prochacková and M. Steklac, *J. Radioanal. Nucl. Chem.*, 1996, **208**, 295–307.
66. J. B. Henry and T. R. Sweet, *Chromatographia*, 1983, **17**, 79–82.
67. P. N. Nesterenko and A. V. Ivanov, *Mendeleev Comm.*, 1994, 174–176.
68. P. N. Nesterenko and A. V. Ivanov, *J. Chromatogr. A*, 1994, **671**, 95–99.
69. J. C. Hsu, C. H. Chang and C. Y. Liu, *Fresenius J. Anal. Chem.*, 1998, **362**, 514–521.
70. K. T. DenBleyker, J. K. Arbogast and T. R. Sweet, *Chromatographia*, 1983, **17**, 449–450.

71. M. Slebioda, Z. Wodecki, A. M. Kolodziejczyk and W. Nowicki, *Chem. Anal.*, 1994, **39**, 149–152.
72. I. N. Voloschik, M. L. Litvina and B. A. Rudenko, *J. Chromatogr. A*, 1994, **671**, 51–54.
73. I. N. Voloschik, M. L. Litvina and B. A. Rudenko, *J. Chromatogr. A*, 1994, **671**, 205–209.
74. S. Srijaranai, N. Ryan, N. Mitchell and J. D. Glennon, Solid phase extraction and chromatography of metal ions using free and immobilised biomimetic chelating agents, in *Recent Developments in Ion Exchange 2*, ed. P. A. Williams and M. J. Hudson, Elsevier Applied Science, London, 1990, pp. 311–315.
75. J. R. Jezorek, K. H. Faltynski and J. W. Finch, *J. Chem. Educ.*, 1986, **63**, 354–357.
76. C. H. Risner and J. R. Jezorek, *Anal. Chim. Acta*, 1986, **186**, 233–245.
77. H. W. Thompson and J. R. Jezorek, *Anal. Chem.*, 1991, **63**, 75–78.
78. S. Watanesk and A. A. Schilt, *Talanta*, 1986, **33**, 895–899.
79. N. Simonzadeh and A. A. Schilt, *Talanta*, 1988, **35**, 187–190.
80. N. Simonzadeh and A. A. Schilt, *J. Coord. Chem.*, 1988, **35**, 187–190.
81. W. Bashir and B. Paull, *J. Chromatogr. A*, 2001, **907**, 191–200.
82. P. N. Nesterenko and P. Jones, *Anal. Comm.*, 1997, **34**, 7–8.
83. P. Nesterenko and P. Jones, *J. Liq. Chromatogr. Relat. Technol.*, 1996, **19**, 1033–1045.
84. W. Bashir and B. Paull, *J. Chromatogr. A*, 2001, **942**, 73–82.
85. A. I. Elefterov, P. N. Nesterenko and O. A. Shpigun, *J. Anal. Chem. (Engl. Transl.)*, 1996, **51**, 887–891.
86. A. I. Elefterov, S. N. Nosal, P. N. Nesterenko and O. A. Shpigun, *Analyst*, 1994, **119**, 1329–1332.
87. I. N. Voloshchik, B. A. Rudenko and M. L. Litvina, *J. Anal. Chem.*, 1994, **49**, 1165–1169.
88. I. N. Voloschik, M. L. Litvina and B. A. Rudenko, *J. Chromatogr. A*, 1995, **706**, 315–319.
89. W. Bashir and B. Paull, *J. Chromatogr. A*, 2001, **910**, 301–309.
90. G. Bonn, S. Reiffenstuhl and P. Jandik, *J. Chromatogr.*, 1990, **499**, 669–676.
91. J. Tria, P. N. Nesterenko and P. R. Haddad, *Chem. Listy*, 2008, **102**, 319–323.
92. A. Haidekker and C. G. Huber, *J. Chromatogr. A*, 2001, **921**, 217–226.
93. E. Sugrue, P. Nesterenko and B. Paull, *J. Sep. Sci.*, 2004, **27**, 921–930.
94. E. Sugrue, P. Nesterenko and B. Paull, *Analyst*, 2003, **128**, 417–420.
95. E. Sugrue, P. N. Nesterenko and B. Paull, *J. Chromatogr. A*, 2005, **1075**, 167–175.
96. R. Garcia-Valls, A. Hrdlicka, J. Perutka, J. Havel, N. V. Deorkar, L. L. Tavlarides, M. Munoz and M. Valiente, *Anal. Chim. Acta*, 2001, **439**, 247–253.
97. N. R. Sumskaya, Y. V. Kholin and V. N. Zaitsev, *Zhurnal Fizicheskoi Khimii*, 1997, **71**, 905–910.

98. P. N. Nesterenko, I. P. Smirnov, G. D. Brykina and T. A. Bolshova, *Vestn. Mosk. Univ. Khim.*, 1991, **32**, 358–362.
99. I. P. Smirnov and P. N. Nesterenko, *Fibre Chemistry*, 1993, **24**, 422–424.
100. C. A. Pohl, J. R. Stillian and P. E. Jackson, *J. Chromatogr. A*, 1997, **789**, 29–41.
101. P. N. Nesterenko and T. A. Bolshova, *Vestn. Mosk. Univ. Khim.*, 1990, **31**, 167–169.
102. P. N. Nesterenko and P. Jones, *J. Chromatogr. A*, 1997, **770**, 129–135.
103. M. G. Kolpachnikova, N. A. Penner and P. N. Nesterenko, *J. Chromatogr. A*, 1998, **826**, 15–23.
104. P. N. Nesterenko and O. A. Shpigun, *Russ. J. Coord. Chem.*, 2002, **28**, 726–735.
105. P. Jones and P. N. Nesterenko, *J. Chromatogr. A*, 2008, **1213**, 45–49.
106. M. J. Shaw, P. N. Nesterenko, G. W. Dicinoski and P. R. Haddad, *J. Chromatogr. A*, 2003, **997**, 3–11.
107. M. J. Shaw, P. N. Nesterenko, G. W. Dicinoski and P. R. Haddad, *Aust. J. Chem.*, 2003, **56**, 201–206.
108. M. A. Rey, C. A. Pohl, J. Jagodzinski, E. Q. Kaiser and J. M. Riviello, *J. Chromatogr. A*, 1998, **804**, 201–209.
109. M. G. Kiseleva, P. A. Kebets and P. N. Nesterenko, *Analyst*, 2001, **126**, 2119–2123.
110. P. A. Kebets, E. P. Nesterenko, P. N. Nesterenko and A. J. Alpert, *Microchim. Acta*, 2004, **146**, 103–110.
111. P. A. Kebets, K. A. Kuz'mina and P. N. Nesterenko, *Russ. J. Phys. Chem.*, 2002, **76**, 1481–1484.
112. W. Bashir, E. Tyrrell, O. Feeney and B. Paull, *J. Chromatogr. A*, 2002, **964**, 113–122.
113. S. Reiffenstuhl and G. Bonn, *J. Chromatogr.*, 1989, **482**, 289–296.
114. L. M. Nair, N. Saari and J. M. Anderson, *J. Chromatogr. A*, 1994, **671**, 43–49.
115. J. Toei, *Fresenius Z. Anal. Chem.*, 1988, **331**, 735–739.
116. P. N. Nesterenko and A. V. Ivanov, *Vestn. Mosk. Univ. Ser. 2 Khim.*, 1992, **33**, 574–578.
117. E. Antico, A. Masana, V. Salvado, M. Hidalgo and M. Valiente, *J. Chromatogr. A*, 1995, **706**, 159–166.
118. S. M. Staroverov and A. Y. Fadeev, *J. Chromatogr.*, 1991, **544**, 77–98.
119. B. Busche, R. Wiacek, J. Davidson, V. Koonsiripaiboon, W. Yantasee, R. S. Addleman and G. E. Fryxell, *Inorg. Chem. Commun.*, 2009, **12**, 312–315.
120. G. V. Kudryavtsev and G. V. Lisichkin, *Zh. Fiz. Khim.*, 1981, **55**, 1352–1354.
121. P. G. Righetti, *Isoelectric Focusing: Theory, Methodology and Applications.*, Elsevier Biomedical Press, Amsterdam, 1983, p. 386.
122. L. C. Sander and S. A. Wise, *CRC Crit. Rev. Anal. Chem.*, 1987, **18**, 299–415.

123. E. Sugrue, P. N. Nesterenko and B. Paull, *Anal. Chim. Acta*, 2005, **553**, 27–35.
124. G. K. Bonn, S. Nathakarnkitkool and P. Jandik, Coordination chromatography of transition metals: efficient and selective separations of cations on iminodiacetate functional groups, in *Advances in Ion Chromatography*, ed. P. Jandik and R. M. Cassidy, Century International, Medfield, MA, 1991, pp. 197–214.
125. B. Paull and P. N. Nesterenko, *Trends Anal. Chem.*, 2005, **24**, 295–303.
126. D. Schaller, E. F. Hilder and P. R. Haddad, *J. Sep. Sci.*, 2006, **29**, 1705–1719.
127. J. Scancar and R. Milacic, *TrAC Trends Anal. Chem.*, 2009, **28**, 1048–1056.
128. D. Ren, N. A. Penner, B. E. Slentz, H. D. Inerowicz, M. Rybalko and F. E. Regnier, *J. Chromatogr. A.*, 2004, **1031**, 87–92.
129. Q. Luo, H. Zou, X. Xiao, Z. Guo, L. Kong and X. Mao, *J. Chromatogr. A.*, 2001, **926**, 255–264.
130. B. Preinerstorfer, D. Lubda, W. Lindner and M. Lämmerhofer, *J. Chromatogr. A.*, 2006, **1106**, 94–105.
131. S. Yamazaki, H. Omori and O. Eon, *J. High Resolut. Chromatogr.*, 1986, **9**, 765–766.
132. T. Yasui, N. Komatsu, K. Egami, H. Yamada and A. Yuchi, *Anal. Sci.*, 2007, **23**, 1011–1014.
133. R. Kocjan, *Analyst*, 1992, **117**, 741–744.
134. R. Kocjan and S. Przeszlakowski, *Sep. Sci. Technol.*, 1989, **24**, 291–301.
135. M. L. Marina, V. Gonzalez and A. R. Rodriguez, *Microchem. J.*, 1986, **33**, 275–294.
136. M. Torre and M. L. Marina, *CRC Crit. Rev. Anal. Chem.*, 1994, **24**, 327–361.
137. H. W. Handley, P. Jones, L. Ebdon and N. W. Barnett, *Anal. Proc.*, 1991, **28**, 37–38.
138. P. Jones and G. Schwedt, *J. Chromatogr.*, 1989, **482**, 325–334.
139. B. Paull, M. Foulkes and P. Jones, *Anal. Proc.*, 1994, **31**, 209–211.
140. B. Paull, M. Foulkes and P. Jones, *Analyst*, 1994, **119**, 937–941.
141. P. Jones, M. Foulkes and B. Paull, *J. Chromatogr. A.*, 1994, **673**, 173–179.
142. O. J. Challenger, S. J. Hill and P. Jones, *J. Chromatogr.*, 1993, **639**, 197–205.
143. P. Jones, O. J. Challenger and S. J. Hill, Recent advances in high-performance chelation ion chromatography for trace-metal determinations, in *Ion Exchange Processes: Advances and Applications*, ed. A. Dyer, M. J. Hudson and P. A. Williams, Royal Society of Chemistry, Cambridge, 1993, pp.279–287.
144. A. Hrdlicka, J. Havel and M. Valiente, *J. High Resolut. Chromatogr.*, 1992, **15**, 423–427.
145. E. A. Gautier, R. T. Gettar, R. E. Servant and D. A. Batistoni, *J. Chromatogr. A.*, 1997, **770**, 75–83.

146. P. Jones, O. J. Challenger, S. J. Hill and N. W. Barnett, *Analyst*, 1992, **117**, 1447–1450.
147. O. J. Challenger, S. J. Hill, P. Jones and N. W. Barnett, *Anal. Proc.*, 1992, **29**, 91–93.
148. B. Paull, P. Nesterenko, M. Nurdin and P. R. Haddad, *Anal. Comm.*, 1998, **35**, 17–20.
149. B. Paull and P. Jones, *Chromatographia*, 1996, **42**, 528–538.
150. M. J. Shaw, J. Cowan and P. Jones, *Anal. Lett.*, 2003, **36**, 423–439.
151. R. C. Sutton, S. J. Hill and P. Jones, *J. Chromatogr. A*, 1996, **739**, 81–86.
152. N. A. Penner, P. N. Nesterenko, M. M. Ilyin, M. P. Tsyurupa and V. A. Davankov, *Chromatographia*, 1999, **50**, 611–620.
153. A. Berthod, M. Kolosky, J. L. Rocca and O. Vittori, *Analysis*, 1979, **7**, 395–400.
154. S. J. Bale, R. M. Smith, S. G. Westcott and M. M. Smith, *Anal. Proc.*, 1988, **25**, 62–63.
155. R. Vespalec, M. Vrchlabsky and M. Cigankova, *Folia Chemia.*, 1985, **26**, 5–34.
156. M. Zenki, *Anal. Chem.*, 1981, **53**, 968–971.
157. J. Toei and N. Baba, *J. Chromatogr.*, 1986, **361**, 368–373.
158. J. Toei, *Analyst*, 1988, **113**, 247–250.
159. J. Toei, *Chromatographia*, 1987, **23**, 583–589.
160. J. Toei, *J. High Resolut. Chromatogr. Chromatogr. Commun.*, 1987, **10**, 111–112.
161. J. I. Toei, *Chromatographia*, 1987, **23**, 355–360.
162. P. K. Dasgupta, K. Soroka and R. S. Vithanage, *J. Liq. Chromatogr.*, 1987, **10**, 3287–3319.
163. H. Wada, T. Matsushita, T. Yasui, A. Yuchi, H. Yamada and G. Nakagawa, *J. Chromatogr. A*, 1993, **657**, 87–93.
164. W. Bashir, S. G. Butler and B. Paull, *Anal. Lett.*, 2001, **34**, 1529–1540.
165. J. M. Gawoski and D. Walsh, *Clin. Chem.*, 1989, **35**, 2140–2141.
166. J. F. van Staden and A. van Rensburg, *Analyst*, 1990, **115**, 605–608.
167. J. Toffaletti and K. Kirvan, *Clin. Chem.*, 1980, **26**, 1562–1565.
168. B. Paull, P. A. Fagan and P. R. Haddad, *Anal. Comm.*, 1996, **33**, 193–196.
169. B. Paull, M. Macka and P. R. Haddad, *J. Chromatogr. A*, 1997, **789**, 329–337.
170. B. Paull, M. Clow and P. R. Haddad, *J. Chromatogr. A*, 1998, **804**, 95–103.
171. K. Ueno, T. Imamura and K. L. Cheng, *Handbook of Organic Analytical Reagents* (second edition), CRC Press, Boca Raton, 2000, p. 615.
172. B. Paull, P. N. Nesterenko and P. R. Haddad, *Anal. Chim. Acta*, 1998, **375**, 117–126.
173. S. Elchuk, K. I. Burns, R. M. Cassidy and C. A. Lucy, *J. Chromatogr.*, 1991, **558**, 197–207.
174. F. Hao, B. Paull and P. R. Haddad, *J. Chromatogr. A*, 1996, **739**, 151–161.
175. M. J. Shaw, P. Jones and S. J. Hill, *Anal. Chim. Acta*, 1999, **401**, 65–71.

176. P. N. Nesterenko, G. Z. Amirova and T. A. Bol'shova, *Anal. Chim. Acta*, 1994, **285**, 161–168.
177. P. N. Nesterenko and G. Z. Amirova, *J. Anal. Chem. (Engl. Transl.)*, 1994, **49**, 447–451.
178. M. J. Shaw, P. Jones and P. N. Nesterenko, *J. Chromatogr. A*, 2002, **953**, 141–150.
179. A. Hulanicki, S. Glab and G. Ackermann, *Pure Appl. Chem.*1983, **55**, 1137–1230.

CHAPTER 4
Elution

4.1 Mobile Phase Parameters Influencing Separation Performance in HPCIC

In HPCIC the mobile phase has several different functions. The first is to ensure that chelation in the stationary phase is the dominant retention and separation mechanism. As non-charged chelating substrates are rarely used as stationary phases in HPCIC (see Chapter 3), unwanted electrostatic interactions between metal cations and charged functional groups are always present, if otherwise not suppressed. Ignoring such electrostatic interactions is a common occurrence in many publications utilising chelating ion exchangers. Often, negatively charged carboxylic, phosphonic or arsenic acid groups attract and retain positively charged metal cations without the formation of any donor–acceptor bonds. Therefore, the presence of chelating groups in the stationary phase does not automatically mean chelation is the sole or indeed dominant retention mechanism. Similar potentially unwanted effects can be seen with the use of chelating substrates having weak amino, phosphine and other basic groups, which in protonated form can retain negatively charged metal species from within the sample, or indeed with some negatively charged complexing agents added to the mobile phase.

4.1.1 Ionic Strength

With the above interactions in mind, to obtain a pure chelation based separation mechanism, the suppression of electrostatic interactions is required. Efficient suppression can be achieved through the addition of strong electrolytes, usually nitrates or perchlorates of sodium or potassium, in the concentration range 0.5–1.0 M, to the mobile phase. The presence of alkali metals in such a large concentration creates an efficient shielding layer of counter-ions around negatively charged functional groups, thus competitively preventing

Elution

Table 4.1 The values of formation constants (log β_1) for the complexes of some metals with IDA obtained at varying ionic strengths in aqueous solutions at 25 °C.[1]

Metal	Ionic strength, I			
	0.1	0.5	1.0	2.0
Ni^{2+}	8.30	8.16	8.16	8.39
Cd^{2+}	5.71	5.56	5.52	5.56
Fe^{2+}	5.80	5.54	5.52	—
UO_2^{2+}	8.96	8.72	8.78	—

significant electrostatic interactions with target metal cations. At the same time, complexation as a site specific multi-point based interaction for divalent and trivalent metal ions is less affected by this increase in electrolyte concentration. This low sensitivity of chelation to such changes of electrolyte concentration can be seen in the relatively small variations in the reported formation constant values obtained for various metal–ligand complexes at different ionic strengths[1] (see Table 4.1).

The effect of the suppression of electrostatic interactions can be clearly seen through changes in the elution order of alkaline-earth metal cations separated on IDA-silica at increasing mobile phase concentrations of KNO_3, whilst maintaining a constant mobile phase pH.[2] Without the presence of KNO_3 in the mobile phase, the retention order on IDA-silica was recorded as $Mg^{2+} < Ca^{2+} < Sr^{2+} < Ba^{2+}$, which corresponds exactly to the standard retention order of these cations observed on a common sulfonated cation exchange resin with a diluted nitric or perchloric acid mobile phase. In other words, with no significant complexation influencing the separation. In terms of eqn (2.17) this means a greater impact of electrostatic interactions on retention of these cations as defined by the selectivity ratio $K_{M,B}$ values rather than by coordination, mostly defined by β_1. A good illustration of the ability of IDA-silica to operate as a simple cation exchanger was obtained by Haidekker and Huber, who used an IDA-silica column for the separation of alkali metals.[3]

However, at increased concentrations of KNO_3 in the mobile phase the retention order changes, in fact there appears an almost complete reversal in the retention order for the alkaline-earth metal cations when using a mobile phase containing 2 M KNO_3 (Figure 4.1). This reversal in elution order is now in close agreement with values of formation constants (as log β_1) of the corresponding iminodiacetates, which are reported to increase in the following order, Ba^{2+} (1.67) < Sr^{2+} (2.23) < Ca^{2+} (2.60) Mg^{2+} < (2.98) (Table 4.2). Thus in the presence of 2 M electrolyte, the ion exchange interactions seen previously between negatively charged carboxyls from within the IDA group and the divalent cations would appear to be significantly suppressed. Such a reversal in elution order is possible for groups of metals displaying the same or similar degree of enthalpy for both electrostatic attraction and complexation based interactions.

It should be noted that the retention order of transition and heavy metals (Mn^{2+}, Co^{2+}, Cd^{2+}, Zn^{2+}, Ni^{2+}, Pb^{2+} and Cu^{2+}) on IDA-silica remains

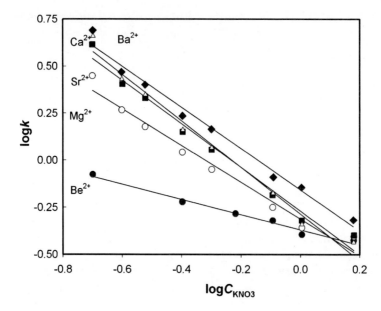

Figure 4.1 A reversal in the retention order of alkaline-earth metals on IDA-silica column with an increase in ionic strength of the mobile phase. Reproduced, with permission, from Bashir and Paull.[2] Copyright (2001) Elsevier.

essentially unchanged for both unsuppressed and suppressed electrostatic interactions elution systems. Obviously, this is due to the dominance of complexation at the surface as the main retention mechanism because of the much higher values of formation constants (log $\beta_1 \sim 4.72$–10.56) for complexes of transition metals with IDA (Table 4.2).

This is not to say that the presence of unsuppressed electrostatic interactions is not having some effect. It is likely that the main effect will be to increase the relative differences between k values, which could be a help or a hindrance on resolution of peaks depending on the metals involved.

An effect of suppression of electrostatic interaction resulting in significant retention order changes was observed for the separation of the lanthanide series cations on an IDA-silica column.[4,5] According to the data shown within Figure 4.2, the retention of lanthanides due to electrostatic interactions on a strong cation exchange resin increases with their atomic number. However, the formation constants for IDA chelates of this series of cations display the exact opposite trend. Therefore, without suppression of electrostatic interactions, through simply using dilute acid mobile phases, IDA-silica retains lanthanides by a combination of the two possible types of interaction, which although both acting to increase retention overall, when combined cause the phase to exhibit very poor separation selectivity, especially for the heavier lanthanides (shown within Figure 4.3). However, addition of KNO_3 to the mobile phase, at a concentration of 0.75 M or above, causes a dramatic improvement in the resolution of these chromatographic peaks, as the coincidental electrostatic interaction effect between trivalent lanthanide cations and IDA groups is suppressed. In the chromatographic example shown, an additional shift of the sorption equilibrium towards chelation was achieved through an increase in column temperature (see Section 4.1.2), resulting finally in excellent separation selectivity (shown later in Figure 4.8) corresponding closely to the difference in formation constants of the individual complexes.

Perchlorates and nitrates of alkali metals are selected as electrolyte additives to the mobile phase in HPCIC due to their weak complexing ability with transition and heavy metal cations. However, there are some exceptions to this rule. Nitrate can form relatively stable associates with the lanthanide series cations. For example, the logarithm of the formation constant for $Eu(NO_3)^{2+}$ is 1.22,[1] and at high concentrations within the mobile phase the formation of nitrate complexes with lanthanides can affect the resultant separation.

The use of a high concentration of chloride salts of alkali metals in the mobile phase for the suppression of electrostatic interactions, instead of nitrates, may also effect on separation selectivity. As shown in Figure 4.4, the presence of a high concentration of KCl in the mobile phase changes the retention order of transition metals, when compared with similar strength KNO_3 based mobile phases. Clearly, the formation of chloride complexes of lead, zinc and cadmium is responsible for these changes in retention selectivity. The addition of chloride to the mobile phase in lower concentrations can therefore be used for the fine optimisation of separation selectivity, by selectively moving chromatographic peaks of these metals into appropriate retention time windows.

One further aspect to consider from the addition of high concentrations of electrolytes to the mobile phase in HPCIC, is the ability of certain alkali metal cations themselves to form relatively stable complexes with the immobilised chelating group, *e.g.* IDA.[1] Because of the high concentration of these electrolytes, this effect can become significant and these cations can begin to compete with divalent alkaline earth and transition metals for the surface sorption sites (see Section 4.2.1.3). In practice this means certain alkali salts, *e.g.* LiCl, have considerably greater effect upon retention (although still relatively small) in HPCIC than others, *e.g.* KNO_3.

The suppression of electrostatic interactions is important not only for chelating ion exchangers with negatively charged functional groups (IDA, phosphonic acid, *etc.*), but also for stationary phases with bonded basic ligands, like oligoethyleneamines, 8-hydroxyquinoline and others. Basic groups in protonated form can operate as anion exchangers and thus retain anions and negatively charged metal complexes from the mobile phase. For example, Kuo and Mottola studied the separation of both inorganic and organic anions, utilising the 8-hydroxyquinolinium ion, both as a counter-ion dissolved in the organic phase during a liquid–liquid extraction, and as a chelating reagent covalently immobilised on the surface of controlled pore glass.[6] In the latter case, the baseline anion exchange chromatographic separation of four arylsulfonates was obtained. The combination of the complexing, hydrophobic and anion exchange properties of the 8-hydroxyquinoline bonded silica have also

Table 4.2 pK_a and logarithms formation constants ($\beta_1 = [ML]/[M][L]$) of metal complexes with common ligands (25 °C, ionic strength 0.1 if otherwise not stated in brackets)[1]

Metal	IDA	Oxalic acid	Tartaric acid	Maleic acid	Diglycolic acid	Dipicolinic acid	Picolinic acid	Chloride	Ethylphosphonic acid
H^+	1.77; 2.62; 9.34	1.2; 3.80;	2.82; 3.97	1.74; 5.81	2.81; 3.91	2.07; 4.66	0.95; 5.21	—	—
Ba^{2+}	1.67	2.33 (18 °C, 0)	1.72	2.30 (0)	2.15	3.44	1.65	−0.44 (1.0)	1.30
Sr^{2+}	2.23	2.54 (18 °C, 0)	1.75	—	2.50	3.85	1.69	−0.22 (1.0)	1.35
Ca^{2+}	2.60	2.46 (37 °C)	1.86	1.60	3.38	4.36 (1.0)	1.80	0.20	1.54
Mg^{2+}	2.98	2.76 (20 °C)	1.44	2.30 (0)	1.70	2.34	2.21	0.20	1.85
Mn^{2+}	4.72	3.00	2.45	1.7	2.53	5.01 (20 °C)	3.57 (20)	0.00 (0.0)	2.51
Cd^{2+}	5.71	2.65 (1.0)	1.6 (0.5)	2.4	3.25	6.36	4.35	1.59	2.94
Fe^{2+}	5.80	3.05	2.24 (20 °C)	—	2.56	5.71 (20 °C)	4.90 (20)	−0.30 (0.0)	—
VO_2^+	6.70	6.11 (0.5)	—	—	5.01	9.00 (1.0)	—	—	—
Co^{2+}	6.97	3.84	2.20	—	2.70	6.36 (20 °C)	5.74 (20)	−0.28 (3.0)	2.27
Zn^{2+}	7.15	4.00	2.69	2.0	3.61	6.35 (20 °C)	5.23	−0.20	2.67
Ni^{2+}	8.30	4.16	2.52	2.0	2.80	6.95 (20 °C)	6.72	−0.64	2.30

Elution

Ion									
Pb²⁺	7.36	4.20	3.12	—	4.41	8.70 (20 °C)	4.58 (20)	—	3.69
Al³⁺	8.18 (0.2)	6.20 (0.2)	2.96 (0.5)	—	3.16 (0.5)	4.87 (0.5)	4.52	−1.00 (1.0)	6.63
Fe³⁺	10.72 (0.5)	7.53 (0.5)	6.49 (20 °C)	—	5.04 (0.5)	10.91 (20 °C)	12.80 (20)	0.80	—
VO²⁺	8.65	5.89 (0.2)	3.9	—	5.01 (0.5)	11.70 (1.0)	6.67	—	—
Sc³⁺	9.80	8.40	—	—	8.30	11.20 (0.5)	—	0.04 (20 °C, 0.7)	—
Cu²⁺	10.56	4.85	3.32	3.40	3.94	9.10	7.87	−0.20 (0.5)	3.59
Hg²⁺	11.80	9.66	7.0	—	6.70 (1.0)	20.28 (20 °C)	7.70 (20)	6.74 (0.5)	—
Bi³⁺	—	—	—	—	7.69 (0.5)	—	—	—	—
Th(IV)	9.69 (1.0)	8.8	—	6.34 (20 °C, 1.0)	8.15	—	5.15	—	—
UO₂²⁺	8.96	6.36 (20 °C)	—	4.46 (1.0)	5.31	—	4.51	−0.30 (1.0)	—
In³⁺	10.14	6.02	4.44	5.0	5.0 (1.0)	—	5.81	2.32 (20 °C, 0.7)	—
Be²⁺	7.70	4.16	1.69 (0.5)	—	—	—	—	−0.36 (0.7)	—
Ga³⁺	12.76	6.45 (20 °C, 1.0)	—	—	—	—	—	0.01 (20 °C, 0.7)	—

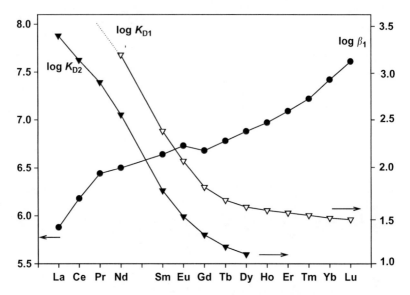

Figure 4.2 Stability constants (as log β_1) for chelates of lanthanide series ions with IDA and their corresponding distribution coefficients reported for the strong cation exchanger Dowex 50 × 4, in 0.1 M (log K_{D1}) or 7 M (log K_{D2}) HNO$_3$–methanol (90:10 v/v) mixture.

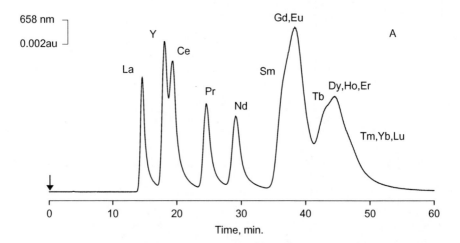

Figure 4.3 Separation of rare-earth elements on IDA-silica column (250 × 4.0 mm i.d., 7 µm. Mobile phase: 2.3×10^{-2} M HNO$_3$; flow rate: 0.8 mL min^{-1}. Sample: 100 µL with concentration of each metal 4 mg L^{-1} in 0.2% HNO$_3$. Spectrophotometric detection at 658 nm after PCR with arsenazo III. Reproduced, with permission, from Nesterenko and Jones.[5] Copyright (1998) Elsevier.

Elution 123

been exploited for the simultaneous chromatographic separation of transition metals (Mn^{2+}, Cd^{2+}, Pb^{2+}, Zn^{2+}, Co^{2+}), aromatic organic substances (*o*-toluidine, phenols, anilines) and inorganic anions (chloride, bromide, nitrate, sulfate) under isocratic conditions.[7] There are also examples of the use of 3-aminopropylsilica,[8,9] oligoethyleneamine bonded silica[10] and diethylenetriamine functionalised polymer resin columns,[11] for the IC separation of inorganic anions.

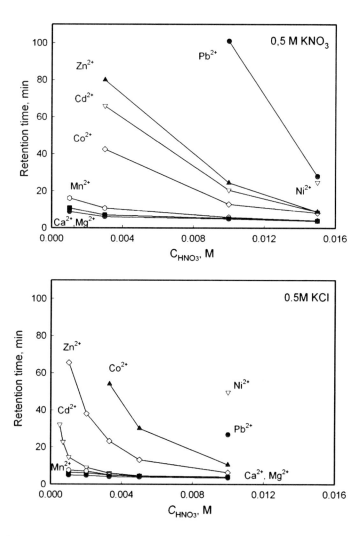

Figure 4.4 Influence of nitric acid concentration on the retention times of alkaline-earth and transition metal ions in the presence of $0.5\,M\,KNO_3$ and $0.5\,M\,KCl$. Column: IDA-silica $250 \times 4.0\,mm$ i.d.; flow rate: $0.7\,mL\,min^{-1}$.

4.1.2 Temperature Effects

A number of studies have been undertaken for the investigation of temperature effects concerning the separation of alkaline-earth metals,[2,12–18] Be^{2+},[19] transition and heavy metals[12,16,18,20] lanthanides[5,21] and other metals both under conventional IC and HPCIC modes. There are various temperature related effects in HPCIC, which can be divided into two groups related to kinetic and thermodynamic changes in the chromatographic system. The theoretical aspects of temperature effects are considered in Chapter 2.

The variation of the column temperature has been mainly used for the improvement of the column efficiency, alteration of the retention times and separation selectivity. In practice the combination of these effects is seen in improved chromatographic separations.

4.1.2.1 Improvement of Separation Efficiency

Rey and Pohl[13] observed a 32–41% increase of peak efficiencies with an increase of the column temperature up to 50 °C for alkaline-earth metal cations separated on an IonPac CS12A column (for details see 5 in Table 3.4) having mixed phosphonic–carboxylic functional groups involving 18 mM methane sulfonic acid as the mobile phase. They also observed a 13–44% decrease of retention factors for the same experiment and associated this effect with changes in hydrated radii of the analytes, which is typical for traditional IC. Intensive investigations of temperature effects for a wider temperature range between 27 °C and 60 °C on the retention of alkaline-earth metal cations on same column has also been performed by Hatsis and Lucy,[14] who confirmed the above-mentioned relationships. However, no significant changes in peak asymmetry at different temperatures were found.

An example of improvement in efficiency and resolution of chromatographic peaks for four metals on a Hamilton PRP-X 800 column, an itaconic acid functionalised chelation ion exchange column (see 8 in Table 3.4), was shown with chromatograms obtained at 19, 40 and 50 °C (Figure 4.5). The peak for cadmium, which is actually retained more strongly at elevated column temperatures, maintains approximately the same peak width in all the presented chromatograms, indicating clearly the increase in separation efficiency obtainable with elevated column temperature.[12]

A further example of the improvement in efficiency possible with an increase in the column temperature has been recently reported by Tria *et al.*[22] The reported study showed dramatic changes in peak shape for aluminium retained on IDA-silica, and increased column efficiency with an increase in the column temperature (Figure 4.6). In fact the study showed the chromatographic peak for aluminium to be tailing significantly at room temperature, 24 °C, yet elute as a fronted peak when the column temperature was raised to 75 °C.

Optimum peak shape and maximum separation efficiency was observed to occur unusually sharply at 71 °C. The changes were reported to be due to at least two kinetic processes occurring within the system. The first is connected

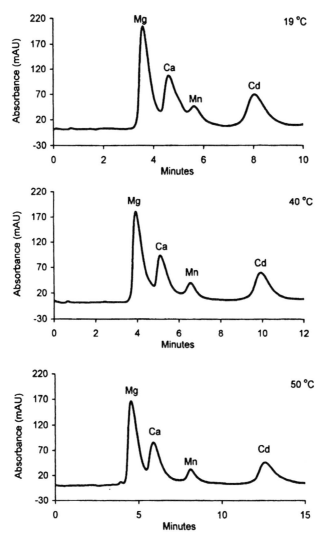

Figure 4.5 Chromatograms showing improvements in separation efficiency and resolution of chromatographic peaks with an increase in the column temperature. Column: Hamilton PRP-X800, 150 × 4.0 mm i.d., 7 μm. Mobile phase: 0.02 M KCl at pH 3.5; flow rate: 1.5 mL min^{-1}. Spectrophotometric detection at 510 nm after PCR with PAR. Reproduced, with permission, from Bashir et al.[12] Copyright (2002) Elsevier.

with changes in complexation rates for kinetically slow aluminium, and the second is due to a simple increase in the diffusion coefficients and decreasing viscosity of the mobile phase according to eqns (2.26–2.28). It should be noted that similar temperature effects are reported by Barron et al.[23] who studied the retention of aluminium on poly-IDA grafted polymeric resin column ProPac IMAC-10 (see properties for adsorbent 4 in Table 3.4).

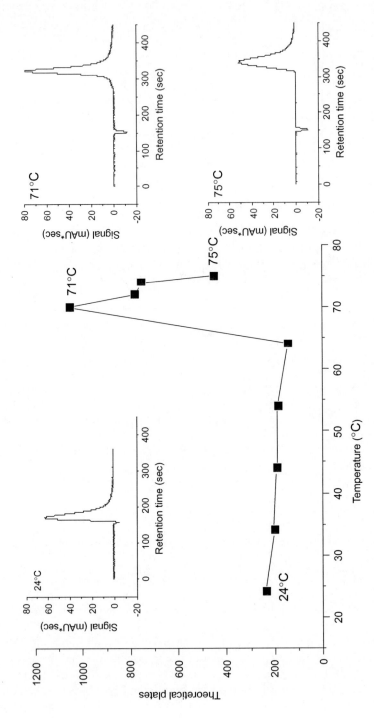

Figure 4.6 The effect of the column temperature on chromatographic peak shape and separation efficiency for aluminium. Column: IDA-silica, 150 × 4.0 mm i.d., 5 μm. Mobile phase: 0.5 M KCl–0.03 M HNO$_3$; flow rate: 0.8 mL min^{-1}. Spectrofluorimetric detection after PCR with lumogallion. Reproduced, with permission, from Tria et al.[22] Copyright (2008) Wiley-VCH Verlag GmbH & Co. kGaA.

The improvements in efficiency with increased temperature may not always be seen in HPCIC. Increased conformational mobility of chelating groups attached to the surface with rising temperature can lead to the formation of complexes with a higher denticity, which exhibit considerably slower dissociation rates.

4.1.2.2 Changes in Retention Mechanism and Separation Selectivity

According to the data on energies for various types of interaction shown in Table 1.1 there must be a substantial difference between heats of adsorption of metal cations for pure electrostatic interactions and those for complexation at the surface. As described in Section 2.4, in the case of conventional ion exchange, the temperature effects are exothermic (negative values of ΔH), while for complexation ion exchange these effects could be both endothermic in the case of bonded ampholyte and acidic ligands and exothermic in the case of basic ligands.

In the absence of complexation, the sorption enthalpies of alkali metal cations calculated using eqn (2.26) for sulfonic cation exchangers, such as IonPac CS10,[15] do not exceed 10 kJ mol^{-1}. Similar values of ΔH for alkali metal cations have been obtained for carboxylic cation exchangers, namely IonPac CS12A, IonPac CS14, PRP-X800, and the Universal Cation column (see properties in Table 3.4),[14,15,17] possessing some complexing properties, and for other chelating ion exchangers, including IDA-silica, lysine and glutamic acid bonded silica.[15,24,25] In each of the above cases, evidently only simple electrostatic interactions were responsible for the observed retention with dilute acid mobile phases.

However, the sorption enthalpies for alkaline-earth metal cations should already reflect the increasing ability of this group of metals to form surface complexes with chelating functional groups. An interesting demonstration of the dual retention mechanism for alkaline-earth metal cations was reported by Kolpachnikova et al.[24] The study demonstrated a convex type dependence of retention for these cations on IDA-silica on column temperature, with a 2 mM perchloric acid mobile phase. The unusual dependence was attributed to the shift from electrostatic interactions (exothermic process) dominating at room temperature, to a dominant chelation based retention mechanism (endothermic process) at high temperatures.

Another good illustration of the opposing thermodynamic effects for simple ion exchange and chelation is given in Table 4.3. A decrease in retention times for alkali-metal cations separated on complexing stationary phases with increasing temperature was observed for an itaconic acid functionalised resin (PRP-X800) column with a 2 mM HNO$_3$ mobile phase. Under these conditions, where there is a dominant simple ion exchange mechanism governing retention, the slopes ($-\Delta H/R$) of van't Hoff plots for magnesium and calcium were positive. When electrostatic interactions were suppressed through the addition

Table 4.3 Slopes ($-\Delta H/R$) and correlation coefficients from van't Hoff plots (eqn. (2.26)) for magnesium and calcium obtained on an itaconic acid functionalised PRP-X800 column.[12]

Metal	Mobile phase					
	0.002 M HNO_3		0.02 M KCl, pH 3.5		0.1 M KCl, pH 5.7	
	Slope	r^2	Slope	r^2	Slope	r^2
Mg^{2+}	0.336	0.319	−0.265	0.952	−0.905	0.969
Ca^{2+}	0.0485	0.326	−0.395	0.967	−0.711	0.990

of 0.02–0.1 M KCl, the relative contribution of chelation in the retention of these metals becomes greater and slopes became negative and more pronounced. Of course, the increasing role of complexation in the retention mechanism caused some changes in the separation selectivity.

A complete reversal in the elution order for the cation pair Mg^{2+}/Ca^{2+} separated on an IDA-silica column under HPCIC conditions with a change in column temperature was noted by Paull and Bashir,[15,20] which was also attributed to a partial change in the retention mechanism. However, due to the combination of exothermic and endothermic effects (see discussion in Section 2.4), the calculated heats of adsorption for alkaline-earth metals on this column do not exceed 1.6–2.3 kJ mol^{-1}.

One more investigation on temperature effects for IonPac CS12A column was performed by Shaw et al.,[16,21] who studied in detail the retention of alkaline-earth, transition, heavy metal cations and lanthanides under suppressed electrostatic interactions. It was found that the retention of all metals increased with an increase of temperature and slopes of van't Hoff plots (see eqn (2.24)) decreased within the group of alkaline-earth metals from Mg^{2+} to Ba^{2+} (Figure 4.7A). The differing responses each cation showed to an increase of temperature was also observed for lanthanide ions, using the same type of the column in 0.01 M HNO_3 to 0.05 M KNO_3 mobile phases (Figure 4.7B). The calculated slopes of log k–1000/T plots varied from −0.62 for La^{3+} to −1.28 for Lu^{3+}, providing a parameter to obtain better separation selectivity and resolution of peaks for the whole group of elements at elevated temperatures.[21] Minimal changes to peak efficiency of lanthanides were noticed with variation of the column temperature.

The advantage of changes in separation selectivity with variation of the column temperature was fully exploited by Nesterenko and Jones for the separation of lanthanides on IDA-silica.[5] The research showed how Y^{3+} exhibited a much steeper slope ($-\Delta H/R$) on the van't Hoff plot than many of the lanthanide ions, and could thus be selectively placed in an elution window between Nd^{3+} and Sm^{3+}, facilitating the isocratic separation of a model mixture of 14 rare earth elements (Figure 4.8). In a similar study, due to the differing sensitivity of beryllium retention to changes in column temperature, compared to the more common alkaline-earth metal cations, it was also possible to optimise the separation of beryllium from these matrix alkaline-earth metals using IDA-silica.[19]

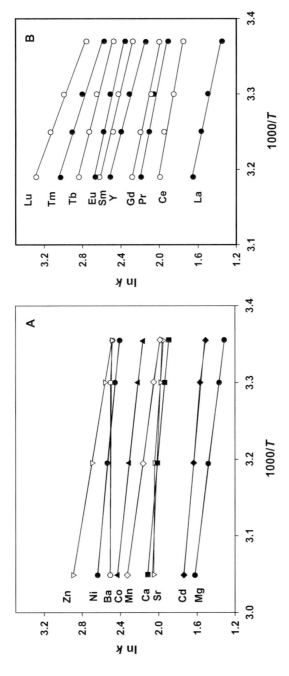

Figure 4.7 Effect of column temperature on the retention of alkaline earth, transition metals and lanthanide ions on IonPac CS12A chelating ion exchanger. Mobile phase: 0.9 mM HNO$_3$–0.2 M KCl (A) and 0.01 M HNO$_3$–0.5 M KNO$_3$ (B). Data sourced from Shaw et al.[16,21]

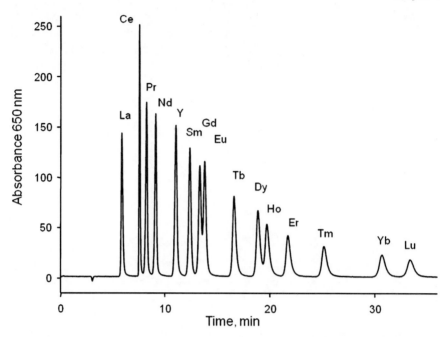

Figure 4.8 Isocratic separation of a model mixture of 14 lanthanides and Y^{3+} on a 150 × 4.0 mm column, packed with 5 µm IDA-silica. Mobile phase: 0.025 M HNO_3 with 0.75 M KNO_3; flow rate: 1.0 mL min^{-1}; column temperature 75 °C; sample volume 20 µL. Sample concentration of each metal was 4 mg L^{-1} in 0.2% HNO_3. Spectrophotometric detection at 650 nm after PCR with arsenazo III. Reproduced, with permission, from Nesterenko and Jones.[72] Copyright (2007) Wiley-VCH Verlag GmbH & Co. kGaA.

A more complex temperature effect could be observed arising from improvements in complexing kinetics occurring between components within the mobile phase system, for example, where a complexing ligand is included within the mobile phase, or within a dynamically modified HPCIC method.

The above temperature effects demonstrate how the variation of column temperature is a useful tool for manipulation of separation selectivity in HPCIC, perhaps more so than alternative modes of LC.

4.1.3 pH of the Mobile Phase

The effect of mobile phase pH on the retention of metal ions in HPCIC depends upon a number of factors, including the existence of pH related secondary equilibria taking place within the mobile phase, which can effect separation selectivity. These reactions may include for example, partial hydrolysis, formation of aqua-complexes and solvates with organic solvents, or most commonly complexation with organic reagents additives to the mobile phase. In HPCIC methods where the basis of obtaining the separation of metals originates purely from a 'formation of substrate-only complexation' mechanism (see Section 4.2.1.1), a decrease in mobile phase pH induces protonation of the ionogenic functional groups within the chelating ion exchanger and thus reduces the effective concentration of the reactive form of the immobilised ligand.

According to eqn (2.32), plots of the logarithm of retention factors, $\log k$, of metal ions *versus* pH of the mobile phase should produce a linear trend with slopes proportional to the number of protons replaced by each metal coordinated to the immobilised ligands. In fact this represents an important difference of HPCIC to simple ion exchange separations, as the definition of ion exchange states an 'equivalent exchange of ions between two and more ionised species located in different phases' (see Chapter 1). Frequently in HPCIC the slopes of $\log k$ – pH plots are not integer numbers, but do have similar values for equally charged metal cations. From a practical point of view, this means no change in separation selectivity under isocratic conditions in mobile phases with different pH or concentrations of strong acids. A typical dependence is shown for lanthanides retained on a mixed phosphonic acid–carboxylic acid functionalised resin within Figure 4.9. The calculated slopes of the $\log k$ – pH dependences are between 1.8 for La^{3+} and 2.1 for Lu^{3+}. Obviously, phosphonic acid functional groups, $-PO_3H_2$, here from an IonPac CS12A column, are responsible for the complexation of trivalent lanthanide cations (Ln^{3+}) with the subsequent release

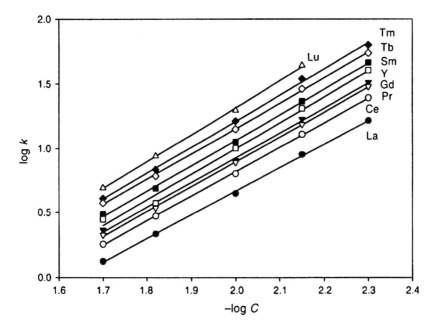

Figure 4.9 Effect of nitric acid concentration on the retention of lanthanides on a IonPac CS12A column. Mobile phase: 0.5 M KNO_3 with various concentration of nitric acid. Reproduced, with permission, from Shaw *et al.*[21] Copyright (2008) CSIRO publishing.

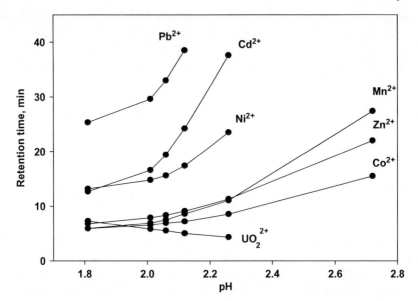

Figure 4.10 Retention of transition metal ions as a function of mobile phase pH. Column: IDA-silica, 250 × 4.0 mm i.d., 7 μm. Mobile phase: 0.02 M oxalic acid; flow rate: 0.7 mL min^{-1}. Reproduced and adapted with permission, from Nesterenko and Jones.[37]

of two protons, according to eqn (4.1):

$$R - P(O)(OH)_2 + Ln^{3+} \leftrightarrow R - PO_3Ln^+ + 2H^+ \quad (4.1)$$

In the presence of competitive complexation within the mobile phase, the effect of pH greatly depends on the acid base properties of added organic reagent/ligand(s). For weakly complexing ligands or dilute concentrations of such, the effect is negligible and a linear dependence for log k – log C is usually observed ($C = [H^+]$). For example, the retention of metal ions on various IDA-functionalised adsorbents in the presence of the weakly complexing maleic acid, obeys the above dependence, in line with that recorded for totally non-complexing mobile phases. However, the use of stronger complexing mobile phase additives, such as oxalic or dipicolinic acids, results in rather complex dependences, as shown in Figure 4.10. The data shown illustrates how the use of the strong complexing ligands within the mobile phase can be used for the further optimisation of separation selectivity through control of pH.

4.1.4 Organic Solvents in HPCIC

There are a number of reasons for the use of polar water-miscible organic solvents (methanol, ethanol, acetonitrile, tetrahydrofuran) as additives to the mobile phase in HPCIC. The most common of these is to achieve an improvement in the solubility of either organic complexing reagent mobile phase additives, or any subsequent metal complexes formed within the mobile phase.[26,27] Usually such organic ligands are rather soluble in aqueous solutions and buffers due to the presence of one or more polar or ionisable functional groups. However, after complexation with metal ions, the precipitation of the resultant complexes can occur, particularly when less polar or neutral complexes, of stoichiometry higher than 1:1 are formed. A typical example is the use of quinaldic acid, which is a known selective reagent for the gravimetric determination of cadmium, copper, zinc and lead in the form of the corresponding quinaldinates. This low solubility of quinaldinates of transition metals restricts the use of aqueous solutions of quinaldic acid within the mobile phase for most modes of cation chromatography. However, the efficient separation of nickel, zinc, cobalt, cadmium and copper was achieved in a reversed-phase HPLC study, based upon on-column formation of complexes utilising a mobile phase of 20% of acetonitrile containing 1 mM quinaldic acid.[28] The improved solubility of the formed complexes was critical to separation efficiency and chromatographic peak shape, and the same principle applies in HPCIC. Additionally, an increase in the column temperature can also improve the solubility of added mobile phase ligands and associated metal complexes (see Section 4.1.2). Organic solvents within the mobile phase may also influence the separation selectivity in HPCIC through a change in the solvation of the metal ions themselves, or through an improvement in the conformation mobility of chelating ligands attached to hydrophobic substrates, thus providing improved accessibility for solute–sorbent interactions.[29]

There have been some studies researching solvent effects in HPCIC, although no comprehensive conclusions can be drawn. For example, in the case of a glutamic acid bonded silica chelating ion exchanger (see Table 3.3), the addition of 5–20% v/v of acetonitrile, methanol or 2-propanol, at varying temperatures did not produce any significant changes in the separation selectivity, but did cause an overall reduction in the retention of alkaline-earth and transition metal cations by 10–15%.[25] However, an improvement in peak shapes for some metal ions was observed, although, this was explained by suppression of hydrophobic interactions between PAR-complexes and PTFE tubing used within the post-column reactor (see Chapter 6). Hatsis and Lucy[14] investigated the effect of the addition of acetonitrile on the separation of alkaline-earth metal cations on an Ionpac CS12A column (see Table 3.4) and also observed a general decrease in the retention of these cations. They also noted a small decrease in peak efficiencies without changes in peak asymmetry.

The use of substantial amounts of organic solvent in the mobile phase may be useful in specific types of applications, such as determination of metals in non-aqueous samples. For example, changes in separation selectivity have been noticed during optimisation of the separation of metal ions present within biofuel samples, on an IDA-silica column. An increase in the mobile phase methanol concentration from 20% to 60%, changed both the elution order (Table 4.4) and column efficiency. Interestingly, in this specific study, the column efficiency increased for iron(II), but decreased for iron(III).[30]

Table 4.4 The effect of the concentration of mobile phase methanol on the separation selectivity and efficiency of IDA-silica (150 × 4.0 mm i.d.) column. Mobile phase: 2.5 mM dipicolinic acid–5 mM HCl; flow rate: 0.8 mL min^{-1}. Data from Dias et al.[30]

Methanol content	Retention time (min)			Theoretical plates		
	Fe^{2+}	Fe^{3+}	Cu^{2+}	Fe^{2+}	Fe^{3+}	Cu^{2+}
20	3.35	4.44	3.88	712	1090	1540
40	3.74	5.23	4.94	1698	899	2040
60	4.31	8.57	9.04	2227	659	1933

Shaw et al.[16] studied the effects of an increase in the concentration of acetonitrile and tetrahydrofuran within a 0.01 M HNO_3 to 0.5 M KNO_3 mobile phase on the retention of rare earth element ions on an Ionpac CG12A column (see properties of the column in Table 3.4). A significant increase in the retention for all cations was observed, while the slopes of log k versus organic solvent volume concentration plots were approximately the same for all metals, bar the heavy lanthanides, Tm^{3+} and Lu^{3+} and Y^{3+}, which displayed considerably steeper slopes. These observations once again provide an additional possibility for modification of separation selectivity, although the exact mechanisms in play here are not completely clear.

Of course, the use of organic solvent additives is possible only with bonded phases in HPCIC, as impregnated/coated resins will suffer from ligand bleed and eventually lose all capacity (see Chapter 3). The same limitations are valid for the use of polar solvents as additives to mobile phases used in extraction chromatography. It should also be noted that the presence of organic solvent within the mobile phase at concentrations of up to 50% and higher, reduces dramatically the solubility of simple inorganic salts, like alkali metal chlorides and phosphates, and so necessary precautions should be undertaken in such cases.[31]

4.1.5 Addition of Oxidising and Reducing Agents

The addition of reducing agents to the mobile phase in HPCIC is an option where samples contain metals ions exhibiting different oxidation states. The most common example would be samples containing both Fe(II) and Fe(III). As Fe(III) can be excessively retained on many chelating resins, its on-column reduction to Fe(II) could be a useful approach to follow. The addition of ascorbic acid to both the sample and to the mobile phase achieves this and only a single peak is eluted. Without such an addition, the oxidation of Fe(II) occurs quite readily in the presence of nitric acid or nitrates within the mobile phase, which can cause significant chromatographic peak distortion and formation of peak 'shoulders'. The above approach can be useful if the chelating column has become contaminated from samples containing high concentrations of Fe(III) and needs subsequent regeneration.

Elution

4.2 Elution Modes

4.2.1 Isocratic Elution

In most modes of chromatography isocratic elution is preferable if suitable separation selectivity is available, mainly for simplicity reasons. In HPCIC this is also the case, as changes in the concentration and composition of the mobile phase can for example cause some serious disturbances in the baseline when using PCR detection, and hence increase the baseline noise, thus decreasing both detection sensitivity and accuracy. When complexation is the main mechanism used to control metal separation then elution can be extremely complex in terms of the number of equilibrium constants involved. Figure 4.11 attempts to give a comprehensive view of how complex the situation for elution can be, where every species concentration is defined by a formation constant. Even in this figure, not all the possible protonated and hydroxyl containing complexes are included. Also, possible metal–metal competition from the background electrolyte is not indicated. If more than one competitive ligand L is present within the mobile phase, for example, up to three is not uncommon (see Figure 4.11), then the metal separation could be characterised by 30 or more formation constants. However, depending on the conditions chosen, such as pH, a number of formation constants will have a very minor or no effect and can be disregarded for the sake of understanding the separation process. Thus, calculations involving many formation constants may not be necessary and useful information on retention processes can be gained using simplified forms of equilibrium equations, while making reasonable assumptions and approximations. There are a number of ways of expressing equilibrium constants and a simplified form has already been used to explain the chelation retention

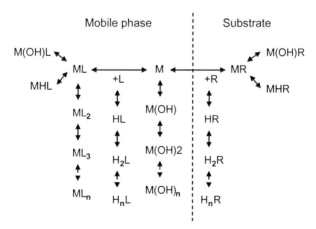

Figure 4.11 A representation of the possible main complexation reactions involved when a ligand (L) and analyte metal (M) in the mobile phase interacts with an immobilised ligand (R) in the stationary phase. Both L and R are completely deprotonated and charges are omitted for clarity.

mechanism in HPCIC in Chapter 2 and elsewhere in this book. Nevertheless, to gain an understanding of the more complex situations to be discussed here, it will be useful to introduce a general equilibrium equation which will encompass all the possible interactions outlined in Figure 4.11. When that equation is established, specific modes of elution can be identified and the equation can be simplified if necessary to explain and predict metal retention. A well established approach for describing complicated complexation reactions is to use equations involving conditional formation constants and so-called alpha coefficients. These have been used for a long time in analytical chemistry pioneered by Ringbom,[32] Schwarzenbach[33] and others in the 1950–1960s. Perhaps the most well known use is in the development of highly specific metal/EDTA titrations involving competitive ligands and pH control as exemplified for use in spreadsheets. A conditional formation constant takes into account the fact that an added component to the system may not all be in the form as defined by the thermodynamic formation constant. Unlike the thermodynamic formation constant (or more correctly the concentration constant measured at a specified ionic strength) the related conditional constant will change with a change in a variety of parameters such as the pH and concentrations of the added components.

First, it is necessary to establish the key formation constant controlling separations in HPCIC. Assuming only a 1:1 ratio of M:R formed in the substrate, which is true for the majority of cases (see Section 3.3.1), it can be defined as:

$$K_f = \frac{[MR]}{[M][\bar{R}]} \quad (4.2)$$

where K_f is the formation constant in terms of molar concentrations. M is the free metal species, *i.e.* not complexed to any ligand except water (the hydroxyl ion is considered a ligand in this context) and $\bar{R}$ is the fully deprotonated complexing or chelating group immobilised on the stationary phase or substrate. Charges will be omitted for clarity in all examples unless important to illustrate a specific point. The conditional constant K_f' related to the above equation can be defined as:

$$K_f' = [\text{total M bound to } \bar{R}]/[\text{total M not bound to } \bar{R}][\text{total R not bound to M}] \quad (4.3)$$

A set of alpha coefficients can now be defined as fractions of the concentration of the three species in eqn (4.3):

$$\alpha_{MR} = [MR]/[\text{total M bound to } \bar{R}] \quad (4.4)$$

$$\alpha_R = [\bar{R}]/[\text{total } \bar{R} \text{ not bound to M}] \quad (4.5)$$

$$\alpha_M = [M]/[\text{total M not bound to } \bar{R}] \quad (4.6)$$

Elution

Thus, we can now express the conditional constant K_f' as:

$$K_f' = K_f' \alpha_M \alpha_R / \alpha_{MR} \tag{4.7}$$

The expression for the conditional constant in eqn (4.7) is fully comprehensive and can incorporate all the complexities outlined in eqns (4.4–4.6), though as already explained, it can be simplified if necessary using appropriate assumptions and approximations.

For this purpose three types or modes of isocratic elution systems can be identified for the separation of metal ions in HPCIC, namely, substrate-only complexation, competitive ligand complexation in the mobile phase and competitive metal complexation in the mobile phase. The last named can only occur in combination with either of the first two.

4.2.1.1 Substrate-only Complexation

This type of elution occurs due to complexation only occurring in the chelating ion exchanger with no ligands in the mobile phase. Concerning eqn (4.3) the [total not bound to M] is much greater in concentration than the other two concentrations and so can be considered a constant. Thus, [total M bound to $\bar{R}$]/[total M not bound to $\bar{R}$] is the distribution ratio D_M and therefore taking into account the phase ratio, the conditional constant K_f' is proportional to the retention factor k. As no complexation is occurring in the mobile phase, $\alpha_M = 1$. However, a small amount of hydrolysis producing $M(OH)_n$, is possible for high charged metals such as Fe^{3+} and Al^{3+}, but at the very low pH values required for elution under substrate-only complexation, hydrolysis can be considered negligible. Furthermore, as the species $M\bar{R}$ is the only one postulated in the substrate under the chosen elution conditions, $\alpha_M = 1$. Therefore under substrate-only complexation:

$$K_f' = K_f \cdot \alpha_R \tag{4.8}$$

where:

$$\alpha_{\bar{R}} = \frac{[\bar{R}]}{[\bar{R}] + [\overline{HR}] + [\overline{H_2R}] + \ldots + [\overline{H_nR}]} \tag{4.9}$$

As R can be increasingly protonated with decreasing pH, α_R and hence K_f' will also decrease. We thus have the same situation as derived in Chapter 2, but expressed in a different way, where a plot of log k against pH will be essentially linear with slope related to the number of protons in the fully protonated form of $\bar{R}$. This linearity will not necessarily hold over a wide range of pH and k, depending on the full expression for α_R, but under conditions normally used for isocratic elution practical measurements, this has shown this to be the case (see Figure 4.9 as an example).

If the species $\overline{\text{MHR}}$ does form to any significant extent, then α_{MR} will not be equal to 1, but essentially the effect of this will be just to produce significant retention at lower pHs than if $\overline{\text{MHR}}$ did not form. It should be emphasised that when considering any form of elution using formation constant and dissociation constant data for the immobilised complexing group $\bar{\text{R}}$ in the substrate, they may not be the same as in free solution (see Section 3.3). These constants reflect the degree of dissociation of the functional groups and upon the degree of metal binding, respectively. The dissociation of part of the functional group changes the effective charge at the surface of the chelating ion exchanger, which should also effect mobility of protons within the rest of the stationary phase. For basic immobilised ligands, such as 3-aminopropylsilica, values of apparent protonation constants calculated for amino-groups from data gained using potentiometric titrations with strong acids, were higher at the beginning of titration and lower at the end.[34] The increasing amount of protonated amino-groups during titration induces the repulsion of protons from the positively charged surface and makes further protonation of the rest of the groups more difficult. A similar sort of effect takes place for the titration of immobilised ligands with transition metal ions in solution.

Taking into account the above theory behind this mode of isocratic elution, some practical limitations should also be mentioned. Under normal circumstances when using bonded phases such as IDA-silica, where the column capacity is quite high, the mobile phase pH range should be limited to greater than 1 and less than 3. pH values up to 4.5 are occasionally necessary when separating the more weakly chelating metals such as the alkaline earths. Lower capacity columns such as the chelating dye coated types will occasionally necessitate pHs of 6 and above. As the substrate is usually a polymer resin there are no problems with stationary phase dissolution. In this case fairly concentrated buffers may be necessary containing mildly complexing ligands to prevent metal hydrolysis at the higher pHs and so strictly cannot be considered as substrate-only complexation.

For all substrates, it is best to avoid using concentrated mineral acids with pHs lower than 1 as it can cause both unwanted stationary phase and equipment stability issues. Also, strong acids can make it difficult to buffer post-column reactions like PAR which operate usually between pH 10 and 11. Additionally, some mineral acids (hydrochloric, phosphoric, sulfuric) can also exhibit complexing properties, significant at higher concentrations, and therefore affect the resultant separation selectivity,[35,36] and so should be considered under the elution mode considered next.

4.2.1.2 Competitive Complexation of Ligands in the Mobile Phase

An alternative elution system is based on the use of a mobile phase containing a complexing ligand (L) to initiate *competitive complexation* within the system, as described in the full form of Figure 4.11. In this case, when considering eqn (4.7), only α_{MR} can be considered equal to 1. Strictly, as mentioned above, this is not really true if the species $\overline{MHR}$ is present, but for the sake of simplicity will not be considered here as it will not affect the outcome to any great extent.

Thus, for this type of competitive complexation:

$$K'_f = K_f \cdot \alpha_M \cdot \alpha_R \tag{4.10}$$

As α_R is the same as in substrate-only complexation, the big change is in α_M. For a fully deprotonated ligand, L, in the mobile phase the alpha coefficient for M becomes:

$$\alpha_M = \frac{[M]}{[M] + [ML] + [ML_2] + \cdots + [ML_n]} \tag{4.11}$$

Similarly as in substrate-only complexation the total concentration of $\bar{R}$ can be considered constant so the distribution ratio will be defined as:

$$D_M = \frac{[\overline{MR}]}{[M] + [ML] + [ML_2] + \cdots + [ML_n]} \tag{4.12}$$

The alpha coefficient for M can be better expressed in terms of L and related formation constants:

$$\alpha_M = \frac{[M]}{[M] + \beta_1[M][L] + \beta_2[M][L]^2 + \cdots + \beta_n[M][L]^n} \tag{4.13}$$

therefore:

$$\alpha_M = \frac{1}{1 + \beta_1[L] + \beta_2[L]^2 + \cdots + \beta_n[L]^n} \tag{4.14}$$

where β_1, β_2 etc. are the cumulative formation constants. It is clear that the addition of the ligand L will decrease both α_M and D_M and hence the retention factor k.

The simplest case is where the ligand L is fully deprotonated throughout the pH range of the mobile phases used with chelating stationary phase. Typically, these ligands would include chloride, bromide, nitrate and sulfate, the last two being only significant for higher charged metals such as the lanthanides, actinides, zirconium and hafnium. In general, the formation constants are relatively small, but will be very useful for fine control of selectivity. As L will not be protonated, once the concentration of L is established in the mobile phase for a particular separation, the change in conditional constant with change in pH will be controlled solely by a change in α_R as in substrate-only complexation. In other words, for a given concentration of L, α_M will remain constant with change in pH.

The most complicated situation is when the ligand L can be protonated in the pH range of interest. So now the concentration of L and hence ML_1, ML_2 etc. will be affected by pH in accordance with the values of the successive acid dissociation constants of L. The α_M will now change with pH, but in the opposite direction to α_R. So, as the pH decreases α_M will increase and α_R will decrease, but unlikely to occur at the same rates. Therefore, plots of pH against log k will not be easy to predict and not necessarily linear as it was shown by Kolpachnikova et al.[24] and Nesterenko and Jones.[37] Furthermore, more than one ligand can be used in the mobile phase to achieve the desired selectivity making the conditional constant equations very complex indeed.

4.2.1.3 Main Types of Competing Ligands

Di- and tri-protic organic acids are the most versatile competing complexing ligands for inclusion within HPCIC mobile phases. The selectivity obtained with these ligands depends on their metal–ligand formation constants, their dissociation constants, and on the pH of the mobile phase. A useful way of comparing the potential effect of a wide range of competing ligands is through the use of reported values of formation constants for some of the common complexing reagents (mainly organic) as presented in Table 4.2. Dissociated organic acids are known to exhibit faster complexation kinetics with metal cations, than those observed for protonated positively charged bases, such as ethylenediamine. This kinetic effect is similar to that described earlier (see Section 3.1) for chelating ion exchangers having negatively charged functional groups. The other important factor to consider is the solubility of the complexes and salts formed within the mobile phase. For example, the direct injection of 0.1 mL samples of seawater onto a chromatographic column of internal diameter 4 mm, equilibrated with 0.02 M oxalic acid, will cause a sharp increase in the column pressure drop, followed by serious chromatographic peak distortion. These effects are due to the precipitation of calcium oxalate at the column inlet. To limit this potential problem, an increase of column temperature and/or the addition of polar organic solvents (methanol, acetonitrile) could be used to improve the solubility of metal complexes with any such mobile phase organic ligands.

Competitive complexation in the mobile phase is frequently used for the optimisation of separation selectivity. These complexing ligands should not form insoluble products with any metal ions present, and ideally also display a different selectivity to such metals than that of the immobilised ligands at the surface of the chelating ion exchanger. Organic acids having acidic (pK_a values < 3) carboxylic groups generally satisfy these demands. Aliphatic acids are frequently used as complexing additives, including oxalic,[35–37,40] maleic,[35,37] tartaric,[35,37,38,41,42] diglycolic[5] and citric acids.[35,38,43–46] Also used are various aromatic acids, such as picolinic,[23,38,47,48] dipicolinic,[38,39,49–51] phthalic[35,52] acids and their derivatives. Some polyampholyte containing mobile phases have also been shown to possess some complexing properties, and have been used for the separation of metals ions with on-column formed pH gradients (see Section 4.2.2.2 and Scheme 4.1). As mentioned previously, some inorganic anions, such as chloride can also be useful in this regard.[36,53]

Sometimes, the addition of the complexing ligands to the mobile phase facilitates photometric detection of separated metals, as they elute in the form of coloured complexes. For example, Slebioda et al. used a 0.1 g L^{-1} solution of 8-hydroxyquinoline in a water–methanol (40:60 v/v) mixed buffered, at pH 8.3, as the mobile phase for the separation of Co^{2+}, Cd^{2+} and Cu^{2+} on an acetylacetone functionalised silica column with direct photometric detection at 390 nm.[26] Toei used 0.1 mM phthalein purple solution in 0.2 M KCl–0.05 M phosphate buffer (pH 5.3) to separate Mg^{2+} and Ca^{2+} within seawater using a TSK-GEL Chelate-5PW column, once again with direct photometric detection, here at 530 nm.[54] Later, a similar approach was used by Paull et al.[27,55] who added phthalein purple to the mobile phase for the dynamic coating of a porous graphitic carbon column, and for the simultaneous provision of sensitive photometric detection. Elefterov et al.[51] separated Co^{2+}, Zn^{2+}, Cu^{2+}, Cd^{2+} and Pb^{2+} on a column packed with IDA-bonded silica and photometrically detected each metal as their corresponding complexes with dipicolinic acid, present as an additive within the mobile phase. The photometric detection of Mo(VI), W(VI), V(v), Re(VII), UO_2^{2+}, Fe^{3+} and Cu^{2+}, in the form of citrate complexes at 310 nm, after separation on an amidoxime bonded silica column has also been reported.[44,56] The use of the photometric organic ligand, trifluoroacetylacetone, within developing solutions has been also reported for the visualisation of resulting chromatograms metals on various complexing stationary phases in a thin-layer variant of chelation ion chromatography.[57,58]

Of course the mobile phase addition of hydrophobic complexing ligands can be used for the stabilisation of an adsorbed layer of the same ligand at the surface of a hydrophobic stationary phase support, resulting in a dynamically modified chelating ion exchangers. Although the background theory is not dealt with here, but covered in detail in Chapter 3, this aspect of competitive complexation should be mentioned for the sake of completeness. Such an approach has been demonstrated using dipicolinic,[59,60] 4-chlorodipicolinic acid,[61,62] quinaldic acid,[60] methylthymol blue,[63] mandelic acid,[64] amongst other reagents. In this case complexation takes place in both within the mobile phase and within the stationary phase, with selectivity similar in both instances. However, complexes of higher stoichiometry may be formed in the mobile phase due to more sterically favourable conditions.

When chelating ion exchangers exhibit relatively weak complexing properties and particularly when the pK_a of functional groups within the immobilised ligand are higher than $\sim 3–4$, the retention of metal ions will only be possible under weakly acidic conditions. In this case the use of buffers is required. Usually, the addition to the mobile phase of mono-carboxylic acids, such as acetic,[65] 3-hydroxypropanoic or lactic acid,[66–71] at concentrations up to 0.05 M, maintains pH and thus improves retention time precision during the separation, and also helps to avoid any undesirable hydrolysis of the metals due to the formation of weakly stable complexes. The logarithms of formation constants (log K_3) for copper complexes, of composition ML_3, are approximately 3.3 and 4.7, for acetate and lactate, respectively.[1] Hence, some effect from use of these weak acid additives on the retention of metal ions might also be expected and can be treated as just additional competing ligands. Inorganic buffers based on phosphate[54] and borate[27] have been also used for keeping pH within a desirable range, although with the former, poor solubility of certain metal-phosphate compounds should be noted.

4.2.1.4 Prediction of Retention and Selectivity Relationships

Under isocratic conditions under suppressed electrostatic interactions (see eqn (2.19)) and in the absence of competitive complexation with components of the mobile phase, the retention of metals can be estimated via values of formation constants as shown by eqn (2.16). A good example of the linear dependence of $\log k - \log \beta_1$ was obtained for IDA-silica with a potassium nitrate–nitric acid mobile phase (Figure 4.12).[72] Obviously, in this case, the ratio of β_1 values for pair of metals can be used for prediction of separation selectivity.

The complexity of competitive complexation, although causing difficulties in assessing and predicting retention and relative retention, still shows the most promise and potential for developing optimum separations to suit a wide

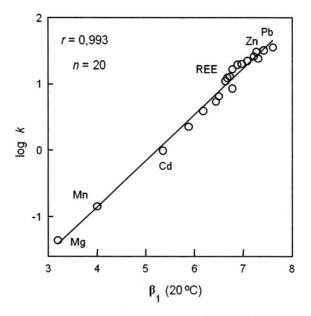

Figure 4.12 Correlation between retention of metal ions on IDA-silica and formation constants of corresponding complexes with IDA measured at 20 °C. Mobile phase: 13.6 mM HNO_3–0.5 M KNO_3, 65 °C. Reproduced, with permission, from Nesterenko and Jones.[72] Copyright (2007) Wiley-VCH Verlag GmbH & Co. KGaA.

Elution 143

variety of sample analyses. In particular it would be very useful indeed to change retention orders to avoid overlap of peaks especially when massive amounts of matrix metals are present within the sample. The equilibrium equations given in this section and elsewhere have been used in a general way to give an overview and understanding of the main types of isocratic elution systems used in HPCIC. To obtain more detailed information on selectivity relationships, which can be used to predict and develop new separation systems, quantitative data needs to be processed based on published formation constants and calculated conditional constants. There are a number of ways this can be done and two examples of very different approaches will be given, namely speciation plots and formation constant plots.

4.2.1.5 Speciation Plots

The conditional formation constants and hence the selectivities of metal separations can be obtained by calculating the alpha coefficients as described in the above equations. This is an onerous task to be done by hand, even when a relatively small number of formation and dissociation constants are involved. As already mentioned, 30 or more constants could be involved when several competitive ligands are used in the mobile phase. When also considering the large number of possible ligands that can be used, computer processing is the only answer. Powerful, fast computer software is now available to process very complex elution systems as encountered in HPCIC. No assumptions or approximations need to be made and all that is required are the values of the formation and dissociation constants and the concentrations of the ligands. Plots of the variation of the concentration of every species against pH can then be obtained. Although conditional constants or distribution ratios will not be given directly, it is relatively straightforward to calculate these from the given data. One such excellent speciation plotting software called MEDUSA can be downloaded as freeware[73] it can process up to 200 equilibrium constants from a maximum of 20 added components (metals and ligands). The input and plotting displays are very 'friendly' and examples of its use are given in Chapter 6, which deals with detection.

Metal speciation models such as MEDUSA have not yet been applied to any significant extent in HPCIC, but the potential is clear. However, there are a number of considerations, which can limit its use, though they can equally apply to other methods. One is the lack of formation constant information on protonated complexes such as $\overline{\text{MHR}}$ involving the immobilised ligand. Without these it is difficult to assess retention selectivity at low pHs. The other thing to watch is the quality of formation constant data, which if not very accurate can give misleading information.

4.2.1.6 Formation Constant Plots

The relationship between formation constants and conditional formation constants is clear, though can be rather complex as already described. However, it is possible that conditional formation constant calculations will not be necessary if trends or relatively major effects in selectivity changes are required. Thus, to understand the possible changes in separation selectivity achieved within HPCIC from the use of mobile phase complexation, the correlation between formation constants of the immobilised ligand and those of the mobile phase complexing ligands can be checked. In some cases complete orthogonality of complexation may exist, allowing considerable scope for selectivity manipulation. Under the separation conditions of HPCIC, the formation of 1:1 surface complexes for divalent metal ions is expected, due to the low flexibility and mobility of grafted chelating ligands. According to the Irving–Williams order,[74] the formation of these complexes should follow, $Ba^{2+} < Sr^{2+} < Ca^{2+} < Mg^{2+} < Mn^{2+} < Fe^{2+} < Co^{2+} < Ni^{2+} < Cu^{2+} > Zn^{2+}$ for relatively acidic mobile phases and multi-dentate attached ligands, including for example, IDA. Nevertheless, occasionally the trend in formation may not be that smooth, with some metals showing significant deviations for a variety of reasons, including changes in crystal field spin states, coordination number, Jahn Teller distortion and so on. Jones and Nesterenko[53] studied the selectivity characteristics of HPCIC when separating a range of metal ions on an IDA-silica column, using a number of complexing mobile phases. Using formation constants from Table 4.2, the corresponding orthogonality plots were obtained for the correlation of complexing properties of IDA and chloride, IDA and picolinic acid, IDA and dipicolinic acid, and IDA and oxalic acids (see Figure 4.13). Consideration of the plots shown in Figure 4.13 revealed those conditions required to alter separation selectivity, through the correct choice of ligand to the mobile phase. Selectivity changes are clear for Hg^{2+}, Fe^{3+}, UO_2^{2+} in the case of oxalic acid additives, Ni^{2+} and Cu^{2+} in the case of picolinic acid, and Hg^{2+}, Cd^{2+} and Pb^{2+} in the case of chloride. However, as discussed in the previous section concerning speciation plots, in practice, the prediction of the separation selectivity is more complicated, as the apparent concentrations of the reactive form of the ligands should also be considered, as the degree of dissociation and total concentration of the ligand in the mobile phase are important parameters.

Nevertheless, the information gained from these plots proved to be very useful and the application of this approach for the adjustment of separation selectivity for seven metal ions on a 150×4.0 mm i.d. column packed with 5 μm IDA-silica is shown in Figure 4.14. The primary separation of seven metals with an oxalic acid containing mobile phase takes approximately 12 min. The presence of two relatively large elution gaps between chromatographic peaks provides the possibility for improvement of the overall separation selectivity. The next step was to add a ligand demonstrating a higher selectivity for Cu^{2+}, here 2 mM picolinic acid. Now the separation of the metal mixture could be performed in just 7 min. However, a further decrease in runtime and an improved resolution of the peaks of Cd^{2+}, Cu^{2+} and Pb^{2+}, could be achieved through the addition of KCl to the mobile phase. The presence of 0.02 M of chloride decreased the retention times of Cd^{2+} and Pb^{2+}, without any significant changes in the retention of other metals and resulted in an optimised 6.5 min baseline separation of all analytes (Figure 4.14c).

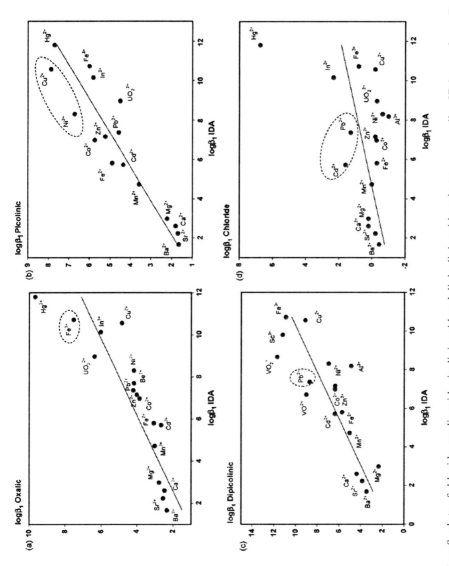

Figure 4.13 Log β_1 plots of chloride, oxalic acid, picolinic acid and dipicolinic acid against corresponding IDA complexes. Reproduced, with permission, from Jones and Nesterenko.[53] Copyright (2008) Elsevier.

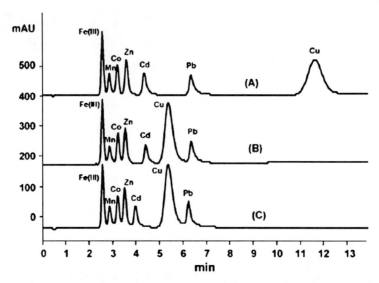

Figure 4.14 Specific changes in separation selectivity of a model mixture of metal ions using multi-component mobile phases, with a 150 × 4.0 mm i.d. IDA-silica column. Mobile phase: (A) 25 mM oxalic acid containing 0.5 M KNO$_3$; (B) 0.025 M oxalic acid containing 0.5 M KNO$_3$ and 2 mM picolinic acid; (C) 0.025 M oxalic acid containing 0.5 M KNO$_3$, 2 mM picolinic acid and 0.02 M KCl. All mobile phases were adjusted to pH 2.40. Spectrophotometric detection at 510 nm after PCR with PAR. Reproduced, with permission, from Jones and Nesterenko.[53] Copyright (2008) Elsevier.

4.2.1.7 Competitive Complexation of Metal Ions in the Mobile Phase

The capacity of a HPCIC column in terms of the apparent concentration of the immobilised ligand on the substrate is very much greater than the concentration of injected metal analytes. Thus, the interaction of each metal analyte with the substrate can be treated independently in terms of their own conditional constants as given in Section 4.1.1. However, if a metal ion is added to the mobile phase at concentration levels similar to the immobilised ligand, then theoretically, the *competitive complexation between various metal ions within the chelating ion exchanger phase* can be used for the manipulation of the elution of metals. This aspect of elution control has been little explored and to avoid the even further complexity of adding it to the isocratic complexation models already discussed, it is better treated here in a simplified form for clarity. The corresponding reaction for substitution of metal ion, M, by a similarly charged metal cation, M′, can be written as:

$$\overline{MR} + M' = \overline{M'R} + M \quad (4.15)$$

Elution

with the equilibrium constant expression:

$$K^M = \frac{[\overline{M'R}][M]}{[M'][\overline{MR}]} = \frac{\beta_1^{M'}}{\beta_1^{M}} \tag{4.16}$$

Clearly, the elution power of cation, M', depends on its formation constant, $\beta_1^{M'}$, and on its concentration within the mobile phase. This type of competitive elution although relatively small should be taken into account in HPCIC, particularly when considering the use of concentrated solutions of alkali metal salts for the suppression of unwanted ion exchange interactions. For example, as previously mentioned, IDA forms relatively stable complexes with Li^+ and Na^+ ions, with log K values of 0.96 and 0.36, respectively, at 20 °C.[1] So, IDA-silica columns should exhibit weaker retention of transition metal cations when used with mobile phases containing lithium salts, compared with the same columns used with sodium and potassium salt containing mobile phases. It would be interesting to consider the use of a background electrolyte containing a dipositive metal ion such as calcium, which should give stronger elution characteristics. However, the large concentrations involved could cause very high baseline absorbances when using certain post-column reagents such as PAR.

4.2.2 Gradient Elution

It is difficult to achieve simultaneous separation of some metal ions under isocratic conditions if such metals form complexes exhibiting significantly different formation constants with the functional groups of the chelating ion exchanger. In this case the separation can be achieved either by addition of a specific additional ligand, or ligands, to the mobile phase as discussed above, and which selectively alters the retention of stronger retained metals, or through the application of gradient elution.

4.2.2.1 Concentration Gradients

Usually chelating ion exchangers exhibit very different selectivity towards various metal cations. As discussed previously, the strength of the interaction between metal ions and chelating ligands, and thus their retention in HPCIC, can be assessed in simple terms by referring to their formation constant values, β_1. These constants, in most instances, are too far apart to obtain an isocratic separation of more than five to six metal ions in a single chromatographic run, using a simple mobile phase based upon a dilute mineral acid solution, with elution according to the simple *substrate-only complexation* type model. Clearly, one solution to this problem is to use concentration gradients.

According to the various possible elution mechanisms within HPCIC (see Sections 4.2.1.1 to 4.2.1.3), concentration gradients need to be formed through variation of the concentration of either non-complexing mineral acids, complexing ligands or metals cations. The profile of these applied concentration gradients is of course very important for final optimisation of the obtained separation. However, the development of concentration gradients of any specific profile is not a trivial task, particularly in the case of using strong acids. If the chelating ion exchanger has weak acid functional groups, as if often the case, changes in the concentration of the strong acid in the mobile phase will obviously cause the protonation of the stationary phase groups, with inevitable pH buffering effects occurring around their corresponding pK_a values. Therefore, in practice there is often a significant difference between the profile of an applied acid concentration gradient at the inlet of the chromatographic column, and the profile of the concentration of acid in the column effluent.

Because of the above difficulties, simple gradients produced through applied step changes in the concentration of acids, have commonly been used for simultaneous separations of cation mixtures containing both weakly and strongly retained components.[66,71,75–77] Actually, in many cases the gradient starts in the mildly alkaline range then changes through to weak and then strong acids. Whatever the starting pH, as a result of the buffering capacity of the chelating ion exchanger, the actual gradient profile formed within the column bed is in fact rather smooth and still acts to provide some beneficial peak shape compression (see Section 4.1.3).

An example of the fast and efficient separation of metal ions, which exhibit wide variations in the formation constants for the stationary phase complexes formed, is presented in Figure 4.15. The formation constants ($\log \beta_1$) of complexes of Cd^{2+} and Cu^{2+} with xylenol orange, used here for the impregnation of PS-DVB particles, are 9.67 and 12.52, respectively.[78] According to eqn (2.20), in the case of isocratic elution, following the *substrate-only complexation phase* model of elution, the separation selectivity for this pair of metals should be:

$$\alpha = \beta_1^{Cu}/\beta_1^{Cd} = 10(\log \beta_1^{Cu} - \log \beta_1^{Cd}) = 10^{2.85}, \sim 708$$

This corresponds to the separation taking approximately 12 h to elute. Strictly, it is the ratio of the conditional constants which should be considered, but the approximate relative ratio should still apply. However, with an applied gradient,[79] all components of the cation mixture could be eluted in only 10 min.

Sometimes the combination of different elution mechanisms can be exploited for the separation of metal cations from complex samples. Such a combined complex gradient elution system was applied for the determination of transition metals in coastal and estuarine seawaters by HPCIC using a 250 × 4.0 mm i.d. IDA-silica column. The elution of an excessive amount of alkaline-earth metals and manganese was achieved first by flushing the column with 0.5 M KCl, adjusted to pH 3.9 with dilute HNO_3. The next step included the selective separation of Cd^{2+}, Co^{2+} and Zn^{2+} with 0.08 M tartaric acid in 20 min. Finally the elution of Ni^{2+} and Cu^{2+} was obtained by washing the column with 0.01 M picolinic acid. The corresponding chromatogram of the model mixture of these metals is shown in Figure 4.16.

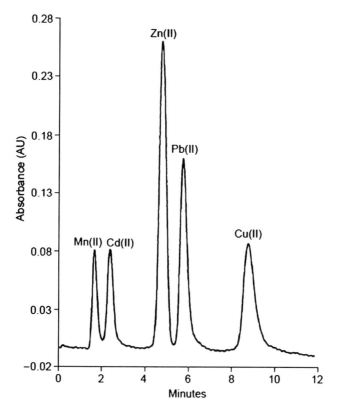

Figure 4.15 Gradient elution of a mixture of 0.5 mg L^{-1} Mn^{2+}, 20 mg L^{-1} Cd^{2+}, 10 mg L^{-1} Zn^{2+}, 10 mg L^{-1} Pb^{2+}, and 1 mg L^{-1} Cu^{2+} on a 100 × 4.6 mm i.d. xylenol orange impregnated 7 μm Hamilton resin column. The gradient was formed by equilibrating the column at pH 3.5, then switching to pH 2.0 for 4 min before injection and switching immediately to 0.1 M HNO$_3$ after injection. Injection volume used was 100 μL with detection at 520 nm with PAR post-column reaction. Reprinted, with permission, from the PhD dissertation of James Cowan.[79]

4.2.2.2 Internal pH Gradients

The above step concentration gradients of inorganic acids in the mobile phase have been used for many years in the HPCIC of metal ions. In many cases where the separation occurred with dilute acids of concentration less than 0.1 M, it could be more appropriate to consider the formation of pH gradients within the chromatographic columns. For example, Challenger et al. separated nine metals on a 100 × 4.6 mm i.d. column packed with micro-particles of PS-DVB resin impregnated with xylenol orange (see details in Table 3.1 and Figure 4.17). The method applied step changes in the concentration of HNO$_3$ in the mobile phase, which also comprised 0.05 M lactic acid and 1 M KNO$_3$.[68] The metal mixture was injected at pH 10, where all metals would be retained on the top of the column. Three further steps of increasing acidification of the

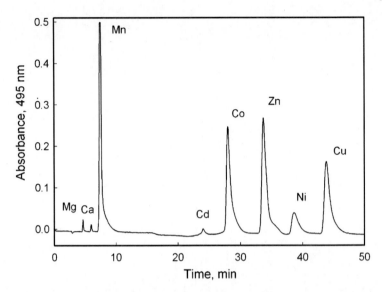

Figure 4.16 Separation of a mixture of alkaline earth and transition metal ions. Three step gradient elution: 0–10 min, 0.5 M KCl–5 × 10^{-4} M HNO$_3$; 10–30 min, 0.08 M tartaric acid; and 30–50 min, 0.01 M picolinic acid. Flow rate: 0.8 mL min^{-1}. Photometric detection 490 nm, PCR with PAR/NH$_4$OH/HNO$_3$ reagent. Sample volume 100 μL; concentrations of metals: 4 mg L^{-1} of Mn^{2+}, Cd^{2+}, Co^{2+} and Zn^{2+}; 9 mg L^{-1} of Mg^{2+}, Ca^{2+} and Cu^{2+}; and 13 mg L^{-1} of Ni^{2+}. Reproduced with permission from Nesterenko and Jones.[38] Copyright (1997) Elsevier.

mobile phase, to pH 6.5, then to 3.0 and finally to pH 0.5, were undertaken, which caused the elution of alkaline-earth metals at the first step, Mn^{2+}, Cd^{2+} and Zn^{2+} at the second stage, and Ni^{2+} and Cu^{2+} at the final step. Xylenol orange has two IDA groups, two phenol groups and one sulfo group, with the following set of pK_a values covering a wide pH range: 12.23, 10.56, 6.74, 2.85, 2.32, <1.5, −1.04, −1.83 and −3.32.[80] The pK_a value of lactic acid used as additive to the mobile phase was 3.86.[1] Altogether, this set of nine ionogenic groups provides an effective polybuffer system and will inevitably result in a relatively smooth pH gradient within the column bed after step increases in the concentration of nitric acid.

Retention of transition metal ions on chelating ion exchangers is strongly dependent upon the mobile phase, so the use of elution with acid concentration gradients is a very useful approach to obtain the simultaneous separation of weakly and strongly retained cations. A change of the mobile phase pH is obviously similar, but as pH is the logarithmic expression for the concentration of H$^+$, large retention changes with relatively small changes in pH, can perhaps appear more dramatic. Because pH is a logarithmic scale, it is much more difficult to create and operate with any particular type of pH gradient. Usually pH gradients can be created externally by mixing two streams of solutions of different concentration according to an exponential function. However, another

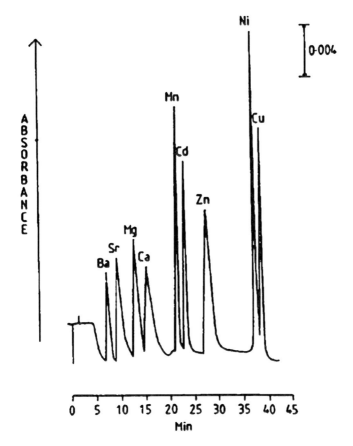

Figure 4.17 Separation of nine metal ions on a 100 × 4.6 mm i.d. column packed with PLRP-S 8 μm PS-DVB resin impregnated with xylenol orange. Mobile phase: 0.05 M lactic acid–1 M KNO_3 adjusted to pH 10 followed with three step acid concentration gradient. Sample concentration: 5 mg L^{-1} of Mg^{2+}, Ca^{2+}, Mn^{2+}, Ni^{2+} and Cu^{2+}; 10 mg L^{-1} of Cd^{2+} and Zn^{2+}; and 20 mg L^{-1} of Ba^{2+} and Sr^{2+}. Detection: photometric at 490 nm after PCR with Zn-EDTA. Reproduced, with permission, from Challenger et al.[68] Copyright (1993) Elsevier.

exciting option is development of a pH gradient directly within the chromatographic column whilst only using isocratic chromatographic equipment.

The use of an internal pH gradient formed within a weak anion exchange chromatographic column has been demonstrated for the separation of proteins and peptides in accordance with their isoelectric points. The other name of this technique is chromatofocusing of proteins. The development of such a pH gradient includes three steps:

1. Equilibration of the weak anion exchange column or, so-called polybuffer ion exchanger, at a pH where the majority of primary and

secondary amino functional groups in the phase of the ion exchanger are partly or fully deprotonated.
2. Passing of a polyampholyte solution through the column, at a pH of three to four units lower than that used for the column equilibration.
3. Re-equilibration of the column with the original buffer used in step 1.

The formation of descending pH gradients is possible due to consecutive titration of the polybasic ion exchanger having a set of amino functional groups with the polyampholyte mobile phase having another set of different pK_a values and a lower pH, adjusted by addition of an acid solution. In practice, the presence of two arrays of functional groups within the polyampholyte molecules in the mobile phase and grafted to the polybuffer ion exchanger matrix cause multiple titrations around each pK_a of the ionogenic groups, monotonically distributed across a broad pH range. This results in an almost linear pH profile along the column and in the effluent. At an originally high pH in the column, the charges of polyampholyte protein molecules from the sample are negative and they are retained by the column packed with weak anion exchanger. The linear pH gradients cause a decrease in the negative charge and acts to reverse the charge of the proteins. The positively charged protein molecules are subsequently repulsed from positively charged functional groups of the anion exchanger, so the elution and separation takes place according to the isoelectric points of the proteins.

Nesterenko and Ivanov[81–83] suggested the use of such a chromatofocusing technique, in particular, the principle of the formation of linear pH gradients within weak anion exchange or weak cation exchange columns, for the separation of metal ions. Instead of electrostatic interactions occurring between protonated groups of a polyamine functionalised adsorbent and negatively charged proteins molecules, the retention of metals on the column was purely due to the formation of complexes with bonded ligands. At high pH, deprotonated amino groups ($-NH_2$, $=NH$) or dissociated carboxylic groups ($-COO^-$) are able to complex transition metal ions which go onto be separated with an applied linear pH gradient according to the 'substrate-only complexation' elution model.

The suitability of different oligoethyleneamine functionalised silicas for use as polybuffer type chelating ion exchangers have been investigated. The ideal polybuffer ion exchanger should have a constant buffering capacity over a broad pH range and chelating groups capable of selectively coordinating metal ions within this pH range. The pH range of constant buffering capacity depends on the number of functional groups with different pK_a values within this range. With a variety of primary and secondary amino groups, with five different pK_a values (9.9, 9.1, 7.9, 4.3 and 2.7)[1] tetrathylenepentamine (tetren) immobilised on a silica surface (see Table 3.2) has a constant buffering capacity over the pH range from 3 to 8, making it an ideal phase for the generation of linear pH gradients.[84] In addition, this chelating ion exchanger adsorbs metal ions over the same pH range (Figure 3.2), with a reasonable intra-group selectivity. At pH 8.5 a variety of metals can be completely adsorbed and retained within the chromatographic column. The separation of four transition metals was possible on the tetren-silica column equilibrated at pH 8.0 with 4 mM histidine buffer, with an internal pH gradient created by passing a diluted solution of Polybuffer 74 at pH 3.5, as shown in Figure 4.18.

The accurate composition of the majority of commercially available polyampholyte buffers is unknown, as are their complexing properties. Thus it is difficult to specify all possible types of interactions with metal ions in this specialist variety of HPCIC. Polyampholyte polybuffer mobile phases, including Polybuffer™ (Pharmacia, Sweden), are composed of concentrated solutions of aliphatic polyaminocarboxylic compounds of molecular mass 300–900, with evenly distributed pK_a and pK_b values.[85] The basic functional groups include primary, secondary and tertiary amino groups, while acidic groups include carboxyls, hydroxyls and in some cases phosphonic groups, possessing the approximate structures shown in Scheme 4.1.

The ratio of the number of basic to acidic groups is approximately 2:1.[85] An interesting property of such polyampholyte buffers is they exhibit very small changes in conductivity with changes in pH. This allows the generation of isoconductive internal linear pH gradients and the use of simple conductivity detection, as shown in Figure 4.18. Obviously aminocarboxylic moieties in polyampholyte buffers could be responsible for complexation with metal ions in the mobile phase, so further investigation of the exact separation mechanism of

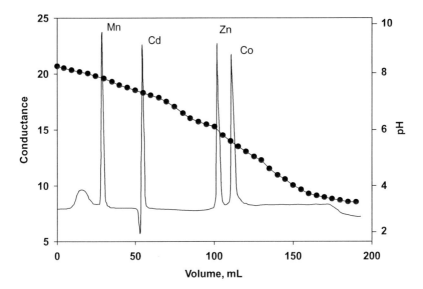

Figure 4.18 Separation of a model mixture of metals using a pH gradient elution. Column: 250 × 4.6 mm i.d. packed with tetraethylenepentamine bonded 10 μm silica particles. Equilibrating buffer: 4 mM histidine pH 8.0; Mobile phase, 1:50 Polybuffer 74 (Pharmacia Fine Chemicals, Sweden); flow rate: 1 mL min^{-1}. Conductivity detection. Reproduced, with permission, from Nesterenko and Ivanov.[82] Copyright (1994) Elsevier.

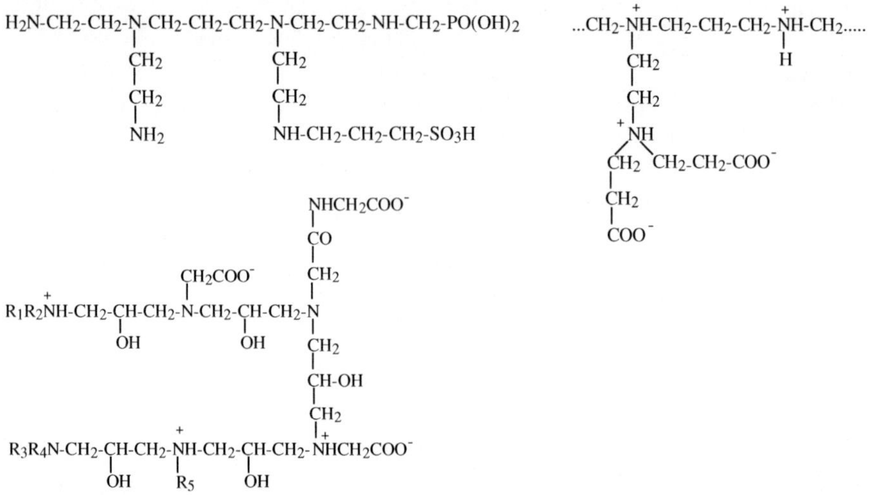

H₂N-CH₂-CH₂-N(-CH₂CH₂NH₂)-CH₂-CH₂-CH₂-N(-CH₂CH₂NH-CH₂CH₂CH₂SO₃H)-CH₂-CH₂-NH-CH₂-PO(OH)₂

...CH₂-N⁺H-CH₂-CH₂-CH₂-N⁺H-CH₂.....
         | |
         CH₂   H
         |
         CH₂
         |
         N⁺H(CH₂CH₂CH₂COO⁻)
         |
         CH₂
         |
         COO⁻

R₁R₂N⁺H-CH₂-CH(OH)-CH₂-N(-CH₂COO⁻)-CH₂-CH(OH)-CH₂-N(-CH₂-CH(OH)-CH₂-N⁺HCH₂COO⁻(-NHCH₂COO⁻)(-CO-))

R₃R₄N⁺-CH₂-CH(OH)-CH₂-NH(R₅)-CH₂-CH(OH)-CH₂-N⁺HCH₂COO⁻

Scheme 4.1 Chemical structures of polyampholyte polybuffer.

metal ions is required. Nevertheless, this approach represents an interesting type of the gradient elution, resulting in impressive separation efficiency.

Clearly internal pH gradients could also be obtained using columns packed with weak cation exchangers. Dissociated carboxylic groups within the chelating ion exchangers can coordinate metal cations resulting in the formation of complexes at high pH, as described in Section 3.2. During the development of an internal descending pH gradient within the column, the complexes are dissociated according to their stability and eluted.

References

1. A. E. Martell and R. M. Smith, *NIST Standard Reference Database 46.* Version 8.0., 2004.
2. W. Bashir and B. Paull, *J. Chromatogr. A*, 2001, **907**, 191–200.
3. A. Haidekker and C. G. Huber, *J. Chromatogr. A*, 2001, **921**, 217–226.
4. P. N. Nesterenko and P. Jones, *Anal. Comm.*, 1997, **34**, 7–8.
5. P. N. Nesterenko and P. Jones, *J. Chromatogr. A*, 1998, **804**, 223–231.
6. M. S. Kuo and H. A. Mottola, *Anal. Chim. Acta*, 1980, **120**, 255–266.
7. H. W. Thompson and J. R. Jezorek, *Anal. Chem.*, 1991, **63**, 75–78.
8. H. J. Cortes, *J. Chromatogr.*, 1982, **234**, 517–520.
9. H. J. Cortes and T. S. Stevens, *J. Chromatogr.*, 1984, **295**, 269–275.
10. G. Domazetis, *Chromatographia*, 1984, **18**, 383–386.
11. J. Li and J. S. Fritz, *J. Chromatogr. A*, 1998, **793**, 231–238.
12. W. Bashir, E. Tyrrell, O. Feeney and B. Paull, *J. Chromatogr. A*, 2002, **964**, 113–122.
13. M. A. Rey and C. A. Pohl, *J. Chromatogr. A*, 1996, **739**, 87–97.
14. P. Hatsis and C. A. Lucy, *Analyst*, 2001, **126**, 2113–2118.

15. B. Paull and W. Bashir, *Analyst*, 2003, **128**, 335–344.
16. M. J. Shaw, P. N. Nesterenko, G. W. Dicinoski and P. R. Haddad, *J. Chromatogr. A*, 2003, **997**, 3–11.
17. P. A. Kebets, K. A. Kuz'mina and P. N. Nesterenko, *Russ. J. Phys. Chem.*, 2002, **76**, 1481–1484.
18. E. Sugrue, P. Nesterenko and B. Paull, *J. Sep. Sci.*, 2004, **27**, 921–930.
19. W. Bashir and B. Paull, *J. Chromatogr. A*, 2001, **910**, 301–309.
20. W. Bashir and B. Paull, *J. Chromatogr. A*, 2002, **942**, 73–82.
21. M. J. Shaw, P. N. Nesterenko, G. W. Dicinoski and P. R. Haddad, *Aust. J. Chem.*, 2003, **56**, 201–206.
22. J. Tria, P. R. Haddad and P. N. Nesterenko, *J. Sep. Sci.*, 2008, **31**, 2231–2238.
23. L. Barron, M. O'Toole, D. Diamond, P. N. Nesterenko and B. Paull, *J. Chromatogr. A*, 2008, **1213**, 31–36.
24. M. G. Kolpachnikova, N. A. Penner and P. N. Nesterenko, *J. Chromatogr. A*, 1998, **826**, 15–23.
25. A. I. Elefterov, M. G. Kolpachnikova, P. N. Nesterenko and O. A. Shpigun, *J. Chromatogr. A*, 1997, **769**, 179–188.
26. M. Slebioda, Z. Wodecki, A. M. Kolodziejczyk and W. Nowicki, *Chem. Anal.*, 1994, **39**, 149–152.
27. B. Paull, M. Macka and P. R. Haddad, *J. Chromatogr. A*, 1997, **789**, 329–337.
28. P. N. Nesterenko and G. Z. Amirova, *J. Anal. Chem. (Engl. Transl.)*, 1994, **49**, 447–451.
29. M. A. Kraus and A. Patchornik, *J. Polym. Sci. Macromol. Rev.*, 1980, **15**, 55–106.
30. J. C. Dias, P. N. Nesterenko, P. R. Haddad, G. W. Dicinoski and L. T. Kubota, *Anal. Methods*, 2010, **2**, DOI:10.1039/COAY00417K.
31. A. P. Schellinger and P. W. Carr, *LC GC North America*, 2004, **22**, 544–548.
32. A. Ringbom, *Complexation in Analytical Chemistry*, Interscience Publishers, N.Y., 1963, p. 395.
33. G. Schwarzenbach and H. Flashka, *Complexometric Titrations*, Methuen, London, 1969, p. 490.
34. G. V. Kudryavtsev and G. V. Lisichkin, *Zh. Fiz. Khim.*, 1981, **55**, 1352–1354.
35. C. H. Risner and J. R. Jezorek, *Anal. Chim. Acta*, 1986, **186**, 233–245.
36. N. Simonzadeh and A. A. Schilt, *Talanta*, 1988, **35**, 187–190.
37. P. Nesterenko and P. Jones, *J. Liq. Chromatogr. Relat. Technol.*, 1996, **19**, 1033–1045.
38. P. N. Nesterenko and P. Jones, *J. Chromatogr. A*, 1997, **770**, 129–135.
39. I. N. Voloschik, M. L. Litvina and B. A. Rudenko, *J. Chromatogr. A*, 1994, **671**, 51–54.
40. I. N. Voloschik, M. L. Litvina and B. A. Rudenko, *J. Chromatogr. A*, 1994, **671**, 205–209.
41. K. H. Faltynski and J. R. Jezorek, *Chromatographia*, 1986, **22**, 5–12.

42. S. Yamazaki, H. Omori and O. Eon, *J. High Res. Chromatogr.*, 1986, **9**, 765–766.
43. G. Bonn, S. Reiffenstuhl and P. Jandik, *J. Chromatogr.*, 1990, **499**, 669–676.
44. P. N. Nesterenko, T. I. Tikhomirova, V. I. Fadeeva, I. B. Yuferova and G. V. Kudryavtsev, *J. Anal. Chem. USSR (Engl. Transl.)*, 1991, **46**, 800–806.
45. P. M. Jonas, D. J. Eve and J. R. Parrish, *Talanta*, 1989, **36**, 1021–1026.
46. I. P. Smirnov and P. N. Nesterenko, *Fibre Chem.*, 1993, **24**, 422–424.
47. P. Jones and P. N. Nesterenko, *J. Chromatogr. A*, 1997, **789**, 413–435.
48. P. Jones and P. N. Nesterenko, Chelation ion chromatography, in *Encyclopedia of Analytical Science*, ed. P. J. Worsfold, A. Townsend and C. E. Poole, Academic Press, Amsterdam, 2004, pp. 467–480.
49. P. N. Nesterenko and O. A. Shpigun, *Russ. J. Coord. Chem.*, 2002, **28**, 726–735.
50. I. N. Voloschik, M. L. Litvina and B. A. Rudenko, *J. Chromatogr. A*, 1995, **706**, 315–319.
51. A. I. Elefterov, S. N. Nosal, P. N. Nesterenko and O. A. Shpigun, *Analyst*, 1994, **119**, 1329–1332.
52. S. Watanesk and A. A. Schilt, *Talanta*, 1986, **33**, 895–899.
53. P. Jones and P. N. Nesterenko, *J. Chromatogr. A*, 2008, **1213**, 45–49.
54. J. Toei, *Fresenius Z. Anal. Chem.*, 1988, **331**, 735–739.
55. B. Paull, P. A. Fagan and P. R. Haddad, *Anal. Comm.*, 1996, **33**, 193–196.
56. P. N. Nesterenko, I. B. Yuferova and G. V. Kudryavtsev, *Zh. Vses. Khim. O va im D. I. Mendeleeva.*, 1989, **34**, 426–427.
57. K. T. DenBleyker and T. R. Sweet, *Chromatographia*, 1980, **13**, 114–118.
58. J. B. Henry and T. R. Sweet, *Chromatographia*, 1983, **17**, 79–82.
59. J. Cowan, M. J. Shaw, E. P. Achterberg, P. Jones and P. N. Nesterenko, *Analyst*, 2000, **125**, 2157–2159.
60. M. J. Shaw, P. Jones and S. J. Hill, *Anal. Chim. Acta*, 1999, **401**, 65–71.
61. M. J. Shaw, P. Jones and P. N. Nesterenko, *J. Chromatogr. A*, 2002, **953**, 141–150.
62. M. J. Shaw, S. J. Hill, P. Jones and P. N. Nesterenko, *Anal. Comm.*, 1999, **36**, 399–401.
63. B. Paull, P. N. Nesterenko and P. R. Haddad, *Anal. Chim. Acta*, 1998, **375**, 117–126.
64. S. Elchuk, K. I. Burns, R. M. Cassidy and C. A. Lucy, *J. Chromatogr.*, 1991, **558**, 197–207.
65. M. J. Shaw, J. Cowan and P. Jones, *Anal. Lett.*, 2003, **36**, 423–439.
66. P. Jones, O. J. Challenger, S. J. Hill and N. W. Barnett, *Analyst*, 1992, **117**, 1447–1450.
67. O. J. Challenger, S. J. Hill, P. Jones and N. W. Barnett, *Anal. Proc.*, 1992, **29**, 91–93.
68. O. J. Challenger, S. J. Hill and P. Jones, *J. Chromatogr.*, 1993, **639**, 197–205.
69. B. Paull, M. Foulkes and P. Jones, *Anal. Proc.*, 1994, **31**, 209–211.
70. P. Jones, M. Foulkes and B. Paull, *J. Chromatogr. A*, 1994, **673**, 173–179.
71. B. Paull, M. Foulkes and P. Jones, *Analyst*, 1994, **119**, 937–941.

72. P. N. Nesterenko and P. Jones, *J. Sep. Sci.*, 2007, **30**, 1773–1793.
73. MEDUSA and HYDRA software for chemical equilibrium calculations. Royal Institute of Technology (KTH). Stockholm, (1999).
74. H. Irving and R. J. P. Williams, *J. Chem. Soc.*, 1953, 3192–3210.
75. P. Jones and G. Schwedt, *J. Chromatogr.*, 1989, **482**, 325–334.
76. C. Pohlandt and J. S. Fritz, *J. Chromatogr.*, 1979, **176**, 189–197.
77. P. Jones, O. J. Challenger, S. J. Hill and N. W. Barnett, *Analyst*, 1992, **117**, 1447–1450.
78. S. Murakami, K. Ogura and T. Yoshino, *Bull. Chem. Soc. Jpn.*, 1980, **53**, 2228–2235.
79. J. Cowan, The deveopment and study of chelating substrates for the separation of metal ions in complex matrices, PhD thesis, University of Plymouth, UK (2002).
80. A. Hulanicki, S. Glab and G. Ackermann, *Pure Appl. Chem.*, 1983, **55**, 1137–1230.
81. P. N. Nesterenko and A. V. Ivanov, *Vestn. Mosk. Univ. Ser. 2 Khim.*, 1992, **33**, 574–578.
82. P. N. Nesterenko and A. V. Ivanov, *J. Chromatogr. A*, 1994, **671**, 95–99.
83. P. N. Nesterenko and A. V. Ivanov, *Mendeleev Commun.*, 1994, 174–176.
84. P. N. Nesterenko, A. V. Ivanov, N. A. Galeva and G. B. C. Seneveratne, *J. Anal. Chem. (Engl. Transl.)*, 1997, **52**, 736–742.
85. P. G. Righetti, *Isoelectric Focusing: Theory, Methodology and Applications*, Elsevier Biomedical Press, Amsterdam, 1983, p. 386.

CHAPTER 5
Liquid–Liquid Chromatographic Methods

5.1 High Speed Counter-current Chromatography of Metal Ions

Counter-current chromatography (CCC) was introduced in early 1970s and was considered for a long time as a simple preparative method for the separation of simple multi-component mixtures. There are several versions of CCC, as shown in Scheme 5.1, which are all based upon the repetitive multiplication of extraction equilibrium between two immiscible phases.

High-performance and high-speed CCC methods are those which are capable of achieving separation efficiencies comparable to those of analogous modes of LC, and are of course based upon essentially the same underlying chromatographic principles. In the CCC of metal ions, a hydrophobic chelating ligand is dissolved within an organic solvent phase, which acts as the stationary phase. In this respect, the technique is clearly analogous to HPCIC, but in this case based on liquid-liquid partition of analytes.

The separation selectivity exhibited by CCC methods is based upon differences in the distribution ratio of the metal (D_M) defined as follows for any extraction based chromatographic system:

$$D_M = C_s/C_m \tag{5.1}$$

where C_s and C_m are concentrations of the solute within either the stationary or mobile phases, respectively. Similarly for all common chromatographic systems, the retention volume (V_R) of the solute within the CCC system can be expressed as follows:

$$V_R = V_m + D_M V_s \tag{5.2}$$

RSC Chromatography Monographs No. 14
High Performance Chelation Ion Chromatography
By Pavel N. Nesterenko, Phil Jones and Brett Paull
© The Royal Society of Chemistry 2011
Published by the Royal Society of Chemistry, www.rsc.org

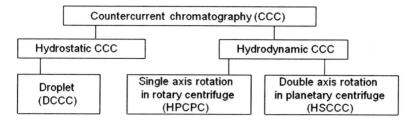

Scheme 5.1 Classification of counter-current chromatography methods.

where V_m and V_s are the volumes of mobile and stationary phases, respectively. According to eqn (5.2), the separation factor (α) of two metal ions eluted in volumes V_R^{M1} and V_R^{M2} is simply defined by the ratio of their partition coefficients for the corresponding extraction system:

$$\alpha = \frac{V_R^{M2} - V_m}{V_R^{M1} - V_m} = \frac{D_{M2}}{D_{M1}} \quad (5.3)$$

All CCC methods can be divided into hydrostatic or droplet CCC (DCCC) and droplet centrifugal partition chromatography (CPC) and hydrodynamic or planetary centrifuge CCC, including high-speed (HSCCC), high-performance centrifugal partition chromatography (HPCPC) and fast centrifugal partition chromatography (FCPC) modes (Scheme 5.1). Original hydrostatic CCC used gravity for passing droplets of mobile phase through a layer of immiscible liquid stationary phase, as shown in Figure 5.1a. Depending upon the relative density of the aqueous and organic phases, droplets either moved up in 'ascending type' or down in 'descending type' of DCCC. In 1982, a faster and technically more advanced type of hydrostatic CCC, also known as high-performance centrifugal partition chromatography (HPCPC) (or fast centrifugal partition chromatography (FCPC), was introduced.[1] The application of centrifugal force facilitated improved control of the separation process with specially constructed columns resulting in more efficient separations. In HPCPC, the analytical scale chromatographic system consists of several hundred to several thousand small 10–15 mm length partition cells placed within a rotor (Figure 5.1b), of total volume equal to between ~50 or 250 mL. The maximum rotation speed applied within HPCPC is approximately 2000 rpm, with a typical flow rate of between 1 and 5 mL min^{-1}. The separation efficiency possible depends on number of partition cells, the flow rate of mobile phase, rotation speed and phase ratio (V_s/V_m). For the separation of metal ions by HPCPC with chelating ligands within the stationary extracting phase, the reported maximum number of theoretical plates per column is typically around 1500, under optimum conditions for the a column with >2000 partition cells.[2]

Until relatively recently, low separation efficiency and long run times seriously limited the applications of CCC. However, significant improvements in instrumental design, such as the work of Ito,[3] who described the use of new

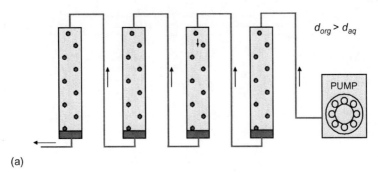

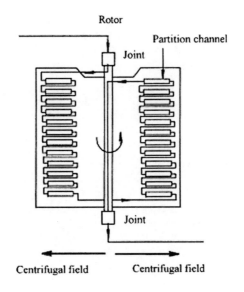

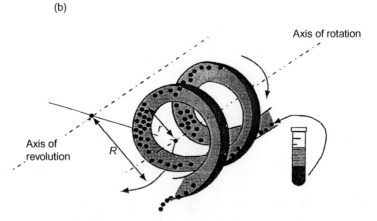

Figure 5.1 Schematic diagrams of instrumental design for (a) hydrostatic and (b) hydrodynamic (centrifugal) variants of counter-current chromatography. Figure (b, bottom) is reproduced, with permission, from Kitazume.[6]

sophisticated planetary centrifuge based CCC systems, have led to improvements in the speed, resolution and efficiency of separations. So-called high-speed CCC (HSCCC) has achieved separation efficiencies for metal cations of 5900 theoretical plates, for a column of 300 m length and 1.07 mm i.d.,[4] with even more impressive efficiencies of up to 12 000 theoretical plates reported for the separation of indolacetic acids.[5,6]

Commercial sources of analytical scale HPCPC chromatographs (EverSeiko Corporation, Tokyo, Japan and Kromaton, Sainte Gemmes sur Loire, France), and HSCCC chromatographs (Tauto Biotech Co., Shanghai, China, and J Coil-Planet Centrifuge, Renesas Eastern Japan Semiconductors Inc., Tokyo, Japan) are now available. Importantly, various sources of CCC chromatographic equipment are also now available fully compatible with, or come with actual built-in post column reactors, allowing in both instances the selective and sensitive detection of separated metal species.

5.1.1 Extracting Ligands and Metal Ion Selectivity

Counter-current chromatography employs a two-phase liquid system for the separation of metals, one of which contains complexing ligands to provide the desired separation selectivity, in a similar way as HPCIC. An important advantage of HSCCC is an absence of the solid stationary phase support, with the ligands being covalently bound or physically adsorbed. This simplicity permits the use of chelating ligands of great diversity, and often highly complex structure. This approach also means surface induced steric and structural effects upon metal–ligand complexation are avoided, with more predictable complexes of known stoichiometry being formed. However, the choice of ligand in HSCCC of metal ions requires careful consideration, as it needs to be soluble at reasonable concentrations within the non-polar organic solvent which composes the stationary phase, and also should exhibit different selectively toward the metals requiring separation. The wide variety of chelating ligands used in HSCCC for metal ion separations include, carboxylic, dialkylphosphoric, dialkylphosphonic, dialkylphosphinic and dithiophosphinic acids, acylpyrazolones, carbamoylphosphonates, bridge substituted tetraalkylmalonamides, crown ethers, oximes and many other similar compounds, as detailed within Table 5.1.

In 1957 Peppard et al.[7] first suggested dialkylphosphates as extractant ligands and later in 1963, Pierce and Peck[8] demonstrated the outstanding selectivity of di-(2-ethylhexyl)phosphoric acid (DEHPA, Table 5.1) for the extraction of lanthanides. Since that time various dialkyl derivatives of phosphoric, phosphonic and phosphinic acids have been studied, but DEHPA probably remains the most popular extracting ligand for the REEs. Due to hydrogen bonding, dialkylphosphoric acids form dimers in non-polar solvents, which interact with trivalent cations to form complexes according to the following equilibrium:

$$M^{3+}_{aq} + 3(HL)_{2,org} = M(HL_2)_{3,org} + 3H^+_{aq} \qquad (5.4)$$

Table 5.1 Structures of extracting ligands applied in counter-current chromatography of metal ions.

Structure	Name
$R_1O\text{-}P(=O)(OH)\text{-}OR_2$	di-(2-ethylhexyl)phosphoric acid (DEHPA) $R_1, R_2 = -CH_2CH(C_2H_5)C_4H_9$ di-2-methylnonylphosphoric acid (DMNPA) $R_1, R_2 = -CH_2CH(CH_3)C_7H_{15}$
$R_1\text{-}P(=O)(OH)\text{-}R_2$	Bis(2,4,4-trimethylpentyl)phosphinic acid (Cyanex 272) $R_1, R_2 = -CH_2CH(CH_3)CH_2C(CH_3)_3$ bis(2-ethylhexyl)phosphinic acid (BEHPA) $R_1, R_2 = -CH_2CH(C_2H_5)C_4H_9$
$R_1O\text{-}P(=O)(OH)\text{-}R_2$	2-ethylhexylphosphonic acid mono-2-ethylhexyl ester (PC-88A, EHPA) $R_1, R_2 = -CH_2CH(C_2H_5)C_4H_9$
$R_1\text{-}P(=S)(SH)\text{-}R_2$	bis(2,4,4-trimethyl)dithiophosphinic acid (Cyanex 301) $R_1, R_2 = -CH_2CH(CH_3)CH_2C(CH_3)_3$ Dichlorophenyldithiophosphinic acid $R_1, R_2 = -p\text{-}ClPh$ (DCPDTPI)
(sec-octylphenoxy-acetic acid structure)	*sec*-Octylphenoxy-acetic acid (CA12)
(1-phenyl-3-methyl-pyrazol-5-one structure with R_1)	1-phenyl-3-methyl-4-capryloylpyrazol-5-one (HPMCP) $R_1 = -n\text{-}C_7H_{15}$ 1-phenyl-3-methyl-4-benzoylpyrazol-5-one (HPMBP) $R_1 = -C_6H_5$
(malonamide structure with R_1, R_2, R_3)	N,N-dimethyl-N,N-dioctylhexylethoxy-malonamide (DMDOHEMA) $R_1 = -n\text{-}C_8H_{17}$; $R_2 = -CH_3$; $R_3 = -C_2H_4OC_6H_{13}$ N,N'-dimethyl-N,N'-dibutyltetradecylmalonamide (DMDBTDMA) $R_1 = -n\text{-}C_4H_9$; $R_2 = -CH_3$; $R_3 = -n\text{-}C_{14}H_{29}$ N,N'-dimethyl-N,N'-dibutyldodecyloxyethylmalonamide (DMDBDDEMA): $R_1 = -C_4H_9$; $R_2 = -CH_3$; $R_3 = -C_2H_4OC_{12}H_{25}$
$R_1O\text{-}P(=O)(OR_1)\text{-}(CH_2)_n\text{-}C(=O)\text{-}N(R_2)_2$	N,N'-diethylcarbamoylmethylenephosphonic acid di-n-hexyl ester (DHDECMP) $R_1 = -(CH_2)_5CH_3$; $R_2 = -C_2H_5$; $n = 1$ diphenyl(dibutylcarbamoylmethyl)phosphine (Ph$_2$Bu$_2$)

Table 5.1 (*Continued*).

Structure	Name
R₁, R₂, R₃, R₄ substituents on O=P–P=O bridge	Tetraphenylmethylenediphosphine dioxide (TPMDPD) $R_1, R_2, R_3, R_4 = Ph$
(hydroxyoxime structure)	5,8-diethyl-7-hydroxydodecan-6-one oxime (LIX 63)
(crown ether structure)	Dicyclohexano-18-crown-6 (DCH18C6)

The affinity of an extracting ligand such as DEHPA to each metal cation can be described by the following equilibrium constant:

$$K_{ex} = \frac{[M(HL_2)_3]_{org}[H^+]^3_{aq}}{[M^{3+}]_{aq}[(HL)_2]^3_{org}} \qquad (5.5)$$

while the ratio of constants K_{ex} for any specific pair of metal ions is equal to the separation factor α (eqn (5.3)), and reflects the selectivity, and hence ability of this system to separate these metal ions.

The retention of a particular metal in CCC, is of course proportional to its specific distribution coefficient, D_M, which is simply defined as:

$$D_M = \frac{[M(HL_2)_3]_{org}}{[M^{3+}]_{aq}}. \qquad (5.6)$$

The selectivity of extraction of REEs with dialkylphosphoric acids is predetermined by the difference in stability constants of the resultant complexes. However, because of the complexity of adducts within the organic phase, no critical values for stability constants for metal complexes with dialkylphosphoric acids have been published, so the selectivity of the corresponding CCC separation is usually correlated with extraction constants,[9] K_{ex} (Table 5.2).

From eqns (5.5) and (5.6) the following relationship can be derived:

$$\log D_M = \log K_{ex} + 3\log[(HL)_2]_{org} - 3\log[H^+] \qquad (5.7)$$

According to eqn (5.7), retention in CCC depends strongly upon pH and the concentration of reagent within the stationary phase. The corresponding plots of $\log D_M$ against pH and $\log[(HL)_2]_{org}$ will give straight lines with slopes equal

Table 5.2 log K_{ex} values for the extraction of REEs and actinides using DEHPA. Adopted from Kolarik[9].

Metal	Toluene, 1 M NaNO$_3$	Toluene, 1 M NaClO$_4$	Kerosene	Metal	Toluene, diluted HCl
La^{3+}	−2.4±0.1			Am(III)	
Ce^{3+}	−1.8±0.1			Cm(III)	−1.4±0.1
Pr^{3+}	−1.5±0.1			Bk(III)	−0.4±0.1
Nd^{3+}	−1.3±0.1			Cf(III)	−0.15±0.1
Pm^{3+}	−0.85±0.1	−1.7±0.1		Fm(III)	0.5±0.1
Sm^{3+}	−0.5±0.1	−0.95±0.1			
Eu^{3+}	−0.1±0.1	−0.8±0.1			
Gd^{3+}	−0.05±0.1				
Tb^{3+}	0.6±0.1	0.4±0.1			
Dy^{3+}	1.0±0.1	0.9±0.1			
Ho^{3+}	1.3±0.1	1.25±0.1			
Er^{3+}	–	1.75±0.1			
Tm^{3+}	2.3±0.1	2.3±0.1			
Yb^{3+}		2.8±0.1			
Lu^{3+}		2.95±0.1			
Y^{3+}			4.4±0.1		
Sc^{3+}	8.8±0.1		11.5±0.1		
Am(III)	−1.8±0.1				
Bi^{3+}	1.45±0.1				
U(VI)	3.55±0.1	3.58±0.1			

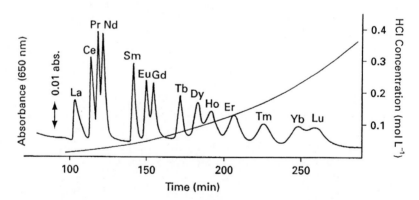

Figure 5.2 Separation of REEs by HSCCC and applied pH gradient elution. Column 300 m × 1.07 mm i.d., comprising three multilayer coils connected in series with a 3 mM solution of DEHPA in *n*-heptane as the stationary phase. Mobile phase: exponential gradient of 0–0.4 M HCl; flow rate: 5 mL min^{-1} at 900 rpm, pressure 2.1 MPa. Sample: 0.1 mL solution containing 25 µg of each component. Detection: spectrophotometric at 650 nm after PCR with arsenazo III. Reproduced, with permission, from Kitazume *et al.*[4] Copyright (1991) Elsevier.

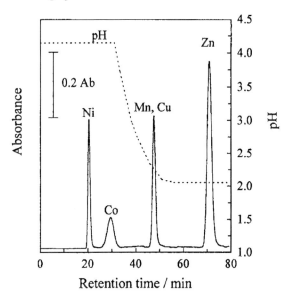

Figure 5.3 Separation of divalent metals by HSCCC and applied pH gradient elution. Column: consisted of three layers of 150 helical turns of a 50 m PTFE tubing (0.5 mm i.d.) connected in series with solution of 0.02 M (EHPA)$_2$ in *n*-hexane as stationary phase. Mobile phase: pH gradient with 0.01 M acetic acid–sodium monochloroacetate (pH 4.15) at starting point and 0.01 M sodium monochloroacetate–sodium dichloroacetate (pH 2.05) at the end of elution; flow rate: 0.3 mL min^{-1} at 1200 rpm. Sample: 0.1 mL of 0.01 mM each of separated metals. Detection: spectrophotometric at 500 nm after PCR with 10^{-4} M solution of PAR in 0.1 M NH$_4$OH–acetone mixture. Reproduced, with permission, from Hosada et al.[10] Copyright (2002) Japanese Society for Analytical Chemistry.

to 3. Thus stepwise or continuous pH gradient elution is often used to obtain the separation of complex mixture of metals, as shown in Figures 5.2 and 5.3, acquired using a DEHPA and EHPA in *n*-heptane stationary phases, correspondingly.[4,10] Obviously, the retention of metal ions in other CCC systems can also be described in a similar way, by taking into consideration the ion charge and the stoichiometry of complexes formed.

Clearly the stability of the associates formed between the extracted metal and ligand within the stationary phase solvent determines the magnitude of K_{ex}. The previous basic complexation equilibrium (eqn (5.4)), in simplified form can be expressed as:

$$M^{3+}_{aq} + 3HL_{org} = ML_{3,org} + 3H^+_{aq} \quad (5.8)$$

With the extraction constant expressed as:

$$K_{ex} = \frac{D_M K_a^3 \beta_3}{D_R^3} \quad (5.9)$$

where D_R is distribution coefficient for reagent, K_a, the dissociation constant of the chelating ligand, and β_3, is the stability constant of the ML$_3$ complex. For complexes of similar structure, the separation factor for a pair of metals (eqn (5.3)) is equal to the ratio of the corresponding stability constants, $\alpha = \beta_{M2}/\beta_{M1}$. A good illustration of this rule is the elution order of lanthanides separated with the DEHPA containing stationary phase, using a gradient of HCl within the mobile phase (Figure 5.2). The elution order from La^{3+} to Lu^{3+} corresponds precisely to the stability of lanthanide complexes with dialkylphosphoric acids.[11]

Therefore, the approximate retention order for various other cations, with the DEHPA containing stationary phase can also be estimated, as shown below, according to these extraction constants, as critically evaluated by Kolarik.[9] Similar selectivity was observed for other reagents, for example EHPA or PC-88A:[10]

$$Ni^{2+} < Co^{2+} < Ba^{2+}, Mg^{2+}, Sr^{2+} < Cu^{2+} < Cd^{2+} < Mn^{2+}$$
$$< Ca^{2+} < Zn^{2+} < REE < UO_2^{2+} < Fe^{3+} < Sc^{3+}.$$

For most of the published data on CCC separations, the retention data shown is in good agreement with this order.[2,4,12–16]

Figure 5.4 shows a typical chromatogram of nickel, cobalt, magnesium, and copper obtained with a 0.2 M DEHPA in *n*-heptane stationary phase and elution with a 7 mM citric acid mobile phase delivered at a flow rate of 5 mL min^{-1}. The retention times of actinides and lanthanides on this stationary phase overlap. It should also be noted that the retention and the separation of alkali metal and alkaline earth cations with such acidic extractants is possible due to extraction of simple salts or ion pairs, formed through electrostatic interactions.[9]

In general, because of similarity in the structures of the functional groups of the ligands listed within Table 5.1, the separation selectivity achieved using dialkylphosphoric, dialkylphosphonic and dialkylphosphinic acids based stationary phases, is also very similar. The main difference between these ligands is their acidity, which increases from dialkylphosphoric (p$K_a \sim 2.75$), to dialkylphosphonic (p$K_a \sim 4.85$), to dialkylphosphinic acids (p$K_a \sim 5.95$).[13] The optimum pH range required for interaction of metal ions with these complexing stationary phases follows a similar order. Of course, when exceeding this range the probability of secondary equilibria (*e.g.* partial hydrolysis) is increased. Minor factors effecting the extraction ability of organo-phosphoric acids include ligand hydrophobicity and dimerisation constants.

The manipulation of separation selectivity through competitive complexation in both stationary and mobile phase has not been widely used in CCC. However, the addition of complexing agents like tartaric or citric acids in concentration up to 0.14 M to the mobile phase has been explored to facilitate elution of strongly retained cations, such as Fe^{3+} [14,15] with alkylphosphoric and alkylphosphinic acid based stationary phases. The use of 0.1 M tartaric

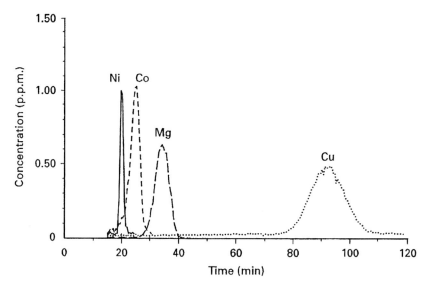

Figure 5.4 Isocratic HSCCC separation of metal cations. HSCCC centrifuge with 10 cm revolution radius; Column: 150 m × 1.6 mm i.d., one multilayer coil, 300 mL capacity. Stationary phase: 0.2 M DEHPA in heptane. Mobile phase: 7 mM citric acid; flow rate: 5.0 mL min^{-1} at 800 rpm, pressure 1 MPa. Sample: 0.1 mL solution containing 10 pg of Ni^{2+}, 20 pg each of Co^{2+} and Mg^{2+}, 40 pg of Cu^{2+}. Detection: direct current plasma atomic emission spectrometry (DCP-AES). Reproduced, with permission, from Kitazume et al.[14] Copyright (1993) American Chemical Society.

acid within the mobile phase, in combination with stepwise pH changes, not only resulted in the early elution of Fe^{3+}, but also resulted in a number of other changes to the separation selectivity of the system.[15]

5.1.2 Efficiency of Metal Separations using HSCCC

The kinetics of complexation reactions within both the stationary and mobile phases is crucial in relation to achievable separation efficiency. For example, the peak efficiency for Fe^{2+} and Fe^{3+}, separated on DEHPA containing stationary phase with a 0.14 M citric acid containing mobile phase (native pH about 1.9) was reported as 3700 and 410 theoretical plates, respectively (separation shown as Figure 5.5). The difference indicates that when Fe^{3+} has to exchange from its relatively stable (log $\beta_2 = 15.94$) citrate complexes within the mobile phase into the corresponding DEHPA complex, ML_3, within the stationary heptane phase, the exchange kinetics of this re-complexation are rather slow, translating into a relatively poor mass transfer term. At the same time, Fe^{2+}, which does not form stable citrate complexes under the same conditions, can be directly transferred into the organic phase, meaning an improved mass transfer term and resulting in the high peak efficiency shown, relative to Fe^{3+}.

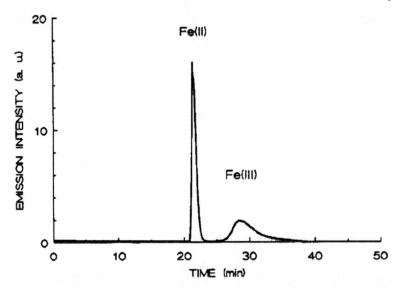

Figure 5.5 Isocratic separation of Fe^{2+} and Fe^{3+} ions by HSCCC. Stationary phase: 0.002 M DEHPA in heptane. Mobile phase: 0.14 M citric acid; flow rate: 4.5 mL min^{-1} at 800 rpm, pressure 1 MPa. Sample: 0.1 mL solution containing 0.5 mg each of metals. Detection: direct current plasma atomic emission spectrometry (DCP-AES). Reproduced, with permission, from Kitazume et al.[14] Copyright (1993) American Chemical Society.

It should also be noted here that in the case of CCC based separations, the formation of fully coordinated complexes is more likely, due to the mobility of both the extracted metal ion and the extracting ligand molecules within the stationary phase. In HPCIC only 1:1 or occasionally 1:2 complexes can be formed due to the fixed position of the chelating molecules upon a surface. The overall lower stoichiometry of complexes seen in HPCIC tends to result in faster complexation kinetics, which is an significant factor, adding to the generally higher levels of efficiency seen with HPCIC as compared to HSCCC.

Another important factor in obtaining improved mass transfer within CCC separations of metal ions, is the charge of the complexing ligand in the stationary phase. As discussed previously in relation to HPCIC (see Chapter 3), neutral (e.g. TBP) or positively charged ligands (e.g. TOA, TOEDA) have tended to provide poorer separation efficiencies for cationic species in comparison with more acidic ligands. The reason for this is the occurrence of long distance repulsion of the cations from the positively charged extracting ligand within the organic phase, or in the case of the neutral ligand, a lack of any electrostatic attraction to increase the rate of ion interaction. Such electrostatic attraction, as expected between negatively charged acidic reagents and cations, contributes significantly to improved phase-transfer kinetics.

Finally, as with many forms of chromatography, miniaturisation of the column used within HSCCC can also increase the separation efficiency

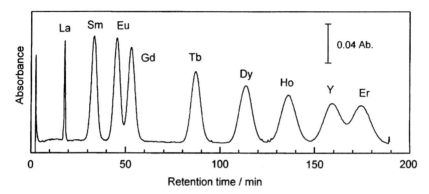

Figure 5.6 Enrichment and chromatographic and separation of selected REEs using HSCCC. Stationary phase: 0.04 M PC-88A in n-hexane. Mobile phase: 0.2 M $Cl_2CHCOONa$, with a pH gradient from 2.16 to pH 1.14 over 240 min; flow rate: 0.5 mL min^{-1} at 1250 rpm. Sample: 100 mL solution containing related metal ions of 5×10^{-8} M each. Detection: spectrophotometric at 650 nm after PCR with arsenazo III. Reproduced, with permission, from Tsuyoshi et al.[17] Copyright (2000) Taylor and Francis.

achievable. For example, Tsuyoshi et al. used a 50 m × 0.5 mm i.d. column, of total volume equal to 10 mL, for the separation of REEs and obtained a 1.5–3.0 fold increase in the number of theoretical plates, as compared with that achieved using a 150 m × 1.5 mm i.d. column of total volume 268 mL, with all other conditions similar.[17] The impressive HSCCC REE chromatogram obtained with the smaller diameter column is shown as Figure 5.6. It should be also underlined that 100 mL of solution containing separated metal ions of 5×10^{-8} M each, was enriched on the same column and used as a sample. This is a good illustration of using the same chelating ion exchange column for both selective preconcentration and efficient separation of metal ions.

5.1.3 Applications

There is a broad spectrum of extraction mechanism and interactions, which can be exploited in CCC. The main application area of CCC is within inorganic chemistry, and then mainly associated with the preparative isolation or separation of lanthanides, actinides and some radiotoxic elements from complex mixtures. Therefore, obviously one of the main areas of application of CCC and its high-performance variants is in the nuclear industry. A number of liquid–liquid extraction technologies have been developed for reprocessing of the spent nuclear fuels, for example, PUREX (plutonium uranium redox extraction), TRUEX (trans uranium extraction), TALSPEAK (trivalent actinide lanthanide separation by phosphorus extractant and aqueous complexes), DIAMEX (DIAMide EXtraction) and SANEX (Selective ActiNide EXtraction) and others.[18] Reprocessing of high level liquid radioactive wastes includes the fractional separation of highly active components and the large scale separation of plutonium and uranium, which is a key stage of reprocessing spent nuclear fuel. Other important elements requiring separation include the so-called minor actinides (Am, Cm and Np) in various oxidation states, lanthanides and selected fission products (Cs^+, Sr^{2+}, etc). Table 5.3 provides summary information on the many applications of high-performance modes of CCC to the separation of the above groups of metals.

A typical chromatogram originating from spent fuel, showing the separation of americium 241, uranium 238 and plutonium 239, with stepwise elution and simultaneous isotope selective detection with a multi-channel alpha ray spectrometer[19] is shown in Figure 5.7. Although the separation efficiency of the separation is rather low, peak resolution is acceptable.

The possibility of using a single column for enrichment of zinc from a large volume of mineral water (20 mL) with following isocratic separation and spectrophotometric detection by PCR with PAR was demonstrated by Hosoda et al.[10] The separation system included PC-88A in n-hexane as the stationary phase and a continuous descending pH gradient elution (see conditions in caption to Figure 5.3). Good linearity was obtained for 10–40 ppb Zn under experimental conditions. The determined concentration of zinc in mineral water (23.1 ppb, RSD 1.7%, $n = 3$) was in good agreement with data from ICP-AES and ICP-MS analysis.

The additional information on applications of CCC for the separation of metals can be found in recent review of Berthod et al.,[31] and in two chapters from the ACS Symposium Series monograph on metal-ion separation.[32,33]

5.2 High-performance Extraction Chromatography

Extraction chromatography of metal ions using chelating ligands, is a mode of partition liquid-liquid chromatography, which can be considered to be closely related to both HSCCC and HPCIC using dynamically modified and ligand impregnated stationary phases.[34] The typical extraction chromatography system consists of particles of a macro-porous inert support material coated with a water immiscible organic solvent layer containing one or more typically hydrophobic ionophores or chelating ligands. This thin layer of organic phase provides the opportunity for more efficient mass transfer (as compared with CCC) due to a short diffusion pathway for metal ions to diffuse into and out of the organic phase, leading to the possibility of achieving relatively high-performance separations, approaching those typical of HPCIC. Obviously the big advantage of extraction chromatography over common liquid–liquid extraction is the absence of the time consuming process of phase separation, and of the convenient use of standard chromatographic equipment.

Originally, extraction chromatography was developed for preparative isolation of some radioactive isotopes and utilised coarse particles of diameter 0.1–0.5 mm and low-pressure LC equipment. However, the current trend in the development of extraction chromatography is to utilise micro-particulate

Table 5.3 Selected CCC separations of metal ions using stationary phases containing chelating ligands.

Separated metals	Stationary phase	Mobile phase	Conditions	Detection	Reference
		Acidic surfactants			
$Ni^{2+} < Co^{2+} < Cu^{2+} < Mn^{2+}$	0.12 M DMNPA in n-heptane	10% ethanol–90% 0.1 M ClCH$_2$COONa, step pH gradient at 180 min	2 mL min^{-1}, 375 min	PCR, PAR, Vis 515 nm	12
$Ni^{2+} < Co^{2+} < Mg^{2+} < Cu^{2+}$	0.2 M DEHPA in n-heptane	7 mM citric acid	5 mL min^{-1}, 110 min	Atomic emission spectroscopy	14
$Fe^{2+} < Fe^{3+}$	2mM DEHPA in n-heptane	0.14 M citric acid	4.5 mL min^{-1}, 40 min	Atomic emission spectroscopy	14
14 REE	3 mM DEHPA in n-heptane	Exponential gradient of HCl from 0 to 0.3 M, (see Figure 5.2)	5 mL min^{-1}, 300 min	PCR, Arsenazo III, Vis 650 nm	4
$Fe^{3+} < Ni^{2+} < Co^{2+}$, $Cu^{2+} < Mn^{2+} < Zn^{2+}$	0.16M BEHPA in n-heptane	0.1 M tartrate, pH 6.25 with step pH changes to 4.0 at 100 min and to 1.90 at 175 min	2 mL min^{-1}, 475 min	PCR, PAR, Vis 515 nm	15
$Ba^{2+} < Sr^{2+} < Ca^{2+} < Mg^{2+}$	0.16M BEHPA in n-heptane	5% ethanol–95% 0.1 M ClCH$_2$COONa, step pH gradient from 6.5 to 5.3 at 180 min	1.5 mL min^{-1}, 340 min	PCR, Phtalein Purple, Vis, 575 nm	13
$La^{3+} < Pr^{3+} < Sm^{3+} < Gd^{3+} < Dy^{3+} < Tm^{3+}$	0.1 M Cyanex 272 in n-heptane	pH gradient from 2.5 to 1.4	1.0 mL min^{-1}, 400 min	PCR, Arsenazo III, Vis 654 nm	20
$Gd^{3+} < Tb^{3+} < Dy^{3+} < Ho^{3+} < Er^{3+} < Tm^{3+} < Yb^{3+} < Lu^{3+}$	0.02 M PC-88A in toluene	0.1 M Cl$_2$CHCOONa, step pH gradient from 2.13 (380 min) to 1.52 (540 min)	5 mL min^{-1} 540 min	PCR, Arsenazo III, Vis 650 nm	16
	0.04 M PC-88A in heptane		0.5 mL min^{-1}, 190 min	PCR, Arsenazo III, Vis 650 nm	17

Table 5.3 (Continued).

Separated metals	Stationary phase	Mobile phase	Conditions	Detection	Reference
$<La^{3+}<Sm^{3+}<Eu^{3+}<Gd^{3+}$ $<Tb^{3+}<Dy^{3+}<Ho^{3+}<Y^{3+}$ $<Er^{3+}$		$0.2\,M\,Cl_2CHCOONa$, pH gradient from 2.16 to 1.14 over 240 min			
$Ni^{2+}<Co^{2+}<Mn^{2+}$, $Cu^{2+}<Zn^{2+}$	0.02 M PC-88A in n-hexane	Continuous pH gradient from 0.01 M CH_3COOH– $ClCH_2COONa$ (pH 4.15) to 0.01 M $ClCH_2COONa$– $Cl_2CHCOONa$ (pH 2.05)	$0.3\,mL\,min^{-1}$, 80 min	PCR, PAR, Vis 500 nm	10
$^{152}Eu^{3+}<{}^{241}Am(III)$	0.1 M Cyanex 301 in n-hexane	$0.1\,M\,NaNO_3$–$0.01\,M$ CH_3COONa, step pH gradient from 3.91 to 1.22 at 80 min	$0.3\,mL\,min^{-1}$, 120 min	Scintillation counter	21
$Eu^{3+}<Am(III)$	0.1 M DCPDTPI in xylene	$0.1\,M\,NaClO_4$ pH 3.3	$0.5\,mL\,min^{-1}$, 120 min	Scintillation counter	22
$Y^{3+}<La^{3+}<Sc^{3+}$	0.03 M CA12 in n-heptane	$0.1\,M\,ClCH_2COONa$ pH 4.6 step pH changes to 4.1 at 280 min and to 2.1 at 485 min	$1.0\,mL\,min^{-1}$, 1000 min	PCR, Arsenazo III, Vis 650 nm	23
Neutral extractants					
$Tb^{3+}<Yb^{3+}$	0.06 M HPMCP in toluene	$0.1\,M\,NaClO_4$ pH 2.85	$1\,mL\,min^{-1}$, 250 min	PCR, Arsenazo III, Vis 654 nm	24

Separation	Extractant	Mobile phase	Flow rate, time	Detection	Ref.
$Pr^{3+} < Eu^{3+} < Yb^{3+}$	0.08 M HPMBP in toluene	0.1 M $NaClO_4$ pH 2.36	1 mL min^{-1}, 500 min	PCR, Arsenazo III, Vis 654 nm	24
$Rb^+, Ca^{2+} < Ba^{2+} < Sr^{2+}$	0.1 M DCH18C6 in chloroform	Step gradient from 5 M HNO_3 to 0.5 M CH_3COOH at 85 min	1 mL min^{-1}, 95 min		25
$Eu^{3+} < Am(III)$	2.5 mM TPMDPD in chloroform	1 M HCl–0.5 M NH_4SCN	1.1 mL min^{-1}; 40 min		26
$Nd^{3+} < Eu^{3+} < Yb^{3+}$	0.2 M LIX 63 in kerosene	0.1 M $NaNO_3$, 0.05 M MES; pH 5.60	3 mL min^{-1}	PCR, Arsenazo III, Vis 650 nm	27
$^{152}Eu^{3+} < ^{241}Am(III)$	0.2 M LIX 63 in n-hexane	0.1 M $NaNO_3$, 0.05 M MES; pH step gradient from 5.60 to 4.60 at 95 min	3 mL min^{-1}, 170 min	Scintillation counter	28
REE	25% DHDECMP in cyclohexane	2.91 M HNO_3	1 mL min^{-1}, 120 min	ICP-MS	29
$^{244}Cm(III) < ^{243}Am(III)$	0.25 M DMDOHEMA in a mixture of dodecane isomers	3.5 M HNO_3	0.5 mL min^{-1}, 100 min	Alpha ray spectrometer	19
$Fe^{2+} < Eu^{3+} < Fe^{3+}$	5 mM TPMDPD in chloroform	0.9 M HCl–0.5 M NH_4SCN step gradient to 2 M HCl at 40 min	1 mL min^{-1}, 54 min		30
Mixed extractants					
$Cs^+ < Sr^{2+}$	0.05 M DCH18C6–3% DEHPA in chloroform	0.001 M $Ba(NO_3)_2$ in 0.1 M HNO_3	1 mL min^{-1}; 50 min		25

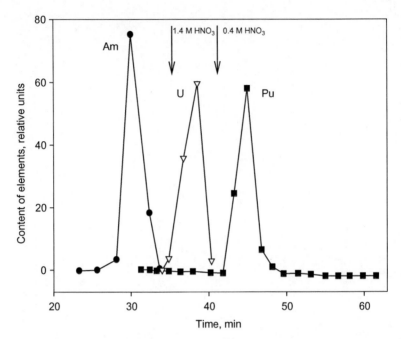

Figure 5.7 Stepwise CCC separation of ^{239}Pu, ^{233}U and ^{241}Am from spent nuclear fuel. Column: 15.6 m length, volume 30.5 mL. Stationary phase: 0.075 M DMDBDDEMA in *n*-dodecane. Mobile phase: 2 M HNO$_3$–0.1 M NaNO$_2$ with step changes of HNO$_3$ to 1.4 M HNO$_3$ at 35 min and to 0.4 M HNO$_3$ at 41 min; flow rate: 0.5 mL min^{-1} at 660 rpm. Sample: 0.2 mL of solution of MOX fuel in DMDOHEMA 12 M HNO$_3$ + 0.9 mL of xylene. Detection: alpha-ray spectrometry. Reproduced, with permission, from Maryutina *et al.*[19] Copyright (2004) Springer.

phases, typically modified reversed-phase materials, as the stationary phase. Where such reversed-phase substrates are high-performance grade materials, then the boundary between modes of HPCIC and extraction chromatography becomes very difficult to define. However, the general distinguishing features of extraction chromatography are the use of often large bulky, very strongly complexing and highly specific, and fully immobilised chelating ligands, in conjunction with a mobile phase which in most cases contains a second, often less specific and weaker chelating or complexing species. Many recent studies using this combination of conditions, no longer use the term extraction chromatography, often instead simple referring to high-performance liquid chromatography on modified supports, although they have clearly evolved from earlier extraction chromatography methods. Despite the slight confusion over the terminology, it is clear that the significant improvements in separation efficiencies achieved from the switch to such high-performance grade substrates has lead to extraction chromatography becoming a more attractive option for metal ion separations.

5.2.1 Stationary Phase Supports

The inert supports and stationary phases used in extraction chromatography must clearly be both mechanically stable and strongly retain the liquid stationary phase. Typical particle sizes (d_P) range between 30 and 150 μm for more traditional approaches and between 5 and 20 μm for standard and high-performance variants of extraction chromatography. A correct porous structure of the inert support is important to provide for the formation of a uniform thin liquid film at the surface, which is required to achieve high-performance separation efficiencies. Micro-pores of diameter less than 2 nm can be filled with occluded organic solvent through capillary action, resulting in a dramatic decrease of intra-particle volume and the interfacial area between organic and water phases, resulting in longer diffusion pathways for extracted metal ions. This effect can significantly decrease the separation efficiency. Inversely, macro-porous supports with large pore diameter > 50 nm avoid the above effect but have a lower specific surface area, and therefore do not always provide a high column loading capacity.

The surface chemical properties of the supporting substrate are also significant. Ideally the support should not interact with the extracting ligand or indeed with the metal ions being separated, but should immobilise the extracting ligand through its preferential partition within the retained organic solvent layer. In addition, the support should be chemically resistant, especially to highly acidic media, typical of those found in many industrial samples. Therefore, a combination of a neutral hydrophobic macro-porous support, coated with a layer of a non-polar low viscosity organic solvent containing the extracting ligand, appears to be the best option to obtain the efficient separations required. Polyfluorocarbons like poly(tetrafluoroethylene) (PTFE) or poly(chlorotrifluoroethylene) (PCTFE, also known under trade name Kel-F) are suitably inert, but suffer from compressibility at high pressures and have a limited surface area range (usually less than $15 \, m^2 \, g^{-1}$). Therefore, PS-DVB based polymers and alkylsilica based phases, although less chemically neutral, have found more application as supports in extraction chromatography.

As with most liquid chromatographic modes, stationary phases with thin and uniform functional interfaces tend to demonstrate better separation efficiencies. Of course, possibly the thinnest non-polar uniform layer achievable is the anchored long-chain alkyl groups found on common ODS reversed-phase particles. There are many different types of commercially available ODS phases which can be successfully modified with various chelating ligands, as shown previously in relation to in dynamic HPCIC (see Section 3.4.2). Under certain conditions, particularly higher temperatures, the bonded alkyl chain layer can justifiably be considered as a pseudo-liquid stationary phase, into which hydrophobic chelating ligands preferentially partition and remain when used in conjunction with aqueous mobile phases. The relative content of monomerically bonded alkyl groups on silica supports depends on both particle size and specific surface area. This can be as high as 18–25 wt% for typical ODS and

25 wt% for triacontyl silica, e.g. Prontosil C30.[35] The melting points of the corresponding hydrocarbons $C_{18}H_{38}$ and $C_{30}H_{62}$ are 29 °C and 68 °C, respectively. Therefore, under these conditions the dense bonded layers in the above phases should indeed exhibit the properties of pseudo-liquids. This 'phase melting' effect has been confirmed using various techniques, including GC, NMR and differential scanning calorimetry. For example, Serpinet et al.[36] reported a reversible phase transition at 26.5–27.0 and 57 °C for a silica surface densely bonded with mono-layers of octadecyldimethylsilane and triacontyldimethylsilane moieties, respectively. This increased alkyl chain mobility at elevated temperatures has also been observed by solid-phase NMR for C_{30} silica.[35] Therefore, both C_{30} and C_{18} phases can be utilised as efficient supports for loading of suitable extracting ligands and subsequent application as stationary phases for the efficient and even high-performance extraction chromatography of metal ions.

There are a number of differences in the properties of stationary phases prepared, as discussed above, by loading the surface of ODS phases with suitable chelating ligands, and those phases formed based upon an immobilised layer of a non-polar organic solvent containing the ligand. The viscosity of the organic solvent layer in the latter cases is significantly less than that of the covalently bonded pseudo-liquid layer in ODS. In practice, this equates to differences in the molecular mobility of the extracting ligand and subsequent differences in the stoichiometry of the complexes formed with extracted metal ions. Therefore, both the selectivity and the separation efficiency of the two related approaches can differ considerably. Additionally, it should be noted that when using ODS phases, the thin layer of the bonded alkyl groups cannot totally shield metal ions and ligands from unwanted interactions with residual silanol groups, which of course can also affect observed selectivity and separation efficiencies. Some comparative studies have been reported, discussing stationary phase effects upon observed selectivity. For example, Kocjan and Przeszlakowski[37] compared bare silica with C_8 and C_{18} bonded silica gels (40–63 µm), modified with the chelating ligand, Nitroso-R-salt, (Aliquat 336, methyltrioctylammonium chloride was also added as an ion-pairing reagent) for the separation of alkaline earth and transition metals and the purification of alkali and alkaline earth metal solutions from trace amounts of heavy metals. Figure 5.8 shows the extraction chromatographic separation achieved using the C_8 bonded silica phase, modified in the above way, separating Ca^{2+}, Cu^{2+}, Fe^{3+} and Co^{2+} using with a perchloric acid mobile phase and a concentration step gradient programme.[38]

The results from the above study showed the importance of having the bonded alkyl layers to maintain a stable ligand loading. However, it is clear that the density of bonded alkyl groups, their structure (polymeric or monomeric alkylsilanes), and the pore size of the base silica can all additionally affect the observed chromatographic performance. Despite these variations, as mentioned above, modified high-performance ODS phases have now become popular in modern applications of extraction chromatography, replacing the use of more traditional supports.

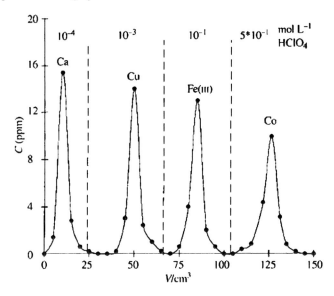

Figure 5.8 Extraction chromatogram of Ca^{2+}, Cu^{2+}, Fe^{3+} and Co^{2+} on 65×13 mm i.d. column packed with 40–63 μm octadecylsilica LiChroprep RP-18 modified with a mixture of Aliquat 336 and nitroso-R-salt. Mobile phase: perchloric acid with step concentration gradient elution; flow rate: 1 mL min^{-1}. Detection: voltammetric in collected fractions. Reproduced, with permission, from Kocjan et al.[38] Copyright (2000) Royal Society of Chemistry.

5.2.2 Ligand Loading Stability and Retention Mechanism

The exact retention mechanism in extraction chromatography is strongly dependent upon on the stability of the adsorbed layer, which in simple terms equates to the distribution ratio, D_R, of the chelating ligand between the mobile and stationary phase. As D_R reflects the probability of chelation occurring at the surface of the stationary phase, low D_R values report a significant proportion of the ligand remaining within the mobile phase, where the formation of the metal complexes of various stoichiometries can take place. In this instance the formed complexes can be retained through a number of stationary phase interactions, including electrostatic (ion exchange) or hydrophobic based retention (reversed-phase). In this case, the system is no longer directly analogous to HPCIC, which by definition would exhibit predominantly stationary phase complexation (even in the case of dynamic HPCIC). However, the diffuse boundary between the dominance of stationary phase complexation, over complexation within the mobile phase, is one which is hard to determine chromatographically and thus their distinction can become somewhat blurred.

In general, D_R values for the distribution of the chelating ligand between phases and, hence, the stability of adsorbed layer on hydrophobic adsorbents, depends foremost upon the hydrophobicity of the ligand. The hydrophobicity of typical ligands used in extraction chromatography (as log P values) varies widely, between 0.76 for 3-oxo-pentanedioicacid bis-diisobutylamide (OPAIBA), and 10.28 for N-methyl-tris(dihexylcarbamoyl-3-methoxy)-pivolamide. For example, Kimura et al. checked the stability of the adsorbed coating of a lipophilic crown ether derivative (log $P = 4.17$) on an ODS phase, with a methanol–water (40:60) mobile phase. The study reported that no decrease in retention times of the separated metal ions was noticed after elution with a volume of mobile phase equivalent to 100 h of continuous flow.[39]

5.2.3 Type of Extracting Reagent

To ensure a stable chromatographic column with domination of chelation occurring within the stationary phase, the distribution ratio of the extracting (chelating) ligand between the immobile organic solvent and the mobile phase should be as high as possible. However, the reagent should also exhibit some sufficient solubility within a polar solvent to enable it to be loaded onto hydrophobic surface of the supporting substrate in the first instance. Other important characteristics include the ability to form relatively stable complexes with a variety of metal ions, and that the kinetics of such complex formation/dissociation are sufficiently rapid, to provide fast enough partition of metal ions between the two phases to achieve acceptable chromatographic efficiency.

The use of both acidic (alkylphosphoric and alkylphosphinic acids, amido carboxylic acid, iminoacids) and neutral ligands (amides, phosphine oxides, crown ethers and other macrocycles) for coating of various supports has been reported. The charge of the extractant is also important in relation to the kinetics of interaction and therefore the separation efficiency achieved. The use of bases is generally less effective. Kimura et al.[39] reported a significantly reduced separation efficiency for an ODS column coated with a lipophilic decyl substituted cryptand, named Kryptofix 222, for the separation of metal cations, as compared with a similarly coated dodecyl-18-crown-6 column. The origin of the very broad peaks observed was attributed to slow complexation kinetics. Relatively broad chromatographic peaks have also been observed during the separation of transition metal cations on ODS columns coated with long alkyl chain derivatives of heterocyclic azo dyes[40] (see Figure 5.9). In both cases the extractants used had some degree of positive charge due to protonated amino groups within the reagent molecule.

Wang et al.[41] used bis(1,1,3,3-tetramethylbutyl)phosphinic acid (HMBP) as the extracting ligand in a study reporting the high-performance extraction chromatography of REEs. The method utilised a HNO_3 generated pH gradient with a HMBP modified 150 × 4.6 mm, 5 μm ODS column (Micro Pak SPC-18). The separation of up to 15 REEs was achieved in a little over 25 min using the developed gradient, shown below as Figure 5.10.

Maheswari et al.[42] investigated amide modified reversed-phased supports (Hypersil C_{18}, 250 × 4.6 mm i.d., 5 μm ODS particles) for the extraction chromatography of lanthanides, uranium and thorium, using α-hydroxy isobutyric acid (α-HIBA) as the mobile phase. Four structurally different amide moieties

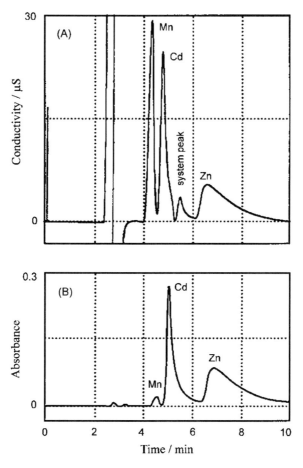

Figure 5.9 Typical extraction chromatograms of metal ions on Unisil Q C8 column (250 × 4.6 mm, 5 μm) dynamically modified with PAOOP. Mobile phase: 1 mM oxalate and 1 mM ethylenediammonium ion, pH 6.5; flow rate: 0.7 mL min^{-1}. Column temperature, 40 °C. Detection: (A) conductometric, (B) spectrophotometric detection at 575–586 nm after. PCR reaction with TAR. Reproduced, with permission, from Yasui et al.[40] Copyright (2007) Japanese Society for Analytical Chemistry.

namely, 4-hydroxy-N,N-dihexylbutyramide (4HHBA), 4-hydroxy-N,N-di-2-ethylhexylhexanamide (4HEHHA), bis(N,N,N',N'-2-ethylhexyl)malonamide (B2EHM) and N-methyl-tris(dihexylcarbamoyl-3-methoxy)pivolamide (MTDCMPA) were synthesised (see structures in Table 5.4) and used to modify the above ODS support. The influence of modifier content, mobile phase concentration and its pH on the retention of metal ions was studied. Under optimised conditions impressive separations were developed for the rapid separation of uranium from thorium, and for a number of the heavier lanthanides. Figure 5.11 shows retention profiles of heavier lanthanides as a function of extracting ligand loading, with a 4HHBA coated ODS column.

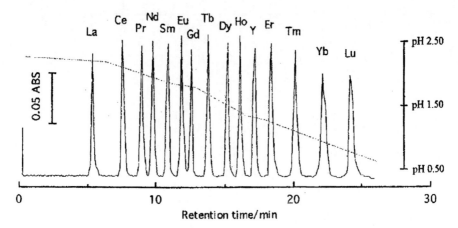

Figure 5.10 Extraction chromatogram of REEs using a HMBP modified ODS column and applied pH gradient. Column: Micro Pak SPC-18 150 × 4.6 mm, 5 μm. Gradient elution: 6.3 mM to 210 mM HNO_3 over 26 min; flow rate: 0.5 mL min^{-1}. Sample: 20 μL containing 5 μg of each REE. Detection: photometric at 655 nm after PCR with arsenazo III. Reproduced, with permission, from Wang et al.[41] Copyright (1999) Chemistry Society of Japan.

Table 5.4 shows the range of chelating ligands applied in extraction chromatography of metal ions and contains calculated log P values showing hydrophobicity of the reagents. Obviously hydrophobicity is a key parameter indicating stability of extraction chromatographic system.

5.2.4 Applications

Most of the past developments in the area of extraction chromatography for metal ion separations have arisen due to its potential in the area of heavy metal separations and within the nuclear industry. Related applications of extraction chromatography have arisen from these developments, some of which have demonstrated high-performance separations, comparable in efficiency to those of HPCIC.

For example, the use of a commercial extraction chromatography resin (TRU resin, Eichrom Tech.) for the separation and determination of trace level actinides in a urine matrix was reported.[46,50,51] The TRU resin is an extraction chromatographic material in which the extractant ligand is octylphenyl-N,N-di-isobutyl carbamoylphosphine oxide (CMPO) dissolved in tri-n-butyl phosphate (TBP) (see Table 5.4). The chromatographic system was coupled on-line with an inductively coupled plasma mass spectrometer (ICP-MS) and applied to the separation of uranium, thorium, plutonium, neptunium, and americium. The capillary column, of 100 cm length and 750 μm i.d., was packed with TRU resin (bed volume 1.7 mL) and generated a column back-pressure of ~10 MPa at a mobile phase linear velocity of 14 cm min^{-1}. The metal ions were loaded

Table 5.4 Structures of complexing extractants used in high-performance extraction chromatography.

Structure of extractant	Name	log P
	4-(2-thiazolylazo)-1-(octyloxy)naphthalene (p-TAOON)[40]	8.51
	2-(2-pyridylazo)-1-(octyloxy)naphthalene (α-PAOON)[40]	8.32
	1-(2-pyridylazo)-2-(octyloxy)naphthalene (β-PAOON); R_1: OC_8H_{17}; R_2: H;[40] 4-(2-pyridylazo)-1-(octyloxy)naphthalene (p-PAOON); R_1: H; R_2: OC_8H_{17};[40]	8.32
	2-(2-pyridylazo)-4-(octyloxy)phenol (PAOOP)[40]	6.67
	bis-2-ethylhexyl succinamic acid (BEHSA), $R = -CH_2CH(C_2H_5)C_4H_9$;[43]	7.00

Table 5.4 (Continued).

Structure of extractant	Name	log P
(R₁R₂N-C(=O)- structure with R₁, R₂)	N,N-dihexyl hexanamide (DHHA); $R_1 = n\text{-}C_6H_{13}$; $R_2 = n\text{-}C_5H_{11}$;[44]	6.39
	N,N-dibutyl heshxanamide (DBHA); $R_1 = n\text{-}C_4H_9$; $R_2 = n\text{-}C_5H_{11}$;[44]	4.43
	N,N-diisooctyl butanamide(DiOBA); $R_1 = i\text{-}C_8H_{17}$; $R_2 = n\text{-}C_3H_7$;[44]	6.50
	N,N-dioctyl hexanamide (DOHA); $R_1 = n\text{-}C_8H_{17}$; $R_2 = n\text{-}C_5H_{11}$;[44]	8.35
	4-hydroxy-N,N-di-2-ethylhexylhexanamide (4HEHHA); $R_1 = -CH_2CH(C_2H_5)C_4H_9$; $R_2 = -C_2H_4CH(OH)C_2H_5$;[42]	7.71
	4-hydroxy-N,N-dihexylbutyramide (4HHBA); $R_1 = -n\text{-}C_6H_{13}$; $R_2 = -n\text{-}C_3H_7OH$;[42]	4.90
(R₁R₂P(=O)OH)	Bis(1,1,3,3-tetramethylbutyl)phosphinic acid (Hmbp); $R_1,R_2 = -C(CH_3)_2CH_2C(CH_3)_3$;[41]	5.81
(R₁O)(R₂O)P(=O)OH	di-(2-ethylhexyl) phosphoric acid (DEHPA); $R_1,R_2 = -CH_2CH(C_2H_5)C_4H_9$[45]	6.07
$C_5H_{11}O-P(=O)(OC_5H_{11})-$	Diamyl amylphosphonate (DP[PP], active ingredient of U/TEVA resin (Eichrom)[46]	7.26
$C_8H_{17}-P(=O)(Ph)-CH_2-C(=O)N(iBu)_2$	Octylphenyl-N,N-diisobutyl carbamoylmethyl-phosphine oxide, active ingredient of TRU resin as a solution in tri-n-butyl-phosphate (Eichrom)[46]	5.71
(R₂N-C(=O)-CH₂-C(=O)-NR₂)	bis(N,N,N',N'-2-ethylhexyl)malonamide (B2EHM) $R = -CH_2CH(C_2H_5)C_4H_9$;[42]	4.91

Structure	Name	log P*
	N-methyl-tris(dihexylcarbamoyl-3-methoxy)-pivolamide (MTDCMPA); R = -n-C₆H₁₃;[42]	10.28
	3-oxo-pentanedioicacid bis-[bis-(2-ethylhexyl)-amide (OPAEHA); R = -CH₂CH(C₂H₅)C₄H₉;[43] 3-oxo-pentanedioicacid bis-diisobutylamide (OPAIBA); R = i-C₄H₉;[43]	8.62 0.76
	N,N,N′,N′-tetraoctyl diglycolamide (TODGA)[47,48]	11.18
	N,N′-dioctylcarbamoylbenzo-18-crown-6 ether[49]	6.56
	Dodecyl crown ethers, R = -C₁₂H₂₅, n = 1–3;[39,49] Dodecanoyloxymethyl crown ethers, R = -CH₂OC(O)C₁₁H₂₃, n = 2, 3;[49]	4.72–4.17 3.70; 3.42

*log P values are estimated with help of KoWin software.

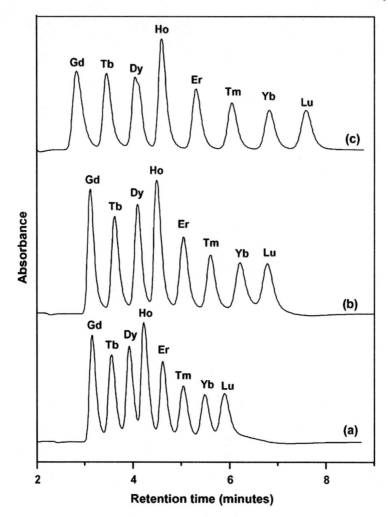

Figure 5.11 Retention profiles of heavier lanthanides obtained on reversed-phase Hypersil C_{18} column (250 × 4.6 mm i.d., 5 µm) dynamically coated with: (a) 0.17 mmol 4HHBA; (b) 0.25 mmol 4HHBA; (c) 0.54 mmol 4HHBA. Mobile phase: 0.1 M α-HIBA, pH 2.80; flow rate: 1 mL min^{-1}. Sample: 20 µL of 10 µg mL^{-1} each of lanthanides. Detection: spectrophotometric at 655 nm after PCR with arsenazo III. Reproduced, with permission, from Mahleswari et al.[42] Copyright (2007) Elsevier.

onto the separation column in 3 M nitric acid, under which conditions the actinide ions were fully retained in the chelating stationary phase, whilst matrix ions were eluted. A 4-step gradient mobile phase programme of HNO_3 and oxalic acid was used to provide the optimised separation of up to five actinides (shown as Figure 5.12) within 25 min. Injection volumes of 400 µL provided detection limits for separated actinides in the region of 50–100 ng L^{-1} with

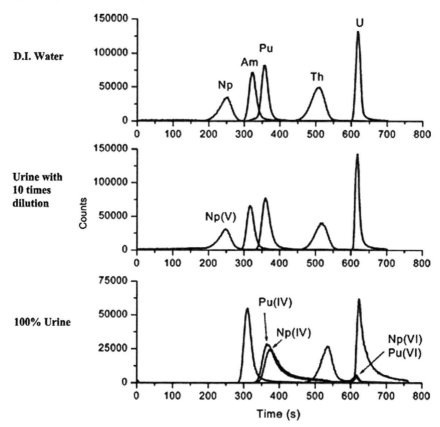

Figure 5.12 Extraction chromatogram of actinides in pure water, diluted urine and 100% urine. Column: 1000 × 0.75 mm i.d. TRU resin, particle size 20–50 μm. Sample volume: 400 μL. Detection: ICP-MS. Reproduced, with permission, from Hang et al.[46] Copyright (2004) Royal Society of Chemistry.

ICP-MS. Figure 5.12 shows overlaid extraction chromatograms for each actinide within water, diluted urine and 100% urine, and illustrates the quality of the separation achieved using this chromatographic approach.

Raju et al.[43] investigated the retention behaviour of uranium and thorium on 3-oxo-pentanedioicacid bis-[bis-(2-ethyl-hexyl)-amide] (OPAEHA), 3-oxo-pentanedioicacid bis-diisobutylamide (OPAIBA) and bis-2-ethylhexyl succinamic acid (BEHSA) modified reversed-phase supports. Using the BEHSA modified support, the isolation and quantitative determination of lanthanides (as a single peak), within uranium matrix was possible. Samples obtained from the pyrochemical reprocessing of molten salts containing lanthanides in the uranium matrix (ratio 1:20 000) were separated and determined within 7 min using the coated support. Figure 5.13 shows the separation of lanthanides, Th(IV) and U(VI) using a BEHSA modified 5 μm Hypersil C_{18} phase, housed within a 250 h 4.6 mm i.d., with a 0.1 M HIBA mobile phase, pH 2.5. The chromatogram

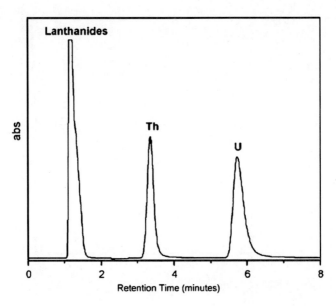

Figure 5.13 Extraction chromatography separation of lanthanides, Th (IV) and U(VI). Column: Hypersil C_{18} 250 × 4.6 mm i.d., 5 μm dynamically coated with 5 mM BEHSA. Mobile phase: 0.1 M α-HIBA, pH 2.5; flow rate: 2 mL min^{-1}. Sample: 20 μL of 100 mg L^{-1} each of U(VI) and Th(IV), and 50 mg L^{-1} of lanthanide mixture. Detection: spectrophotometric at 655 nm after PCR with arsenazo III. Reproduced, with permission, from Raju et al.[43] Copyright (2007) Elsevier.

shown highlights the advantage of chelation as a retention mechanism, as the range of possible and specific selectivities achievable is significantly greater than simple ion exchange based methods.

Sivaraman et al.[45] reported on the separation of lanthanides using a developed chromatographic method based upon the use of di-(2-ethylhexyl) phosphoric acid (HDEHP) coated reversed-phase columns. The developed method utilised α-hydroxy isobutyric acid as a complexing reagent within the mobile phase. Gradient elution was used for the separation of the entire lanthanide series, with an isocratic elution procedure developed for the separation of lighter (La to Gd) as well heavier lanthanides (Lu to Tb). The developed approach was applied to the determination of lanthanides in a fission product mixture, with Figure 5.14 showing an extraction chromatogram of a range of lanthanide metals present in a 'dissolver' solution a thermal reactor fuel.

More recently Husain et al.[47] have reported the extraction chromatography of lanthanides using N,N,N',N'-tetraoctyl diglycolamide (TODGA) as the immobilised complexing ligand on Chromosorb-W (a dimethyl dichlorosilane treated acid washed celite diatomaceous silica, particle size 0.25–0.42 mm). Studies on the uptake of several lanthanides, including La^{3+}, Pr^{3+}, Pm^{3+}, Eu^{3+}, Tb^{3+}, Ho^{3+}, Er^{3+}, Yb^{3+} and Lu^{3+} from nitric acid medium were carried out. In the batch studies, values of the distribution ration (D_M) for each cation

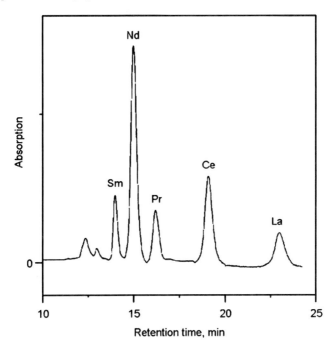

Figure 5.14 Extraction chromatogram of lanthanides present in a dissolver solution of a thermal reactor fuel. Column: LiChrospher RP-18 250 × 4.0 mm i.d., 5 μm dynamically coated with HDEHP. Mobile phase: 0.15 M α-HIBA, pH 3.5; flow rate: 2 mL min^{-1}. Detection: spectrophotometric at 655 nm after PCR with arsenazo III. Reproduced, with permission, from Sivaraman et al.[45] Copyright (2002) Springer.

were shown to increase with the HNO_3 concentration, up to approximately 2.5 M HNO_3, following the order of their ionic potential. Some separations were possible on the modified phase, although efficiency was rather low. Modolo et al.[48] used the same ligand, in a resin containing 30% (w/w) TODGA and 10% (w/w) tributyl-phosphate (TBP), for the uptake of several actinides, including U(VI), Th(IV), Am(III), Cm(III), Cf(III), and selected fission products from HNO_3 solutions of high-level (nuclear) liquid waste. The resin was packed within columns for the analysis of a synthetic PUREX raffinate sample, containing nominal amounts of fission products and trace amounts of Am(III), Cm(III) and Cf(III). The study showed how extraction chromatography using the TODGA and TBP modified stationary phase could be used for the separation of actinides and lanthanides from the bulk of other fission products.

Table 5.5 includes a summary of some metal ion separations achieved using extraction chromatography, together with the mobile phase conditions and detection options used. An update on recent progress and opportunities of extraction chromatography for analytical and preparative-scale separation of metal ions is given by Dietz et al.[52,53]

Table 5.5 Extraction chromatographic separations of metal ions using immobilised chelating ligands.

Stationary phase and column	Separated metals	Mobile phase	Flow rate ($mL\ min^{-1}$)	Detection	Reference
TSKgel ODS-80Tm CTR coated with EHPA or PC-88A	$La^{3+} < Ce^{3+} < Pr^{3+} < Nd^{3+}$	40 wt% acetone – 60% 0.08 mmol/kg PC-88A, pH 3.5	0.5 (30 °C)	PCR, Arsenazo III, Vis 655 nm	54
	$Sm^{3+} < Eu^{3+} < Gd^{3+}$ $La-Tb^{3+} < Dy^{3+} < Ho^{3+}$ $La^{3+}-Ho^{3+} < Yb^{3+} < Lu^{3+}$	As above, pH 3.1 As above, pH 2.4 45 wt% acetone–55% 0.08 mmol/kg PC-88A, pH 2.5			
Micro Pak SPC-18 (150 × 4.6 mm, 5 μm) coated with Hmbp	15 REE from La^{3+} to Lu^{3+}	pH gradient elution programme from 6.3 mM (pH 2.20) to 210 mM (pH 0.75) HNO_3 over 26 min	0.5	PCR, Arsenazo III, Vis 655 nm	41
Lichrospher C_{18} (250 × 4.0 mm, 5 μm) coated with DEHPA	14 REE from Lu^{3+} to La^{3+}	Gradient elution from 0.07 M to 0.3 M α-HIBA (pH 3.5)	1.5	PCR, Arsenazo III, Vis 655 nm	45
Unisil Q C_8 (250 × 4.6 mm, 5 μm) coated with PAOOP	$Mn^{2+} < Cd^{2+} < Zn^{2+}$	1 mM oxalic acid – 1 mM ethylenediamine, pH 6.5	0.7 (40 °C)	PCR, PAR or TAR, cond.	40

Column	Analytes	Mobile phase	Flow (mL/min)	Detection	Ref.
Hypersil C_{18} (250 × 4.6 mm, 5 μm) coated with BEHSA	Lanthanides < Th(IV) < U(VI)	0.1 M α-HIBA, pH 2.5	2.0	PCR, Arsenazo III, Vis 655 nm	43
Hypersil C_{18} (250 × 4.6 mm, 5 μm) coated with N,N-dialkylamides	Lanthanides < Th(IV) < U(VI)	0.2 M α-HIBA, pH 4.25	2.0	PCR, Arsenazo III, Vis 655 nm	44
Hypersil C_{18} (250 × 4.6 mm, 5 μm) coated with 4HHBAA	14 REE from La^{3+} to Lu^{3+}	0.1 M mandelic acid, pH 2.5	2.0	PCR, Arsenazo III, Vis 655 nm	42
	Th(IV) < U(VI)	0.1 M α-HIBA, pH 3.0		PCR, 8-HQS, Fluor., λ_{ex} 405 nm, λ_{em} 525 nm	
YMC-Pack A-311 C_{18} (100 × 6 mm, 5 μm) coated with dodecyl-18-crown-6 ether	$Li^+ < \cap$ $Na^+ < Cs^+ < Rb^+ < K^+$; $Mg^{2+}, Ca^{2+} < Sr^{2+} < Ba^{2+}$	Methanol–water (50:50)	1.0	Conductivity	39,49
TRU resin (300, 1000 or 1500 × 0.75 mm, 20–50 μm)	$^{237}Np < ^{241}Am < ^{239}Pu$ $< ^{232}Th < ^{238}U$	Three-step gradient elution programme: sample loading – 3 M HNO_3; 0 min – 0.8 M HCl, 48 mM oxalic acid, 60 mM TMAOH, 30 mM KOH; 6 min – 10 mM HCl, 48 mM oxalic acid, 60 mM TMAOH, 30 mM KOH;	0.25–0.75	ICP-MS	50

Table 5.5 (Continued).

Stationary phase and column	Separated metals	Mobile phase	Flow rate (mL min^{-1})	Detection	Reference
U/TEVA resin (300 × 4.0 mm, 100–150 μm) TRU resin (300 × 4.0 mm, 50–100 μm)	^{237}Np < ^{241}Am < ^{239}Pu < ^{232}Th < ^{238}U	Three-step gradient elution programme: sample loading – 3 M HNO$_3$; 0 min – 1 M HCl, 32 mM oxalic acid, 40 mM TMAOH, 20 mM KOH; 6 min – 10 mM HCl, 32 mM oxalic acid, 40 mM TMAOH, 20 mM KOH	1.5, 25 °C	ICP-MS	46,51
Develosil ODS-5 (150 × 4.0 mm, 5 μm) coated with N-n-dodecyliminodiacetic acid	Ca^{2+} < Sr^{2+} < Mg^{2+} < Ba^{2+}	0.05 M sodium tartrate, pH 5.5	0.5	PCR, PP, Vis, 570 nm	55

References

1. W. Murayama, T. Kobayashi, Y. Kosuge, H. Yano, Y. Nunogaki and K. Nunogaki, *J. Chromatogr.*, 1982, **239**, 643–649.
2. Y. Nagaosa and T. Wang, *J. Chem. Technol. Biotechnol.*, 2004, **79**, 39–43.
3. Y. Ito, *Sep. Purif. Rev.*, 2005, **34**, 131–154.
4. E. Kitazume, M. Bhatnagar and Y. Ito, *J. Chromatogr.*, 1991, **538**, 133–140.
5. H. Oka, Y. Ikai, N. Kawamura, J. Hayakawa, K. Harada, H. Murata, M. Suzuki and Y. Ito, *Anal. Chem.*, 1991, **63**, 2861–2865.
6. E. Kitazume, Ion analysis. High-speed countercurrent chromatography, in *Encyclopedia of Separation Science* ed., I. D. Wilson, E. R. Adlard, M. Cooke, C. F. Poole, Academic Press London 2000, pp. 3141–3148.
7. D. F. Peppard, G. W. Mason, J. L. Maier and W. J. Driscoll, *J. Inorg. Nucl. Chem.*, 1957, **4**, 334–343.
8. T. B. Pierce and P. F. Peck, *Analyst*, 1963, **88**, 217–221.
9. Z. Kolarik, *Pure Appl. Chem.*, 1982, **54**, 2593–2674.
10. A. Hosoda, A. Tsuyoshi and K. Akiba, *Anal. Sci.*, 2002, **18**, 897–901.
11. N. N. Guseva, E. V. Sklenskaya, M. K. Karapetyants and A. I. Mikhailichenko, *Radiokhimiya*, 1972, **14**, 132–134.
12. T. Wang and Y. Nagaosa, *Anal. Lett.*, 2003, **36**, 441–450.
13. T. Wang and Y. Nagaosa, *J. Liq. Chromatogr. Relat. Technol.*, 2003, **26**, 629–640.
14. E. Kitazume, N. Sato, Y. Saito and Y. Ito, *Anal. Chem.*, 1993, **65**, 2225–2228.
15. T. Wang, M. Xue and Y. Nagaosa, *J. Liq. Chromatogr. Relat. Technol.*, 2005, **28**, 2085–2095.
16. S. Nakamura, H. Hashimoto and K. Akiba, *J. Chromatogr. A.*, 1997, **789**, 381–387.
17. A. Tsuyoshi, H. Ogawa, K. Akiba, H. Hoshi and E. Kitazume, *J. Liq. Chromatogr. Relat. Technol.*, 2000, **23**, 1995–2008.
18. E. P. Horwitz and W. W. Schulz, Solvent extraction in the treatment of acidic high level liquid waste: Where do we stand? in *Metal-Ion Separation and Preconcentration. Progress and Opportunities*, ed. A. H. Bond, M. L. Dietz and R. D. Rogers, ACS Symp. Ser., 1999, **716**, pp. 20–50.
19. T. A. Maryutina, M. N. Litvina, D. A. Malikov, B. Y. Spivakov, B. F. Myasoedov, M. Lecomte, C. Hill and C. Madic, *Radiochemistry*, 2004, **46**, 596–602.
20. S. Muralidharan, R. Cai and H. Freiser, *J. Liq. Chromatogr.*, 1990, **13**, 3651–3672.
21. H. Hoshi, A. Tsuyoshi and K. Akiba, *Solvent Extr. Res. Dev. Jpn.*, 2001, **8**, 159–164.
22. J. F. Wu, Y. R. Jin, Q. H. Xu, S. L. Wang and L. X. Zhang, *Chin. J. Anal. Chem.*, 2006, **34**, 1311–1314.
23. Q. K. Shang, D. Q. Li and J. X. Qi, *J. Solid State Chem.*, 2003, **171**, 358–361.
24. G. X. Ma, H. Freiser and S. Muralidharan, *Anal. Chem.*, 1997, **69**, 2835–2841.

25. T. A. Maryutina, P. S. Fedotov and B. Y. Spivakov, Application of countercurrent chromatography in inorganic analysis, in *Countercurrent Chromatography*, ed. J. M. Menet and D. Thiebault. Marcel Dekker, New York, 1999, Ch. 6, pp. 171–221.
26. M. K. Chmutova, T. A. Maryutina, B. Y. Spivakov and B. F. Myasoedov, *Soviet Radiochem.*, 1992, **34**, 684–689.
27. H. Hoshi, S. Nakamura and K. Akiba, *J. Liq. Chromatogr. Relat. Technol.*, 1999, **22**, 1319–1330.
28. H. Hoshi, A. Tsuyoshi and K. Akiba, *J. Radioanal. Nucl. Chem.*, 2001, **249**, 547–550.
29. Y. R. Jin, L. Z. Zhang, S. J. Han, L. X. Zhang, J. M. Zhang, G. Q. Zhou and H. B. Dong, *J. Chromatogr. A.*, 2000, **888**, 137–144.
30. P. S. Fedotov, T. A. Maryutina, A. A. Pichugin and B. Y. Spivakov, *Zh. Neorg. Khim.*, 1993, **38**, 1878–1884.
31. A. Berthod, T. Maryutina, B. Spivakov, O. Shpigun and I. A. Sutherland, *Pure Appl. Chem.*, 2009, **81**, 355–387.
32. B. Y. Spivakov, T. A. Maryutina, P. S. Fedotov and S. N. Ignatova, Different two-phase liquid systems for inorganic separations by countercurrent chromatography, in *Metal-ion Separation and Preconcentration. Progress and Opportunities*, ed. A. H. Bond, M. L. Dietz and R. D. Rogers, ACS Symp. Ser., 1999, **716**, 333–346.
33. S. Muralidharan and H. Freiser, Fundamental aspects of metal-ion separations by centrifugal partition chromatography, in *Metal-ion Separation and Preconcentration. Progress and Opportunity*, ed. A. H. Bond, M. L. Dietz and R. D. Rogers, ACS Symp. Ser., 1999, **716**, 347–389.
34. T. Braun and G. Ghersini, (eds.), Extraction Chromatography, Elsevier Amsterdam, 1975, pp. 566.
35. K. Albert, *TrAC Trend. Anal. Chem.*, 1998, **17**, 648–658.
36. D. Morel, S. Soleiman and J. Serpinet, *Chromatographia*, 1996, **42**, 451–461.
37. R. Kocjan and S. Przeszlakowski, *Sep. Sci. Technol.*, 1989, **24**, 291–301.
38. R. Kocjan, R. Swieboda and I. Sowa, *Analyst*, 2000, **125**, 297–300.
39. K. Kimura, E. Hayata and T. Shono, *J. Chem. Soc. Chem. Commun.*, 1984, 271–272.
40. T. Yasui, N. Komatsu, K. Egami, H. Yamada and A. Yuchi, *Anal. Sci.*, 2007, **23**, 1011–1014.
41. Q. Q. Wang, B. L. Huang, K. Tsunoda and H. Akaiwa, *Bull. Chem. Soc. Jpn.*, 1999, **72**, 2693–2697.
42. M. A. Maheswari, D. Prabhakaran, M. S. Subramanian, N. Sivaraman, T. G. Srinivasan and P. R. V. Rao, *Talanta*, 2007, **72**, 730–740.
43. C. S. K. Raju, M. S. Subramanian, N. Sivaraman, T. G. Srinivasan and P. R. V. Rao, *J. Chromatogr. A.*, 2007, **1156**, 340–347.
44. V. Vidyalakshmi, M. S. Subramanian, N. Sivaraman, T. G. Srinivasan and P. R. V. Rao, *J. Liq. Chromatogr. Relat. Technol.*, 2004, **27**, 2269–2291.
45. N. Sivaraman, R. Kumar, S. Subramaniam and P. R. V. Rao, *J. Radioanal. Nucl. Chem.*, 2002, **252**, 491–495.

46. W. Hang, L. W. Zhu, W. W. Zhong and C. Mahan, *J. Anal. Atom. Spectrom.*, 2004, **19**, 966–972.
47. M. Husain, S. A. Ansari, P. K. Mohapatra, R. K. Gupta, V. S. Parmar and V. K. Manchanda, *Desalination*, 2008, **229**, 294–301.
48. G. Modolo, H. Asp, C. Schreinemachers and H. Vijgen, *Radiochim. Acta*, 2007, **95**, 391–397.
49. K. Kimura, H. Harino, E. Hayata and T. Shono, *Anal. Chem.*, 1986, **58**, 2233–2237.
50. D. S. Peterson and V. M. Montoya, *J. Chromatogr. Sci.*, 2009, **47**, 545–548.
51. D. S. Peterson, A. A. Plionis and E. R. Gonzales, *J. Sep. Sci.*, 2007, **30**, 1575–1582.
52. M. L. Dietz, E. P. Horwitz and A. H. Bond, Extraction chromatography: progress and opportunities, in *Metal-ion Separation and Preconcentration. Progress and Opportunities,* ed. A. H. Bond, M. L. Dietz and R. D. Rogers, ACS Symp. Ser., 1999, **716**, 234–250.
53. M. L. Dietz, Recent progress in the development of extraction chromatographic methods for radionuclide separation and preconcentration, in *Radioanalytical Methods in Interdisciplinary Research – Fundamentals in Cutting-Edge Applications*, ed. C. A. Laue and K.L Nash, ACS Symp. Ser., 2004, **868**, 161–176.
54. A. Tsuyoshi and K. Akiba, *Anal. Sci.*, 2000, **16**, 843–846.
55. S. Yamazaki, H. Omori and O. Eon, *J. High Res. Chromatogr.*, 1986, **9**, 765–766.

CHAPTER 6
Detection

6.1 Background

Since its beginnings in the 1970s, detection in HPLC has always been more of a problem than in gas chromatography. This is not too surprising as the range of analytes is wider and more varied in liquid chromatography and it does not possess the major advantage of having an inert mobile phase. The difficulty of obtaining good sensitive detection is particularly true for metal species, which in general do not have the right physical chemical characteristics for exploitation by the most common detection techniques such as UV–visible absorption. There is one exception however, involving a small group of metals, namely the alkaline and alkaline earths, and that is suppressed conductivity detection.[1,2] As a rule, direct conductivity without suppression of background conductivity of the mobile phase is not too sensitive, so suppression of mobile phase H^+ ions to non-conducting water and conversion of sample anions to highly conducting OH^- ions is used in IC for the detection of these metal ions. For all other metal species, suppressed conductivity detection is not suitable due to hydrolysis and precipitation in the pH range used in suppressors. Furthermore, as the high ionic strength of most mobile phases in HPCIC rules out the use of suppressed conductivity, it will not be considered here.

A considerable amount of effort was put into the development of spectrophotometric detectors for use in HPLC, producing exceptionally low noise and drift characteristics. It is not unusual for UV–visible spectrophotometers now to have short term noise levels of around ±0.00001 absorbance units (0.01 mAU), available at relatively low cost. Thus, relatively sensitive direct photometric detection of metals is possible in cases where the mobile phase contains additives of complexing reagents and the separated metals are eluted as reasonably strong UV-absorbing complexes. Typical examples are complexes of metals with picolinic,[3] dipicolinic,[4,5] quinaldic[6] and citric acid.[7–9] As an illustration, the molar extinction coefficients of metal picolinates at different

Table 6.1 Molar extinction coefficents of metal picolinates in aqueous solution at pH 7.0.

Molar extinction coefficients ($L\,mol^{-1}\,cm^{-1}$)	Metal ion					Ligand
	Co^{2+}	Fe^{3+}	Pb^{2+}	Cu^{2+}	Ni^{2+}	
$\varepsilon^{235}\,10^3$	3.1	5.0	4.0	6.7	2.2	0.05
$\varepsilon^{280}\,10^3$	1.1	1.0	0.8	0.8	1.0	0.003
$\varepsilon^{310}\,10^3$	1.5	3.3	—	0.15	0.2	—

wavelengths are presented in Table 6.1. Direct detection of the UV-absorbance of these complexes allows relatively sensitive detection of many metals with LODs between 10 and 400 μg L^{-1}.[4]

Nevertheless, for the highest sensitivity detection of metal analytes, the molar extinction coefficients need to be much greater than this. Furthermore, it would be better to have the monitoring wavelength in the visible region where interference from highly absorbing trace organic impurities is much less. There is one approach which can achieve both these aims and that is the use of post-column reactions.

6.2 Post-column Reactions for the Photometric Detection of Metal Ions

With the post-column reaction (PCR) detection approach, analytes with weak or no absorbance or fluorescence, can be converted into a new species where these characteristics are much stronger. This can actually be done pre-column or post-column, however to keep the separated species in an unchanged form as much as possible during separation then PCR provide the best option. Typically, to carry out a PCR, a reagent is mixed with the column effluent at a mixer and reacts with the eluted analytes in a flow reactor before the detector to form a species with a much higher molar extinction coefficient and/or fluorescence with high quantum efficiency. Clearly, the main aim of a PCR system is to achieve the lowest possible detection limits for separated metals. Another important aim is to provide more selective detection of specific metal ions if necessary, e.g. in case the complete separation of the desired metal from others is not possible. These goals can be achieved by careful consideration of the following factors:

1. Selection of the detection type and PCR providing the required sensitivity and selectivity at minimal level of background signal.
2. Optimisation of post-column reactor design, including mixing and reactor construction (reactor types, sizes and materials), producing completeness of the selected PCR in the minimum time. Adjusting the temperature of the reaction might be an essential part of the optimisation.

3. Selection of the reagent delivery system, providing minimum fluctuations of flow and mixing inaccuracy, which will provide a stable baseline with the lowest pulsations and noise.

Post-column techniques have attracted a lot of attention from chromatographers as a powerful tool for the enhancement of detection sensitivity and sometimes enhanced specificity of detection of various classes of compounds. According to Fritz,[10] the first application of PCR detection for the determination of transition metals was reported in 1971 by Sickafoose, as part of his PhD thesis at Iowa State University (USA). Highly specific post-column reactions have also been applied widely in flow-injection analysis (FIA) usually without any separation. The development and application of PCR for the detection of metals in IC is reviewed by Cassidy and Karcher,[11] de Jong et al.,[12] Jansen and Frei[13] and Dasgupta.[14,15]

6.2.1 Construction of Post-column Reactors

The typical construction of a post-column reactor consists of four major components: a reactant delivery system, usually a low pressure pump or pneumatic (gas over-pressure) device, a T or Y junction for mixing the two streams, a flow reaction chamber or coil, and a thermostatically controlled heating chamber if necessary.

6.2.1.1 Reagent Delivery Systems

There are two main ways in which the post-column reagent can be delivered to the reactor, by a pneumatic system or mechanical pump. The pneumatic system involves an inert gas over-pressure reagent bottle forcing the reagent out through a valve to maintain a steady flow. They are not too common and mainly found in commercial instruments. The typical operating pressure for the pneumatic system is 0.69 MPa.

The principal types of mechanical pumps are reciprocating, peristaltic, syringe and more recently rather special ones like the Milligat type (Global FIA, USA), discussed later. An important requirement is that the flow path should be as metal free as possible to avoid spurious contamination. Syringe types are rarely used basically because of the limited capacity for the delivered solution of reagent and the main choice is essentially between reciprocating and peristaltic pumps. Any good quality reciprocating pump can be used, such as those made to deliver mobile phases to columns at high pressure, though this is the most expensive option. Cheaper pumps are available dedicated for post-column use, but must have efficient pulse dampening systems. Those with PEEK pump heads are best so no metal parts are involved.

The use of peristaltic pumps is increasingly common, as high quality versions are available at a reasonable price. They are cheap to maintain, the main consumable being replacement of very low cost pump tubing. Multi-roller pumps with six or more rollers are best as the rapid pump pulses are more easily suppressed. One thing to watch out for is the maximum pressure at which the pump can deliver the reagent. It is unusual to find a peristaltic pump which can deliver more than 0.3 MPa, depending on the flow rate. There is generally no problem at the commonly used reagent flow rate of 0.5 mL min^{-1}, combined with reactor coils up to 2 m in length and 0.5 mm internal diameter. Much higher flow rates and/or longer reactor coils could cause problems if the back-pressure is too great.

All pump systems will produce flow variations to a greater or lesser extent depending on design and type of pulse dampener and will impact on detector baseline noise. Although the pneumatic delivery systems are pulse free it does not mean that there is no affect on baseline noise. Small changes in temperature and flow resistance of the connecting tubing can produce baseline variations of the same order as the peak widths, affecting detection limits. See later for detailed discussion on pump noise causes and its suppression in Sections 6.3.2 and 6.3.3.

6.2.1.2 Mixers

The post-column reagent is delivered to the reactor via a mixer, usually a junction of PEEK construction. Efficient mixing is important, particularly if there is a significant difference in viscosity between the reagent and mobile phase. With this in mind, mixers with all manner of designs have been studied, mostly of the T and Y type with different angles of mixing. Other types with more sophisticated mixing arrangements have also been developed, such as the cyclone and rotating flow mixers, all with very low dead volumes to keep peak dispersion to a minimum.[11-13] In practice, for the IC columns in present use, there seems to be very little difference between the T and Y types and low dead volume T junctions are the most popular, including their use in commercial systems. For best mixing the outlet to the reactor should be connected to the 'stem' of the T or Y junction.

6.2.1.3 Flow Reaction Chambers

Flow reaction chambers or reactors can be membrane, packed-bed or open tubular. In general, according to Katz and Scott,[16] band broadening in PCR depends strongly on the volume of the reaction chamber, so the choice of the reactor depends on the type of reaction used and on the conditions required for the completeness of the PCR reaction.

The PCR systems involving packed bed reactors, usually short columns packed with inert glass particles of diameter 20–30 µm, have some drawbacks connected with heat dissipation affecting the peak shape. Also, due to the very high surface area of the particles, it is relatively difficult to avoid substantial adsorption of bulky and 'sticky' molecules of metallochromic reagents and their complexes with metals.[13,17] As a result, peak distortion may occur and decrease the detection sensitivity. Significant band broadening due to resistance to mass transfer can occur in particle packed columns. This effect can be minimised by using smaller diameter particles, which results in a significant increase in backpressure for this type of reactors and requires usage of high-pressure pumps for reagent delivery. Obviously, open-tubular reactors are much simpler in construction and can operate at low pressures, normally between 0.1–0.3 MPa. Therefore, they have been used practically in all separations of metals with PCR detection.

Open-tubular reactors consist of knitted polymer tubing up to 8–10 m long and internal diameter 0.1–0.5 mm. The length of tubing depends on the kinetics of the PCR and can be decreased if the reaction chamber is heated. The knitting pattern is of a special kind in the form of intertwining loops. The loops are so designed as to continually change the direction of the flow in tight curves. By this means peak dispersion is considerably reduced. Although the knitting pattern looks complicated it is quite easy to do and home-made reactors work very well. Home-made knitted loop reactors are best made in the linear form, whereas commercial ones are available in cylindrical form for compactness (see pictures, Figure 6.1). PTFE knitted coils in a variety of lengths with internal diameters of 0.25, 0.50 and 0.75 mm, are available from VICI-Jour (Switzerland).

The length of the knitted coil reactor depends on the kinetics of the PCR and on the temperature of the reactor. Usually, the shorter coil provides less peak broadening and the better resolution of chromatographic peaks. It is recommended also to operate at linear velocity of flow higher than 10 cm/sec in tubing reactor to have stable radial mixing. This equates to a minimum flow rate of about 1 mL min^{-1} for tubing with an internal diameter of 0.5 mm. Using the knitted coils, there is very little reduction in peak efficiency with tubing lengths up to 2 m (internal diameter of 0.5 mm) at a combined flow rate of 1.5 mL min^{-1}. For PAR, the authors have settled for tubing lengths of 1 metre (0.5 mm i.d.), as reactions with the most common metals are very fast at room temperature, apart from nickel, which is a little slower. Keeping the tubing length as short as possible is important as columns containing 3 μm particle size are becoming increasingly common, putting greater demands on the efficiency of post-column fittings.

6.2.1.4 Effect of Reactor Tubing Material and Reagent Composition on Peak Shapes

The type of tubing used in open tubular systems is usually Teflon (PTFE) based and is normally considered to be essentially inert in terms of interactions with the post-column reactants and products. However, this is not strictly true. When considering the PAR post-column reaction in particular, the authors have found a small amount of peak distortion for certain metals specifically related to the type of tubing used. For example, using PTFE tubing and a PAR

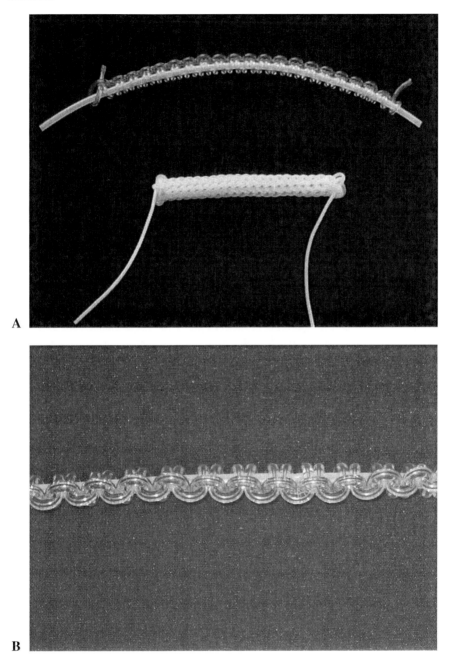

Figure 6.1 Knitted loop reactor coils. (A) Linear and tubular constructions, and (B) magnified portion of a linear knitted loop.

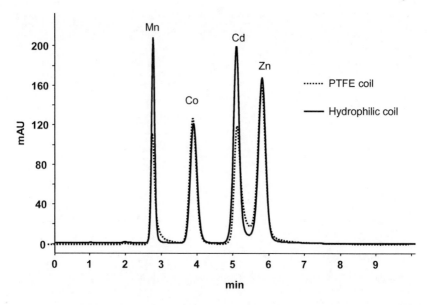

Figure 6.2 Effect of reactor coil material on manganese and cadmium peak shapes. Column: 50 × 4.0 mm i.d. packed with 5 μm IDA bonded silica. Mobile phase: 0.01 M nitric acid. PCR reagent: 5.0×10^{-5} M PAR in 0.4 M ammonia adjusted to pH 10.5 with nitric acid. Reactor coils were of the knitted loop type, 2 m long by 0.5 mm i.d. Sample: 100 μL containing 0.25 mg L^{-1} Mn^{2+}, 0.5 mg L^{-1} Co^{2+}, 1.0 mg L^{-1} Cd^{2+} and 0.5 mg L^{-1} Zn^{2+}.

PCR, there is definite increase in tailing or asymmetry for Mn, Cd and Pb compared to the other common metal ions. The cause of this is not quite clear, but PTFE is quite hydrophobic and may attract some of the less soluble metal/PAR species in the reaction chamber. In other words some 'stickyness' may be present between the reactants and the polymer walls as mentioned for packed bed reactors, but at a lower level. It has been found that the type of tubing and the reagent composition can affect peak shapes for the three metals. Figures 6.2 and 6.3 clearly show that the use of a more hydrophilic polymer tubing gives improved peak shapes for Mn^{2+}, Cd^{2+} and Pb^{2+} and consequently greater responses. However, improved peak shapes for these metals can be obtained using PTFE tubing if a surface active agent is present, but nevertheless it is still not quite as good as an hydrophilic coil. It is also interesting that changing the buffer from ammonia to a Goods type can improve peak shapes of some metals when using PTFE tubing. Interactions with the walls of the reactor tubing could also occur for other metallochromic dyes and their complexes, possibly related to solubility and size. Some similar peak distortions have been observed for Nitro PAPS, an analogue of PAR.

In some cases, a small percentage of a polar organic solvent such as methanol or acetonitrile or a non-ionic surfactant (see Section 6.2.3) can minimise the adsorption of complexes on the internal walls of the post-column reactor and

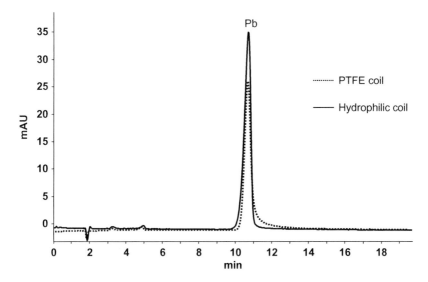

Figure 6.3 Effect of reactor coil material on lead peak shape. Column: 100 × 4.0 mm i.d. packed with 5 µm IDA bonded silica. Mobile phase: 0.015 M nitric acid in 0.5 M KNO$_3$. Sample: 2 mg L^{-1} Pb^{2+}. Photometric detection after PCR with reagent and coils as in Figure 6.2.

improve peak shape, peak efficiency and decrease limits of detection. It should be noted that in spite of good aqueous solubility, the hydrophobicity of the PCR reagents used for the detection of metals can be sufficient for significant adsorption on PTFE tubing walls. The experimentally obtained logP values for Br-PADAP and 1,10-phenanthroline (see structure in Table 6.2) are 4.23 and 2.29, respectively.[18] If the detected metals form complexes of composition ML$_2$ or ML$_3$ the hydrophobicity can be even higher.

6.2.1.5 Commercial PCR Systems

At present, PCRs suitable for the detection of metals are available from the leading producers of IC equipment, Dionex (USA) and Metrohm (Switzerland). The sophisticated PCR detection systems designed by Pickering Laboratories (USA) for the detection of organic and inorganic molecules can be used as well. Early versions of commercially produced equipment for post-column detection are reviewed by Weinberger and Femia.[19]

Dionex produces two PC-10 Postcolumn Pneumatic Delivery Systems designed for use with 2 and 4 mm i.d. columns, which provides a pulseless flow of reagent. The maximum operating gas pressure is 0.7 MPa. The post-column reactors are equipped with mixing tee and Teflon knitted coil reactors of volume 125 and 375 µl, correspondingly for 2 and 4 mm i.d. columns. Metrohm has a different PCR system comprising a high speed multi-roller peristaltic pump with an IC post-column reactor containing T mixer, soft polymer

Table 6.2 Structures of most common reagents for PCR.

General formula	Substitute groups
(pyridyl-azo-phenol structure with R_1, R_2)	**PAR**: $R_1 = H$; $R_2 = -OH$; **Br–PADAP**: $R_1 = -Br$; $R_2 = -N(C_2H_5)_2$; **Nitro–PAPS**: $R_1 = -NO_2$; $R_2 = -N(C_3H_7)(C_3H_6SO_3Na)$
(triphenylmethane structure with R_1–R_5)	**o-CPC**: $R_1 = -COOH$; $R_2 = -CH_2N(CH_2COOH)_2$; $R_3 = -CH_3$; $R_4, R_5 = -H$; **XO**: $R_1 = -SO_3H$; $R_2 = -CH_2N(CH_2COOH)_2$; $R_3 = -CH_3$; $R_4, R_5 = -H$; **PCV**: $R_1 = -SO_3H$; $R_2 = -OH$; $R_3, R_4, R_5 = -H$; **CAS**: $R_1 = -Cl$; $R_2 = -COOH$; $R_3 = -CH_3$; $R_4 = Cl$, $R_5 = -SO_3H$
(azo-naphthalene disulfonic acid structure with R_1, R_2, R_3)	**Arsenazo I**: $R_1, R_3 = -H$; $R_2 = -AsO_3H_2$; $R_3 = -H$; **Arsenazo III**: $R_1 = -N=N-C_6H_5AsO_3H_2$; $R_2 = -AsO_3H_2$; $R_3 = -H$; **Chlorophosphonazo III**: $R_1 = -N=N-p-Cl-C_6H_5PO_3H_2$; $R_2 = -PO_3H_2$; $R_3 = -Cl$; **Tiron**
(Tiron structure: benzene-1,2-diol-3,5-disulfonic acid)	
(8-hydroxyquinoline structure with R_1)	**8–HQ**: $R_1 = H$; **8–HQS**: $R_1 = -SO_3H$;
(diphenylthiocarbazone structure with R_1)	**Dithizones**: $R_1 = -H$, $-COOH$, $-SO_3H$
(dithiocarbamate structure with R_1, R_2)	**Dithicarbamates**: $R_1, R_2 = -CH_2CH_2OH$ $R_1-R_2 = -CH_2CH_2CH_2CH(COOH)-$
(Lumogallion structure)	**Lumogallion**

Detection

dampener and knitted coil in one small box. The soft silicon tubing dampener placed before the mixing tee effectively reduces the fast low-pressure pulsations from the peristaltic pump.

Some components of post-column detections systems are produced by other companies. A low pulsation high rotation pump head Milligat pump (Global FIA, USA) represent a very good alternative to both pneumatic delivery system and peristaltic pump. The special construction of the pump allows delivery of the reagent at flow rates from $5\,\text{nL}\,\text{min}^{-1}$ to $10\,\text{mL}\,\text{min}^{-1}$ with a maximum operating pressure of 0.7 MPa, which is nearly two times higher than that produced by common peristaltic pumps.

6.2.2 Reagents for Post-column Derivitisation

A classic PCR example, first used in the late 1940s, is the detection of amino acids using ninhydrin[20] after separation on ion exchange columns. For the PCR detection of metal ions by IC, a colorimetric reaction is also the one most commonly employed. The reagent is usually a complexing dye, producing an intensely absorbing metal complex at a different wavelength to that of the dye. Although many hundreds of colorimetric reagents were synthesised and studied for the determination of trace metals, particularly in the 1950s and 1960s, only a small number were found to be suitable for use as PCR reagents. Essentially, the main objective of this huge volume of work on photometric reagents was to find ones selective for particular metals. To this end a particular metallochromic reagent was used in combination with auxiliary complexing agents and in many cases with solvent extraction. Dithizone is one such reagent, which attracted a considerable amount of attention, where, for example, lead could be determined with a high degree of selectivity by extraction from a cyanide solution into chloroform.[21] The classic book by Sandell and Onishi discusses the properties of all the photometric (colorimetric and fluorimetric) reagents of any consequence up until 1978.[22] More recent information can be found in the *Handbook of Organic Analytical Reagents* by Ueno et al.[23]

In general, the ideal reagent for a PCR should fulfil a number of requirements, including sensitivity, contrastness of the reaction, specificity, reaction velocity, good solubility of both reagent and products of the PCR and sufficient long term stability in solution. For use as a PCR reagent, specificity is not a major issue. The chromatography column automatically produces a high degree of selectivity by separating the metal species into individual peaks. So, in contrast, the PCR colorimetric reagent needs to react with a wide range of metals. Occasionally, however, as stated earlier, some specificity may be required when analyte peaks overlap. In addition the post-column reagent should produce complexes with very high molar absorptivities and have a wavelength maximum well removed from that of the reagent. The wavelength maxima of the different metal complexes should be reasonably close together so that one compromise detection wavelength needs to be used (see Figure 6.4). Furthermore, as IC separations involve mostly aqueous based mobile phases, the reagents and

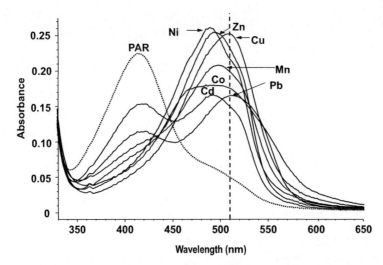

Figure 6.4 The spectra of seven metal–PAR complexes at pH 10. Dotted line is the most commonly used compromise monitoring wavelength.

complexes must be water soluble and fast reacting to avoid long delay times. Taking all these factors into account there are not that many photometric reagents which fit these properties. Thus, although 30 or so colorimetric reagents have been selected for study for PCR of metals[10,17] fewer than 10 appear to be all that are required to cover the majority of metals amenable to IC separation. The structures of the most commonly used reagents are shown in Table 6.2.

Very early on in the development of IC of metals one colorimetric reagent stood out as satisfying these criteria for a wide range of common metals. This was 4-(2-pyridyl)-azoresorcinol (PAR) and after fundamental work by Fritz and co-workers[24] in the 1970s, soon became established as the one of the best and most commonly used PCR reagent for metal ion detection. PAR is particularly suited to the detection of most of the first row transition elements and the so called heavy metals such as Cd^{2+} and Pb^{2+}. Figure 6.4 shows the spectra of seven metal/PAR complexes compared with free PAR. It can be seen that a reasonable compromise monitoring wavelength is about 500 nm. Slightly longer wavelengths are sometimes used to reduce the background absorbance.

Since PAR has been found to produce very good results in PCR, there have been a number of studies on analogues involving the pyridylazo- (PAN,[25] Br-PADAP[26]) or thyazolilazo- (TAR[25]) grouping, mainly to further increase sensitivity. The relatively higher hydrophobicity of PAN and Br-PADAP restricts their effective use without additives of surfactants or organic solvents in the PCR. Also, the complexes of some of the metals with these organic reagents have absorption maxima at significantly different wavelengths (see Table 6.3), which makes it difficult to use a single compromise wavelength for

Table 6.3 Key detection parameters for the formation of coloured metal complexes with different reagents (adapted from Dugay et al.,[27] updated and modified).

Metal cation	Arsenazo III ($\lambda = 516–536\,nm$)		PAR ($\lambda = 410–440\,nm$)		Br-PADAP ($\lambda_{reag} = 450\,nm$)		PCV ($\lambda_{reag} = 450\,nm$)	
	λ_{max}, nm	$\varepsilon \cdot 10^4$ ($L\,mol^{-1}\,cm^{-1}$)	λ_{max}, nm	$\varepsilon \cdot 10^4$ ($L\,mol^{-1}\,cm^{-1}$)	λ_{max}, nm	$\varepsilon \cdot 10^4$ ($L\,mol^{-1}\,cm^{-1}$)	λ_{max}, nm	$\varepsilon \cdot 10^4$ ($L\,mol^{-1}\,cm^{-1}$)
Alkaline earth	650	2.80 (Ca)	—	—	—	—	640	—
Cd^{2+}	540	0.12	495	5.80	548	7.31	710	1.2
Co^{2+}	620	0.9	510	5.60	586	9.10	720	1.1
Cu^{2+}	610	1.03	510	5.89	555	11.0	635	2.08
Fe^{2+}	610	0.32	500	5.00	554, 742	8.06, 3.41	595	6.55
Hg^{2+}	510	0.58	500	6.80	558	5.78	—	—
Be^{2+}	600	3.12	—	—	—	—	680	1.5
Mn^{2+}	—	—	496	8.65	555	1.40	640	—
Ni^{2+}	620	0.25	494	7.30	523	10.1	650	—
Pb^{2+}	665	1.0	522	5.02	567	4.97	580	5.06
Zn^{2+}	520	0.25	495	6.34	554	11.9	665	2.59
Al^{3+}	585	0.37	510	3.64	—	—	580	6.80
Bi^{3+}	610	1.75	515	1.07	583	4.90	610	2.90
Cr^{3+}	610	2.60	550	1.20	595	7.93	590	2.63
Fe^{3+}	515	0.32	536	6.04	590	3.80	605	6.45
Ga^{3+}	502	0.30	504	10.2	552	7.20	580	7.31
In^{3+}	—	—	510	8.64	558	5.90	620	2.90
Tl^{3+}	—	—	520	1.94	557	5.25	—	—
Sc^{3+}	660	0.56	520	1.47	—	—	620	4.90
Sb^{3+}	—	—	540	1.20	595	6.0	630	1.65
REE	650	6.00	510	6.00	—	17.6 for Gd^{3+}	607	1.9 for Y^{3+}
Hf(IV)	665	9.50	510	3.75	582	13.30	—	—
Sn(IV)	—	—	—	—	—	—	555	6.50
Th(IV)	660	5.12	500	3.89	580	16.6	660	6.00
Ti(IV)	420	0.14	—	—	—	—	575	3.82

Table 6.3 (Continued).

Metal cation	Arsenazo III ($\lambda = 516-536\,nm$)		PAR ($\lambda = 410-440\,nm$)		Br-PADAP ($\lambda_{reag} = 450\,nm$)		PCV ($\lambda_{reag} = 450\,nm$)	
	λ_{max}, nm	$\varepsilon \cdot 10^4\,(L\,mol^{-1}\,cm^{-1})$	λ_{max}, nm	$\varepsilon \cdot 10^4\,(L\,mol^{-1}\,cm^{-1})$	λ_{max}, nm	$\varepsilon \cdot 10^4\,(L\,mol^{-1}\,cm^{-1})$	λ_{max}, nm	$\varepsilon \cdot 10^4\,(L\,mol^{-1}\,cm^{-1})$
Zr(IV)	665	12.0	520	1.50	585	15.4	598	3.31
V(V)	510	0.11	550	3.60	550	1.97	595	1.51
U(VI)	665	5.34	530	3.95	584	7.40	605	3.29
Np(IV)	665	12.5	510	4.29	—	—	—	—

the sensitive photometric detection for a group of separated metals. A new generation of water soluble and more sensitive pyridylazo reagents (nitro-PAPs) has appeared and their applications will be discussed later.

It is claimed that PAR can be used for the detection of over 40 metals.[27] However, for the higher charged or strongly hydrolysing metal ions, such as the lanthanides and actinides, better sensitivity is obtained using other reagents, which can also be used in acid conditions, reducing the risk of precipitation due to hydrolysis. Bisazo derivatives of chromotropic acid including arsenazo I, arsenazo III and chlorophosphonazo III give even better sensitivity for the lanthanides and also for U(VI) and Th(IV), though arsenazo III is the one preferred by most workers.[28,29]

The triphenylmethane based dyes such as xylenol orange (XO), phthalein purple (PP) and methylthymol blue (MTB) having IDA chelating groups in their molecules is another popular group of universal photometric reagents reagents in PCRs, which can be effectively used for the detection of alkaline-earth, transition and heavy metal ions and lanthanides.[30] The molecules of the triphenylmethane dyes pyrocatechol violet (PCV) and chrome azurol S (CAS) contain the different chelating groups pyrocatechol and salicylic acid, respectively, so these reagents have been used particularly for the PCR detection of other acid hydrolysing metals such as iron(III), aluminium, beryllium, vanadium etc.[31] Tiron (4,5-dihydroxy-1,3-benzene disulfonic acid, disodium salt) exhibits a selectivity similar to PCV but the reaction is less sensitive.

Some sulfur containing reagents such as dithizone and the dithiocarbamates have good kinetics of interaction and reasonable molar extinction coefficients ($\varepsilon \sim 12\,000$) for complexes with transition and heavy metal ions. The existing problem of poor solubility of the reagents in water has been solved by the development of hydrophilic analogues containing carboxylic, sulfo and hydroxyl substitutes in molecules. However, these reagents decompose under storage in polyethylene containers.

When considering the reagents for the fluorescent detection of metals by PCR only 8-hydroxyquinoline-5-sulfonic acid (8-HQS) stands out as being particularly good for groups of metal ions. However, other reagents like lumogallion can provide very sensitive detection of a single element, *e.g.* aluminium and zinc. The possible solution of the problem is the use of a fluorescent reagent for multielement detection through displacement reactions (Table 6.5).

6.2.2.1 Stability Constants, Sensitivities and Detection Limits for Colorimetric Post-column Reaction Reagents

A number of reviews have compared the different PCRs.[2,14] Table 6.3 presents the key data for the most common metallochromic reagents used for the detection of metals by using PCR systems.

It should be pointed out that a lot of the data in the literature comparing stability constants, sensitivities and limits of detection (LOD) for trace metal determination with colorimetric reagents can be contradictory and very inaccurate and should be treated with great caution.[32] This is especially true for PAR, now the most commonly used post-column reagent for metals in IC. The results of stability constants determinations, particularly in earlier work, varied wildly between different research groups and probably reflects the low purity of the PAR reagent available at the time. It was not uncommon to see differences of many orders of magnitude between reports. Also, measurements at very different ionic strengths will produce significant differences. Among other things, accurate stability constants can be used for the construction of metal speciation plots for a particular post-column reagent system. These plots, easily produced using the latest speciation software, can give very useful information on the reagent conditions, such as pH, type of mobile phase and buffer and concentration necessary for optimum sensitivity and also to improve selectivity for certain metals. Fortunately, more recent work has produced more reliable stability constant data. The large database of critically assessed stability constant data published by NIST quotes a number, mostly of the ML type.[33] Several workers have given data for ML_2, the species producing the highest molar extinction coefficient and the stability constants for Cd^{2+}, Zn^{2+} and Pb^{2+} in particular are used here for constructing metal speciation plots.[34–36]

Metal speciation plots can be very useful for investigating and optimising PCRs. A number of metal speciation software programmes are available commercially, but the authors have found the freeware, MEDUSA (Making Equilibrium Diagrams Using Sophisticated Algorithms) and HYDRA programmes, devised by a Swedish academic group, particularly good.[37] Now available in 32 bit form, it is very powerful with a 'friendly' data input system based on the Periodic Table and versatile graph plotting routines.

The reported sensitivities in terms of molar extinction coefficients and the related limits of detection can also vary widely. Not only because of the use of impure PAR, but also because the composition and pH of the solutions used for the measurements can be very different and not at optimum. The solutions usually contain other complexing agents, such as in the buffer, which could compete with the PAR, thus suppressing the complete formation of the PAR/metal complex at a particular pH. Therefore, depending on the type and concentration of the other complexing agents present, different molar extinction coefficients and LODs will be measured. This of course, affects not only static measurements, but also the PCR under dynamic conditions. One example concerns ammonia, which is commonly used to buffer the post-column solutions to around pH 10. Ammonia forms relatively strong complexes with certain metals such as Zn^{2+}, Cd^{2+} and Cu^{2+} and will compete with the PAR complexes if the ammonia concentration is high. The metal speciation plots for Zn^{2+} and Cd^{2+} (Figures 6.5 and 6.6) illustrate the problems as discussed above.

At pH 10 both Zn^{2+} and Cd^{2+} are little affected by low levels of ammonia, but as the concentration is increased Cd^{2+} becomes especially affected, where the sensitivity could be reduced by more than a factor of ten going from low to

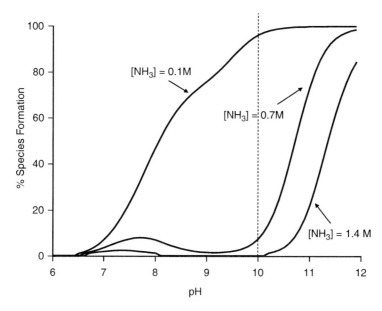

Figure 6.5 Metal speciation plot showing the effect of ammonia concentration on the formation of Cd(PAR)$_2^{2-}$ the major absorbing species. Less important species removed for clarity. Dotted line shows typical pH used in PAR post-column reagents. PAR concentration: 5.0×10^{-5} M; Cd^{2+} concentration: 5.0×10^{-6} M.

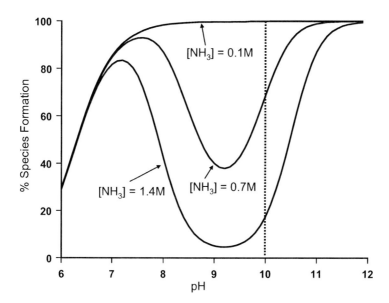

Figure 6.6 Metal speciation plot showing the effect of ammonia concentration on the formation of Zn(PAR)$_2^{2-}$ the major absorbing species. Less important species removed for clarity. Dotted line shows typical pH used in PAR post-column reagents. PAR concentration: 5.0×10^{-5} M; Zn^{2+} concentration: 5.0×10^{-6} M.

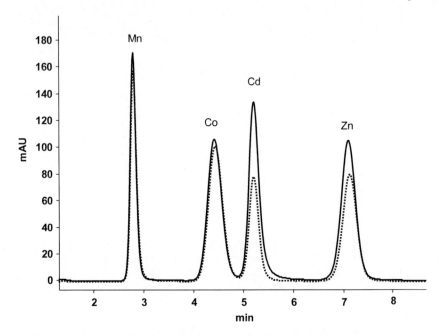

Figure 6.7 Chromatogram showing the effect of ammonia concentration on peak heights of Cd^{2+} and Zn^{2+}. Column: 50×4.0 mm i.d. packed with 5 μm IDA bonded silica. Mobile phase: 0.01 M nitric acid. Post-column reagent: dotted line, 1 M ammonia; solid line, 0.4 M ammonia in PAR solution; concentration, 1.5×10^{-4} M. Sample: 100 μL containing 0.25 mg L^{-1} Mn^{2+}, 0.5 mg L^{-1} Co^{2+}, 1.0 mg L^{-1} Cd^{2+} and 0.5 mg L^{-1} Zn^{2+}.

high concentrations of ammonia. Zn^{2+} is less affected as the log K values for the stability constants for $Zn(PAR)_2^{2-}$ complexes are significantly higher than those for Cd^{2+}: 21 and 18, respectively. The chromatogram in Figure 6.7 clearly shows the effect of decreased ammonia on the improvement in sensitivity for Zn^{2+} and Cd^{2+}. These plots also illustrate the importance of measurement at a particular pH even in the absence of ammonia. One report of the molar extinction coefficient for the Cd^{2+}/PAR complex is very low.[38] However, when examining the original publication it is found that the measurement pH is between 6 and 8, which as can be seen from the speciation plot, is nowhere near the optimum pH for complete complex formation of ML_2. Zinc on the other hand still shows complete formation at pH 8. It should be emphasised that speciation plots are only valid under the specified conditions. Other speciation plots show that substantially increasing the concentration of PAR will reduce the ammonia effect, but at the same time significantly increase the background absorbance and therefore the noise, affecting LODs.

Another example is where relatively strong complexing agents are used in the mobile phases for metal separation by IC. Dipicolinic acid is used in some mobile phases to give a good separation of metals on a CS5 mixed cation/anion exchange column such as sold by the Dionex Corporation.[39–41] Unfortunately,

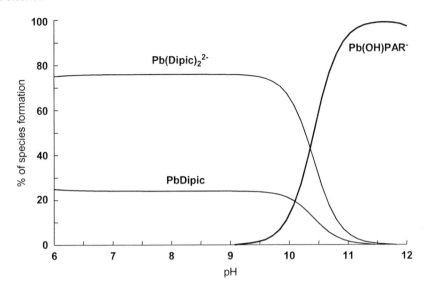

Figure 6.8 Metal speciation plot showing the effect of dipicolinic acid concentration on the formation of Pb(OH)PAR, the major absorbing species. Less important species removed for clarity. Dotted line shows typical pH used in PAR post-column reagents. PAR concentration, 5.0×10^{-5} M, Pb^{2+} concentration, 5.0×10^{-6} M, dipicolinic acid concentration, 2.0×10^{-3} M.

dipicolinic acid is quite a strong complexing agent and even at the millimolar level reduces the sensitivity for most metals with PAR, but the reduction is particularly great for Pb^{2+} and Cd^{2+}. Figure 6.8 shows the speciation plot for lead containing dipicolinic acid and PAR at typical levels used in the mobile phase and post-column reaction. The main Pb/PAR species is Pb(OH)PAR$^-$. The expected species, Pb(PAR)$_2^{2-}$, does not actually form in spite of being reported as present in the earlier literature.[36] The formation of M(OH)L complexes with PAR has also been reported for the lanthanides.[42] It is clear that not only does the sensitivity for lead suffer from just reacting with one PAR molecule, but also from the fact that only a small fraction is present at the most commonly used pH of 10. The speciation model indicates that a higher pH will substantially improve sensitivity for lead. The chromatogram in Figure 6.9, comparing runs at pH 10 and 11, indeed shows this is the case with a considerable increase in sensitivity for lead at the higher pH. There were also significant increases for most other metals indicating the strong competition from the dipicolinic acid in the mobile phase if the pH is too low. Other commonly used mobile phase complexing acids such as oxalic and citric acid will also have an effect on sensitivity, though not as strong as dipicolinic acid and an increase in pH may be required. However, it should be pointed out that operating a PAR PCR at a pH much higher than 10.5 will produce a higher background absorbance, which can affect baseline noise levels.

This situation of competing complexation from certain buffers and mobile phases with metal/PAR complexes must also be true when using other

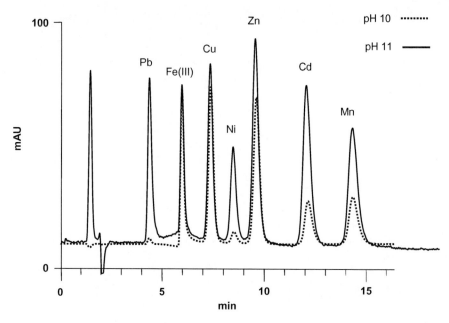

Figure 6.9 Chromatogram of a seven-metal mixture showing the effect of raising the pH to reduce the suppression by dipicolinic acid in the mobile phase. Column: IonPac CS5A, 250 × 4.0 mm i.d. Mobile phase: 4 mM dipicolinic acid and 3 mM K_2SO_4 adjusted to pH 4.3 with KOH. Post-column reagent, 1.5×10^{-4} M PAR in 0.4 M ammonia adjusted to the required pH with nitric acid. Sample: 100 µL containing 0.4 mg L^{-1} of all metals except Pb^{2+} at 3 mg L^{-1}.

photometric reagents. Thus, the composition and conditions of each reagent mixture should be carefully considered to obtain optimum results. If reasonably accurate stability constants are known for the main components then metal speciation plots as described would be very useful in this regard.

For reported limits of detection (LOD) the situation is even more difficult for direct comparison. Not only are there problems of reagent purity, competing buffers and pHs not at optimum, but also lack of knowledge of system noise levels. The LOD is defined in terms of the baseline noise as well as sensitivity (see detailed discussion later) and can vary from instrument to instrument and analytical methodology. Although the reported sensitivities and LODs can be a useful guide, they can not easily be translated from one system to another. For example, the LODs found under static absorbance measurements will be very different to those values found under dynamic flow during chromatography and PCR detection, as they will be affected by additional factors. These will include background pump noise levels, column efficiencies and the retention factor, k. Clearly, the longer the retention time and/or the lower the efficiency, the wider the peak and therefore it will be closer to the background noise. Thus, the limits of detection shown in Table 6.4 should be treated with the

Table 6.4 PCRs for the photometric detection of separated metal ions in IC.

No.	PCR reagent	Reaction conditions	λ (nm)	Detected ions and $LODs^a$ ($\mu g\,L^{-1}$)	Reference
1	PAR	1.0×10^{-4} M PAR, 0.125 M $Na_2B_4O_7$, 0.2 M NaOH, pH 10.5	520	Transition, heavy metals and lanthanides	43,44
		2.0×10^{-4} M PAR, 2 M NH_4OH, 0.02% Triton X-100, pH 10	490	Fe^{3+}, Cu^{2+}, Pb^{2+}, Ni^{2+}, Zn^{2+}, Co^{2+}, Cd^{2+}, Fe^{2+}, Mn^{2+}	45
		PAR, 50 $\mu g\,L^{-1}$, 2 M NH_4OH, 1 M ammonium acetate	520	Lanthanides (11–23)	46
2	PAR-ZnEDTA	1.2×10^{-4} M PAR, 0.2 mM ZnEDTA, 2 M NH_4OH	520	Alkaline earth metals	43
		2.5×10^{-4} M ZnEDTA-PAR, 2 M NH_4OH, 0.02% Triton X-100, pH 11	490	Fe^{3+}, Cu^{2+}, Pb^{2+}, Ni^{2+}, Zn^{2+}, Co^{2+}, Cd^{2+}, Fe^{2+}, Ca^{2+}, Mn^{2+}, Mg^{2+}, Sr^{2+}, Ba^{2+}	45
3	PAN	8.0×10^{-3} M PAN in isopropanol–water (40:60)	552	Cd^{2+}, Zn^{2+}, Fe^{3+}, Pb^{2+}, Cu^{2+}, Co^{2+}, Mn^{2+}	25
4	5-Br-PADAP	3.0×10^{-4} M 5-Br-PADAP, 0.95 % (w/v) Triton X-100, 66 mM glycine, 67 mM NaOH and 0.12 M NaCl	565	Cu^{2+} (1.5), Ni^{2+} (5.0), Zn^{2+} (0.5), Co^{2+} (5.0), Cd^{2+} (5.0), Mn^{2+} (3.0), Hg^{2+} (20) and Pb^{2+} (200)	26,47
5	Nitro-Paps	6.0×10^{-5} M nitro-PAPS, 0.15 M CHES, 0.1 M NaOH, pH 9.4	574	Pb^{2+} (4.8), Cd^{2+} (1.9), Mn^{2+} (1.5)	48
6	Xylenol orange	5.0×10^{-5} M xylenol orange, 4 mM CTAB, 2 M NH_4OH, 4 M NH_4Cl, pH 8.7, 40% methanol	618	Ce^{3+} (3.0), Nd^{3+} (4.0), Er^{3+} (30), Lu^{3+} (60)	30
		3.3×10^{-4} M xylenol orange, 0.2 M ammonium acetate	570	Mn^{2+} (1000), Co^{2+} (200), Zn^{2+} (500) and Ni^{2+} (2000)	49

Table 6.4 (Continued).

No.	PCR reagent	Reaction conditions	λ (nm)	Detected ions and LODs^a ($\mu g\ L^{-1}$)	Reference
		1.2×10^{-4} M XO, 1.2 mM CPB, 1 M urotropine buffer, pH 6.3, 4% ethanol	598	Dy^{3+} (1.1), Tb^{3+} (0.8), Gd^{3+} (0.7), Eu^{3+} (0.7), Sm^{3+} (0.8), Nd^{3+} (1.0), Pr^{3+} (1.1)	50
7	Phthalein purple (PP)	4.0×10^{-4} M PP, 0.25 M H_3BO_3, pH 10.5	510	Alkaline-earth metals	51,52
8	Arsenazo I	1.6×10^{-4} M arsenazo I, 1 M THAM, pH 9.0	590	Mg^{2+}, Ca^{2+}	25
		1.3×10^{-4} M arsenazo I, 1 M NH_4OH	600	Lanthanides from La^{3+} (130) to Lu^{3+} (175)	53
9	Arsenazo III	1.5×10^{-4} M arsenazo III, 1 M HNO_3	654	$U(VI)$; $Th(IV)$; $Bi(III)$, $Hf(IV)$, $Zr(IV)$, $Sn(IV)$	31,43
		1.5×10^{-4} M arsenazo III, 0.5 M acetic acid	658	Lanthanides	54–56
10	Chlorphosphonazo III	6.24×10^{-5} M chlorophosphonazo III, 0.02 M HNO_3	660	Lanthanides	57,58
11	Pyrocatechol violet (PCV)	1.0×10^{-4} M PCV, 0.5 % w/v Triton X-100, 0.2 M imidazole, pH 6.9	580	Al^{3+} (28)	43,59
		1.04×10^{-4} M PCV, 1.5 M hexamine	585	$V(IV)$, $V(V)$	31
12	Dithizone	4.0×10^{-5} M dithizone, 0.01 M CTAHS, pH 2.0	500	Cu^{2+} (58), Hg^{2+} (41), $MeHg^+$ (79), $PhHg^+$ (220)	60
		Dithizone solution in acetone–water (80:20), pH 4.8	590	Cd^{2+} (5), Co^{2+} (4), Cu^{2+} (1), Ni^{2+} (8) and Zn^{2+} (6) and Pb^{2+}	61

	Reagent	Conditions	λ (nm)	Analyte	Ref.
13	Sulfonated or carboxylated dithizone	6.5×10^{-4} M dithizone derivative, 0.5 % (w/v) Triton X-100, 0.05 M NaOH	570	Hg^{2+} (4.0), $MeHg^+$ (8.5), $PhHg^+$ (11.5)	62
14	Eriochrome black T	2.0×10^{-4} M eriochrome black T, 4 mM CTAB, pH 8.7 adj. NaOH	650	20–70 for indirect detection for lanthanides from La^{3+} to Lu^{3+}	63
15	*meso*-tetrakis (4-*N*-trimethyl-aminophenyl)-porphine (TTMAPP)	Four reaction coils: (**1**) 6.0×10^{-6} M EDTA in 0.1 M CH_3COOH, 0.2 M NaAc, 0.12 M NaOH; (**2**) 6.0×10^{-6} M Cu^{2+}, 4.4×10^{-6} M TTMAPP in 0.1 M HNO_3; (**3**) 0.5% ascorbic acid in 25 mM acetic acid, 0.8 M NaOAc; (**4**) 1.9 M HNO_3	433	Lanthanides (10)	64
16	Chrome azurol S (CAS)	2.6×10^{-4} M CAS, 2% (w/v) Triton X-100, 0.05 M MES, pH 6.0	590	Be^{2+} (3.0), Al^{3+} (128)	59, 65
		0.008 % CAS, 1 M hexamine and 0.01 M EDTA, pH 6.0.	560	Be^{2+} (35)	66
		0.01 % CAS, 0.01 % CTAB, 0.05 % Triton X-100, 0.4 M HMTA, pH 5.9	610	Fe^{2+} (2.0), Al^{3+} (1.0)	67
17	Eriochrome cyanine R (ECR)	2.5×10^{-4} M ECR, 1.0mM CTAB 0.2 M hexamine, pH 6.1	580	Al^{3+} (16)	59
18	Tiron	3.0×10^{-4} M tiron in 3 M ammonium acetate	310	Al^{3+} (20)	68
19	Calmagite	0.004% (w/v) calmagite in 0.3 M aqueous ammonia	610	Mg^{2+}, Mn^{2+}, Co^{2+}, Zn^{2+}, Cu^{2+}	69
20	1,5-Diphenylcarbazide	2.0×10^{-3} M 1,5-diphenylcarbazide	520	Cr(VI) (0.5)	70

Table 6.4 (*Continued*).

No.	PCR reagent	Reaction conditions	λ (nm)	Detected ions and LODs^a ($\mu g\ L^{-1}$)	Reference
21	1-Carboxy-N,N-cyclotetramethylene-dithiocarbamate	0.01 M aqueous solution of 1-carboxy-N,N-cyclotetramethylene-dithiocarbamate	320	Ni^{2+} (10), Co^{2+} (25), Cu^{2+} (42.5), Pb^{2+}, Fe^{2+}	71
22	Bis(2-hydroxyethyl)-dithiocarbamate (BHEDTC)	0.1% aqueous solution of BHEDTC	405	Cu^{2+}, Zn^{2+}, Ni^{2+}, Co^{2+}, Fe^{2+}	72,73
23	1,10-Phenanthroline	0.014 M 1,10-phenanthroline, 1.34 M ammonium acetate buffer, pH 5.1	508	Fe^{2+} (45), Fe^{3+} (87)	74
24	8-Hydroxyquinoline	5.4×10^{-3} M 8-hydroxyquinoline in acetone, 0.5 M NH_4OH	405	Ca^{2+}, Cd^{2+}, Mn^{2+}, Pb^{2+}, Fe^{3+}, Cu^{2+}	7
25	2-Mercaptopropionic acid	17.9 mM 2-mercaptopropionic acid in 1 M sodium acetate buffer, pH 4.0	365	Mo(vi)	72

Table 6.5 PCRs for the fluorescent detection of separated metal ions in LC.

PCR reagent	PCR conditions	Wavelength λ_{ex}	Wavelength λ_{em}	Detected ions and LODs[a], ($\mu g\ L^{-1}$)	Reference
8-Hydroxy-quinoline-5-sulphuric acid (8-HQS)	2.0×10^{-3} M 8-HQS, pH 8.3	360	512	Al^{3+} (1.0)	103
	4.0×10^{-3} M 8-HQS, 2.0×10^{-3} M CTAB, 1.0 M sodium acetate buffer, pH 4.4	395	500	Al^{3+} (0.5)	110
	2.0×10^{-3} M 8-HQS, 0.5 M ethanolamine, pH 8.2	405	525	$Ba^{2+}, Mg^{2+}, Sr^{2+}, Ca^{2+}$	111
	2.0×10^{-3} M 8-HQS, 0.05 M Tris buffer, pH 9.2	390	510	Mg^{2+} (0.6), Ca^{2+} (40)	112
	1.0×10^{-3} M 8-HQS, 1.2×10^{-3} M CTAB	389	529	Ga^{3+} (1.5), In^{3+} (1.0)	113
	1.0×10^{-3} M 8-HQS, 0.3 M Bicine buffer, pH 9.2	362	filter	Zn^{2+} (0.1), Cd^{2+}	102
	1.0×10^{-3} M 8-HQS, 4.0×10^{-3} M CTAC, 0.01 M MOPS buffer, pH 9.0	400	520	$Al^{3+}, Cd^{2+}, Ga^{3+}, In^{3+}$ (0.5–1.6)	114
8-HQS–MgEDTA	Two-coil system: **(1)** reaction with MgEDTA in borate buffer at pH 10; **(2)** with 1.95×10^{-3} M 8-HQS	390	510	$Al^{3+}, Ba^{2+}, Ca^{2+}, Cd^{2+}, Co^{2+}, Fe^{2+}, In^{3+}, La^{3+}, Mg^{2+}, Mn^{2+}, Ni^{2+}, Pb^{2+}, Sr^{2+}, Zn^{2+}$ (7.9–191)	109
8-HQS–MgCDTA	Two-coil system: **(1)** reaction with 1.0×10^{-3} M Mg-CDTA at pH 6; **(2)** with 1.0×10^{-3} M 8-HQS in 0.6 M bicine, final pH 8.0	360	500	$Ba^{2+}, Mg^{2+}, Sr^{2+}, Ca^{2+}, Mn^{2+}, Ni^{2+}, Cu^{2+}, Co^{2+}, Zn^{2+}, Pb^{2+}$ (~100)	84
	Two-coil system: **(1)** reaction with 1.0×10^{-3} M Mg-CDTA in 0.25 M MOPSO at pH 6.3; **(2)** with 1.0×10^{-3} M 8-HQS in 0.6 M 2-amino-2-methyl-1-propanol, final pH 8.0	360	500	Lanthanides (~100)	84

Table 6.5 (*Continued*).

PCR reagent	PCR conditions	Wavelength		Detected ions and LODs^a, ($\mu g\ L^{-1}$)	Reference
		λ_{ex}	λ_{em}		
2,2'-pyridyl-benzimidazole (PBI)–ZnEDTA	1.5×10^{-3} M PBI, 6.25×10^{-4} M ZnEDTA, glycine/NaOH buffer, pH 11.5	365	435	Mn^{2+} (1600), Mg^{2+} (850), Ca^{2+} (950), Cu^{2+} (4600), Cd^{2+} (2950), Zn^{2+} (1350)	115
Lumogallion	3.0×10^{-5} M Lumogallion, 0.25 M MES, pH 6.05	500	590	Al^{3+} (0.01)	83

6.2.3 Recent Developments in High-sensitivity Reagents for Post-column Reactions

A priori, the molar extinction coefficient (ε) defines the fundamental sensitivity of photometric detection involving the formation of complexes of metals with corresponding organic reagents. The frequently used PAR gives molar absorptivities of about 6×10^4 L mol^{-1} cm^{-1} to 7×10^4 L mol^{-1} cm^{-1}. As mentioned earlier, due to the success of PAR many studies have been carried out on analogues containing the pyridylazo unit, mainly to improve sensitivity.[22,38,75–79] It has been found that a significant improvement in sensitivity is obtained when the *para*-hydroxy group is replaced with a tertiary amino group (R_2 in Table 6.2). Two of the most successful ones to date are 5-Br-PADAP and nitro-PAPS, which produce higher molar absorptivities of approximately 8×10^4 and 1.2×10^5 L mol^{-1} cm^{-1}, respectively, and could be potentially more suitable in future. They are available commercially and have been used in PCR detection[47,48] and a significant improvement in sensitivity of detection was noted. This can be seen in Figure 6.10 which compares the sensitivity of detection of PAR and nitro-PAPS for selected metals when no complexing agents are present in the buffer or mobile phase.[80] The peak heights are significantly greater for all the metals except Mn^{2+}. This is because the peak for nitro-PAPS shows a lot of tailing for Mn^{2+} reducing the peak height. At the moment, the relatively low solubility of 5-Br-PADAP and the very high cost of nitro-PAPS still affect the wide use of these reagents.

There is one characteristic of these two new reagents that should be noted, and that is the stability constants for the metal complexes are lower than those with PAR. Studies have shown that there is a direct relationship between the pK_a value of the *ortho*-hydroxy group on the benzene ring and the strength of the complexes, that is, the value of the stability constants.[77] This is not too surprising, as normally the pK_a value of the ligating group is related to the strength of the conjugate base and hence the strength of the complex. As expected, adding or changing substituents to the benzene ring will affect the *ortho*-hydroxy pK_a value. When an amine type group is substituted for the *para*-hydroxy group as in 5-Br-PADAP and nitro-PAPS, the pK_a value of the *ortho*-hydroxy group is reduced and so also are the metal complex stability constants.[42,76] One disadvantage of this is that the formation of the metal complexes in the PCR will be even more suppressed by competition from any complexing agents in the buffers and mobile phases. Therefore, even more care is needed in choice of buffer and mobile phase composition. For example, nitro-PAPS cannot really be used with dipicolinic acid mobile phases as the suppression is so great.[80] Nevertheless, one benefit of the reduced pK_a values is that the pH for optimum formation of the metal complexes is lower than that of PAR which will help with the choice of different less complexing buffers and reduce the tendency of the more acid hydrolysing metals to precipitate.

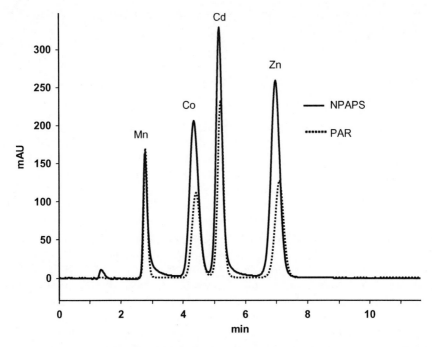

Figure 6.10 Chromatogram showing the difference in sensitivity between PAR and nitro-PAPS with no complexing agents in the buffer or mobile phase. Column: 50 × 4.0 mm i.d. packed with 5 μm IDA bonded silica. Mobile phase: 0.01 M nitric acid. PCR reagents: 0.15 mM PAR in 0.3 M CAPS buffer, pH 10, monitored at 510 nm and 0.17 mM nitro-PAPS in 0.1 M tetraborate buffer, pH 9.2 monitored at 570 nm. Sample: 100 μL containing 0.25 mg L^{-1} Mn^{2+}, 0.5 mg L^{-1} Co^{2+}, 1.0 mg L^{-1} Cd^{2+} and 0.5 mg L^{-1} Zn^{2+}.

Molar extinction coefficients tend to increase with the area of the conjugated system absorbing the incident radiation so, potentially, even more sensitive reagents could be the porphyrins. For example, TTMAPP, gives a molar extinction coefficient for the copper complex of 4.8×10^5 L mol^{-1} cm^{-1};[48] however, the first attempt involving its application as a PCR reagent did not provide any significant improvements in sensitivity of detection (see Table 6.4, row 15). A big problem with using larger organic molecules, such as the porphyrins, with higher molar absorptivities, is that the reactions with the metals become significantly slower, requiring longer reaction coils and/or higher reaction temperatures. Furthermore, the aqueous solubilities of the large metal complexes tend to be lower leading to an increased 'stickyness' on the walls of the reaction tubing as noted earlier.

In spite of all the studies attempting to produce better colorimetric reagents than PAR, it still remains the best all-round PCR reagent for the common metals. It may not be as sensitive as some, but its chelating strength is still the highest and is yet to be surpassed in its ability to resist competition from other complexing agents.

6.2.3.1 Choice of Buffer for Post-column Reactions

It will be useful at this point to discuss some of the factors which help with the best choice of buffer for a particular PCR. It has already been noted that one of the most commonly used buffers with the PAR reagent is ammonia, usually used close to pH 10. In spite of the fact it can easily be demonstrated how the concentration of ammonia can affect sensitivity, it is remarkable that buffers containing very high concentrations are still being reported. Some are as high as 2.5 M ammonia and sometimes even higher.[81,82] Of course, dilution with the mobile phase will reduce this concentration to around 1 M, but as Figure 6.4 shows this level seriously suppresses Cd^{2+} in particular. It is also puzzling why acetic acid is used in some systems to adjust the pH to around 10, as the buffer action will be weaker and furthermore acetate is also a complexing agent. Increasing the pH of the ammonia buffer will improve things, but the buffer capacity becomes quite weak at pH 11 and the problem of higher baseline noise will become increasingly apparent. The authors have found that a concentration of 0.4 M ammonia (0.13 M after dilution in the PCR) adjusted to pH 10.5 with nitric acid gives good buffering action for most separation systems with little affect on sensitivity. Alternative buffers are available in the alkaline pH range and Goods buffers such as Bicine, CAPS, MOPSO, CHES or MES can be suitable[14,48,65,83,84] as they are essentially non-complexing.[85] The CAPS buffering capacity is not that great, however, and a high concentration is necessary, almost at saturation level for some strong acid mobile phases, which means a much higher cost than ammonia based buffers. Borate is also a useful non-complexing buffer, but is best used at its optimum around pH 9, as it becomes significantly weaker at pH 10 and above. The limited solubility of borate is also a problem, with a realistic maximum of 0.1 M or around 0.03 M in the mixed PCR reagent, again impacting on buffer capacity. CAPS and borate, although useful for some PAR systems as long as buffering capacity is not a problem, are particularly useful for the weaker complexing nitro-PAPS where the optimum pH is around 9. Another important consideration to take into account is the level of metal impurities in the buffers, as they are used at quite high concentrations. These metal impurities will increase the absorbance background of the PCR reagent and thus increasing detector noise. Fortunately, ammonia and nitric can be obtained with very low levels of impurities at quite low cost. For other buffers, the purest grades possible need to be purchased giving even higher costs. The buffers used for the reagents involving the detection of the more acid hydrolysing metals such as Al^{3+}, Fe^{3+}, Be^{2+}, Zr(IV), lanthanides and actinides, *etc.* appear to be less of a problem at the lower pHs involved. Hydroxylamine is non-complexing and a useful buffer around pH 6, for example with PCV for determining Al^{3+} and Fe^{3+}. Acetate buffers for use with arsenazo(III) do not appear to pose any problems since acetate is too weak to interfere with the strongly complexing lanthanides and actinides.

6.2.3.2 Effect of Added Surfactants

The positive effect of adding surfactants for photometric measurements has been known for some time. Enhancement of the analytical response of many colorimetric and fluorescent reactions is found if micelles are formed in the solution.[22] This is thought to be due to the environment being very different inside a micelle compared to the bulk solution. Other benefits of added surfactants are increased solubilisation of organic reagents with poor aqueous solubility and surface modification of the surface of the reaction coil, minimising the interaction with metal complexes.

Gautier *et al.* investigated the effect of additives of cationic surfactants (CPC and CTAB, *cmc* for both 9.0×10^{-4} M), anionic surfactant (SDS, *cmc* 8.0×10^{-3} M) and non-ionic surfactant Triton X-100 with *cmc* 2.0×10^{-4} M on a PCR reaction of separated transition metal ions with xylenol orange[30] and eriochrome black T.[63] They noted the almost three times absorbance increase for PCR reaction of xylenol orange with lanthanides in presence of cationic surfactants CPC at pH 8.7 and attributed this effect to the formation of ternary complex with a stoichiometric ratio of lanthanide–xylenol orange–CPC = 1:2:4. Some spectral shift for absorbance maximum from 604 to 618 nm was also noticed due to changes of the micro-environment of the complex due to electrostatic interactions between micelles of the cationic surfactant and anionic dye and due to hydrophobic interactions between alkyl chains of the surfactant and the organic dye. In the case of eriochrome black T,[63] no changes in sensitivity were observed with additives of the anionic surfactant SDS and non-ionic Triton X-100, but a decrease in sensitivity of detection was noticed with addition of cationic surfactants such as CPC and CTAB. The authors also noted some positive effects of the presence of surfactants in a PCR mixture such as improvement of baseline drift and decrease of background noise of PCR.

The addition of 0.5–2.0% w/v of non-ionic surfactant Triton X-100 has used for the solubilisation of such organic reagents as dithizone derivatives,[60,62] CAS,[59,65,67] PCV[43,59] and 5-Br-PADAP[26,47] in PCR.

6.3 Practical Methods for Improving Limits of Detection in Liquid Chromatography

6.3.1 The Detection Limit

The sensitivity and LOD are important parameters which indicate the potential of an analytical system for determining low levels of traces of metals. Unfortunately, the general treatment of these parameters in the literature can be vague and the terms can even be used interchangeably. The statement in the book by Grob in 1985 gives a good idea of the situation at the time:[86]

The word 'sensitivity' is like beauty, truth or happiness. It describes a quantity that is always good, but is not that well defined and is used in different ways. The term is supposed to mean the ratio of signal to sample size. It is also used when LOD is meant.

Well, not much has changed since then, but it is becoming more accepted that the term sensitivity should be restricted to mean the ratio of signal to analyte amount, under the specified conditions, or as defined by Rubinson and Rubinson, the slope of the working curve.[87] For absorbance measurements under static conditions then the slope would be related to the molar extinction coefficient, again under specified conditions such as the measuring wavelength. A species with a higher molar extinction coefficient would be expected to have a higher sensitivity and hence a lower LOD. However, in many cases this is not necessarily so as the LOD will be affected by other factors, one major one being baseline noise. So although sensitivity in terms of the slope of the working curve is an important parameter in defining LODs, these two terms represent different concepts and should not be interchanged. For example, quite often, changing the analytical method to improve the sensitivity may not result in a lower LOD, as the noise levels could increase correspondingly even resulting in worse LODs. Thus, under practical conditions the LOD is the preferred means for comparing the ability to determine the lowest concentrations. It will be found that the LOD is a very important quoted parameter in any quantitative trace analytical method. It not only gives vital information on the lower limit of quantification (LOQ), but also can be used to compare different instrumental techniques, so that a decision can be made on choice of method for a particular sample analysis. Of course, it is important to know how the LOD and LOQ are defined for each system before a judgement can be made.

Unfortunately, there is no universally accepted descriptive definition or calculation of the LOD and LOQ. One descriptive term for the LOD is the lowest amount of an analyte in a sample which can be detected but not necessarily quantified as an exact value.[88] The LOD is called by a variety of other names, such as miniumum detectable quantity (MDQ), detection limit (DL) and detectivity. A descriptive term for the LOQ is the lowest amount of an analyte in a sample that can be determined as an exact value. LOQ is also called by other names including the minimum level of quantification (ML) and depending on the validating organisation or laboratory, can be defined as anywhere between six and 10 times the LOD. The most common way of obtaining a LOD, is from signal-to-noise (S/N) calculations. The analyte response above the background noise is considered as the signal. As the analyte concentration is decreased the analyte signal will eventually merge into the background noise and be 'lost' or become 'invisible'. Most techniques use a statistical approach to calculating the LOD based on the standard deviation of the noise and/or signal. It is not within the scope of this book to discuss the plethora of guidelines for defining and calculating LOD and LOQ by the various organisations such as IUPAC, ISO, AOAC and ASTM *etc*. There is abundant literature on this subject, but if the reader needs a succinct account of

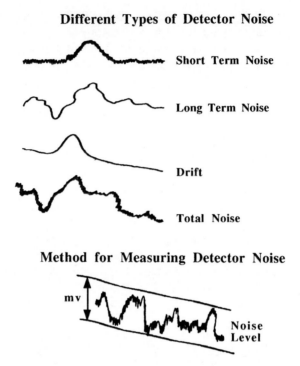

Figure 6.11 Representation of different types of baseline noise and its measurement for determining the limit of detection. Reproduced, with permission, from Scott.[90]

the different guidelines between these organisations, then the chapter in the book by Miller is a good start.[89]

Rather than use a statistical approach, the common method in chromatography is to assess and measure the average peak to peak baseline noise just before and after the analyte signal (Figure 6.11). The LOD is then defined as twice, or sometimes more conservatively as three times this figure. The LOQ would correspondingly be six to 10 times the noise measurement. Although some people would prefer a more statistically exact method, the concept of the LOD being two or three times the measured baseline noise is widely accepted and used by chromatographers.[86,89–91]

6.3.2 Types of Baseline Noise

The nature of the baseline noise on a chromatogram can vary depending on, among other things, the instrumentation and reagents used for a particular determination. A useful classification is based on the time scale of the noise in relation to the peak width, namely, short term noise, long term noise and drift.[90,92]

Detection

The difference between each type, is somewhat arbitrary, but serves to cover most situations. Short term noise involves relatively fast even fluctuations, where the time between the variations is much shorter than the peak base width. Typically, this type of noise could be from the detector electronics and/ or fast pump fluctuations. Long term noise involves slower variations in the baseline of the same time scale as the base peak width. They can be random or more cyclic such as from slow reciprocating pump heads. This is the most serious type, as most noise reduction procedures would not be able to improve the signal to noise ratio without seriously distorting the signal. Finally, drift is a slow unidirectional change in the baseline, which is more or less linear, though can be curved. Unless severe, drift should not interfere with LOD calculations. Both long term noise and drift are sometimes called baseline wander. An idealised example of the three types of noise and the effect on an analyte peak is that shown by R.P.W. Scott in Figure 6.11.[90] It can be seen that the long term noise has masked the analyte peak and has essentially become 'invisible'. The same figure also shows how baseline noise can be measured, which obviously must include the long term noise. An example from a real chromatogram is shown in Figure 6.12 produced during a study of uranium determination by HPCIC. The short term noise is due to pump fluctuations, where the

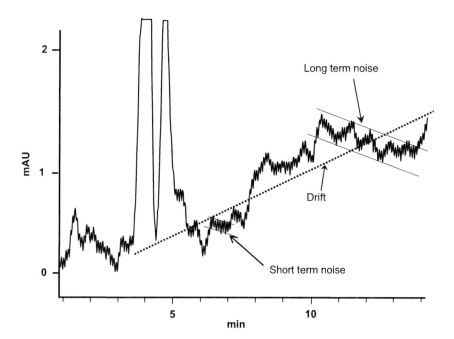

Figure 6.12 The main types of baseline noise on a real chromatogram. Column: 100×4.6 mm i.d., PRP-1, 7 μm PS-DVB. Mobile phase: 1 M KNO_3 in 0.5 M nitric acid containing 5.0×10^{-4} M dipicolinic acid Sample: 100 μL containing 0.5 mg L^{-1} U(VI). Detection: photometric at 654 nm after PCR with arsenazo III.[80]

reciprocating cycle is a lot faster than the peak width. The long term noise is of the same order as the peak width and the drift, though not quite linear is on a clear upward trend. Like the idealised example, the analyte peak, which is an injection of $1\,\mathrm{mg\,L^{-1}}$ U(VI) with a retention time of 8 min, is lost in the total noise. Using noise reduction methods, discussed in the next section, a clear U(VI) peak can be produced.

6.3.3 Noise Reduction Methods

A large number of applications now require the determination of very low levels of components in a wide variety of samples. As stated earlier, instrument manufacturers have put a tremendous effort into reducing the noise and drift of detectors including spectrophotometric detectors, the ones most commonly used in LC. Nevertheless, these low noise levels can be difficult to achieve in practice as external factors can increase the noise when the detectors are used in conjunction with other equipment in HPLC systems. Because of this a number of noise reduction methods have been developed to try and improve S/N ratios.

6.3.3.1 Mechanical Methods

Reciprocating and peristaltic pumps will produce pulsations which may be seen at high sensitivity detection settings, particularly with certain PCRs. The high precision design of pump heads and multi-roller peristaltic manifolds can keep these pulsations to a minimum. The addition of pulse dampeners can further reduce the flow fluctuations. The construction of dampeners depends on the level of pressure at which pulsations take place. HPLC high pressure dampeners are incorporated in-line between the pump and the injector. Most commonly they involve a long coil or chamber with a slight elasticity in the walls or coil tension. The energy of the pump pulses is absorbed and released out of phase with the frequency of the pump strokes, thus reducing or dampening the amplitude of the pulses. Because of the relatively slow frequency of most pump piston reciprocations, it is difficult to design a pulse dampener to completely suppress short term pump pulses and they have little effect on pump variations which produce long term noise. Low pressure dampeners, for example, for peristaltic pumps, have a similar construction but from softer coil materials.

6.3.3.2 Electronic Methods

The electronic circuits of detectors will produce noise involving rapid changes in voltage or current. Electronic noise will also be produced by the light collecting devices in spectrophotometers, with diodes producing greater voltage fluctuations than photomultipliers.

Many detectors have built-in electronic filters, which can control the time constant or rise time of the signals at the output of the detector, before being sent to the computing integrator or recorder. The level of the time constant can be chosen by the operator. The time constant is only useful for suppressing fast or short term noise. As the electronic suppression affects both the baseline noise and the peaks, the rise time has to be chosen with care otherwise serious peak distortion could result.

6.3.3.3 Software Methods

Computer programs have been developed as part of integration software, which can suppress noise either during a chromatographic run, *i.e.* in real time, or post-run. In real time this can be done by digital sampling of the noise followed by averaging to smooth out the baseline. Baseline noise suppression post-run can be achieved by using a number of special algorithms involving averaging or filtering protocols. The most common ones are, moving average, Savitsky–Golay and Olympic processes. For example, Gettar et al.[30,63] applied a Savitzky–Golay algorithm to a registered chromatogram and got a 5-10 fold decrease in noise. Also, a 6-32 fold enhancement of the S/N ratio was noted after a similar smoothing procedure on a chromatogram of the lanthanides detected by PCR reaction with xylenol orange.[30] It is likely the big improvement in S/N ratios is due to suppression of fast short term noise. For both real time and post run methods the degree of noise suppression is under control of the operator and like electronic suppression can seriously distort peaks if overused. As the degree of noise suppression is increased, peaks will broaden and resolution will suffer. For long term noise, the degree of peak distortion in terms of broadening and reduction in height will offset any reduction in noise, so lower LODs will not be achieved.

6.3.3.4 Noise Suppression Using a Multiple Wavelength Approach

Although the systems discussed above can suppress a significant amount of the short term noise, they are virtually useless at suppressing long term noise of the same order as the peak widths in the chromatogram. In future, even short term pump noise will become increasingly difficult to suppress as faster and faster LC systems are developed. This is because peak widths early on in a chromatogram could become similar to the width of pulses from higher frequency piston pump cycles, which are gradually replacing the slower piston types.

The situation is made significantly worse when PCR detection is involved, where the mobile phases and/or reagents have different absorbances. Not only is the noise from the two pumps or solvent delivery systems additive, but also the noise is amplified by the absorbance differences. For example, two pumps producing a combined rhythmic flow variation of ±0.2% will give a peak to peak baseline noise of 0.4 mAU, when the absorbance difference between the mobile phase and PCR reagent is 100 mAU. This is about 40 times the electronic noise levels of typical high quality UV–visible absorbance detectors.

In some ways this is the best case scenario as in reality, even the best pumps get noisier with age, particularly in combination with slightly worn check valves. Furthermore, for most PCRs, the post-column reagent is usually delivered with a low cost pump or pneumatic based solvent delivery system, which will produce larger flow variations than high quality HPLC pumps.

The increased pump noise seen in many systems using PCRs will obviously seriously worsen the LOD. Since much of the noise will be long term and of the same order as peak widths, current noise reduction systems, as explained above, will not be of much help. A radical new approach to reducing pump noise has just been introduced by a chromatography company (JPP Chromatography Ltd, UK) using a multiple wavelength procedure.[93]

The approach is based upon a small electronic piece of equipment called a Signal Extractor, containing specially programmed microprocessors specifically designed to extract a much cleaner analyte signal where noise is present, which is mainly due to pump variations, even when the variations are very large. It works in real time, processing the signals from a dual or multiple wavelength detector, removing the pump noise electronically, whilst leaving the analyte signal unchanged. Unlike present noise reduction methods, such as rise-time filters and computer smoothing algorithms, mentioned above, the Signal Extractor has no effect on peak width or symmetry. It is claimed that the extractor can virtually eliminate pump noise whatever the frequency, including long term drift or baseline wandering. The dual wavelength approach, however, does not reduce fast electronic noise. If this is significant, a small amount of noise filtering or smoothing can be used in combination, but only enough to suppress the electronic noise without distorting peak shapes. The chromatograms shown in Figure 6.13 is the same as the one in Figure 6.12 shown earlier, but subjected to different noise reduction procedures. The one processed with a moving average algorithm post-run (A), successfully removes the fast pump noise but has no effect on the long term noise. Chromatogram (B), processed by the Signal Extractor in real time, removes the fast pump noise and most of the long term noise producing an easily measurable uranium peak.

Another even more dramatic example of signal extraction considerably reducing short and long term noise can be seen for the HPCIC separation of three common metals at very low µg L^{-1} levels (Figure 6.14). The noise is so large there is no indication that any metal peaks are present in the unprocessed chromatogram (A). When the noise reduction is carried out it can be seen that the LODs for all three metals are less than 1 µg L^{-1}. It should be appreciated that even though the baseline noise level looks very bad, the chromatogram is run at very high detector amplification with an absorbance scale (2 mAU full scale) not normally seen in research and commercial publications. In fact it is rare to see any published chromatograms with less than 50 mAU full scale deflection. The ability to extract the signal from noise caused mainly by pump variations without affecting peak shape is very useful. It allows the use of PCR where the background absorbance is unusually high such as the PAR-ZnEDTA system or the PAR system run at a pH above 10.5. Even the

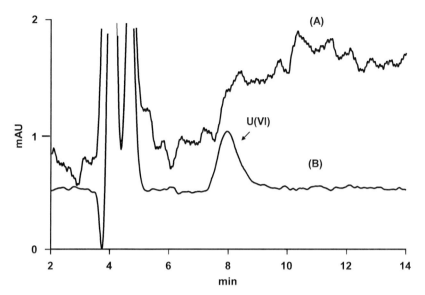

Figure 6.13 The same chromatogram as Figure 6.12 subjected to different noise reduction strategies: (A) after being processed by a moving average smoothing algorithm, post-run; (B) obtained when using the signal extraction device in real time.

best quality pumps with very low pulsations will still produce significant baseline variations at these exceptionally high sensitivity detector settings.

Finally, it should be pointed out that when working at very high detector sensitivities, it tends to be forgotten that small temperature variations of the post-column reagent entering the detector cell can affect the baseline absorbance. Normally, these variations would be masked by the pump noise. Figure 6.15 shows a baseline after signal extraction where the temperature cycling of the laboratory air conditioning unit close by is clearly being followed by the detector baseline.

6.3.3.5 On-line Purification of the Mobile Phase

The possible contamination of the mobile phase can produce a detrimental effect on the sensitivity of PCR detection in HPCIC and increase baseline noise. It is especially important for ultra-trace analysis. The inclusion of a metal trap column before the injection valve can help to solve this problem. A significant improvement in the baseline for the fluorescent detection of aluminium after PCR with Lumogallion has been reported[59,83] using a trap column packed with Diphonix resin (EiChrom Technologies Inc, USA). Diphonix resin has both phosphonic and sulfonic acid functional groups[94] providing efficient removal of aluminium from acid mobile phases. Metal impurities can be also removed

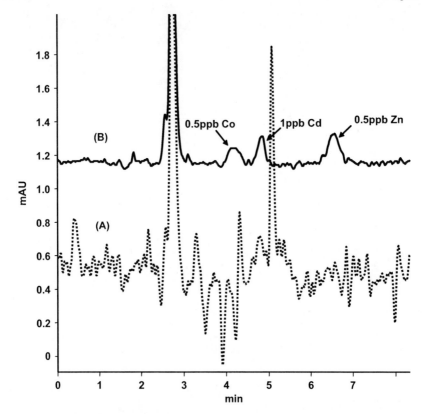

Figure 6.14 Chromatogram showing the effect of multiple wavelength signal extraction on baseline noise. (A) no noise reduction, (B) with signal extraction. Column: 50 × 4 mm i.d. containing 5 μm IDA bonded silica. Mobile phase: 0.01 M nitric acid. Sample volume: 100 μL. Detection: photometric 510 nm after PCR with 5.0×10^{-5} M PAR in 0.4 M ammonia adjusted to pH 10.5 with nitric acid.

from the mobile phase by passing it through a column packed with Kelex-100 (7-(4-ethyl-1-methyloctyl)-8-quinolinol) impregnated ODS.[95]

6.4 Other Types of Post-column Reaction Detection

6.4.1 Displacement Reactions

It is not correct, as stated in some reports, that the alkaline-earth metals do not react with PAR; it is just that Mg^{2+}, Ca^{2+}, Sr^{2+} and Ba^{2+} tend to have much lower sensitivities in PAR colorimetric reactions mainly due to the fact that they only react with one PAR molecule and also give very incomplete reactions at the pHs used in the PCRs. This is not a disadvantage for many applications as large amounts of Mg^{2+} and Ca^{2+} can be present in the samples and interfere

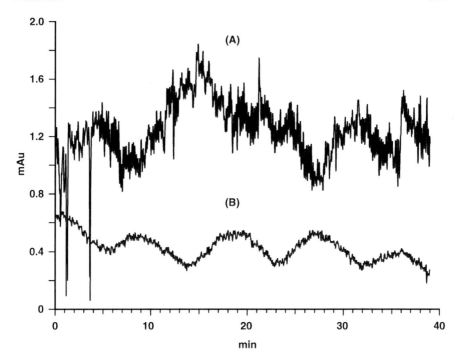

Figure 6.15 Effect of the temperature cycling of a nearby air conditioning unit on baseline absorbance. (A) without smoothing, (B) with signal extraction. PAR post-column reagent.

$Ca^{2+} + PAR^{2-} = Ca(PAR)$ — Only partial reaction at pH 10

$ZnEDTA^{2-} + Ca^{2+} = CaEDTA^{2-} + Zn^{2+}$ — Complete displacement

$Zn^{2+} + 2PAR^{2-} = Zn(PAR)_2^{2-}$ — Essentially complete reaction

Scheme 6.1 Replacement of zinc from ZnEDTA by an alkaline-earth metal.

with the determination of other metals. However, for certain specialised applications, such as their determination in saturated brines by HPCIC,[96,97] it would be useful to have as high a sensitivity as possible. One way of achieving this is to use displacement reactions. The most commonly used displacement reaction involves the addition of ZnEDTA to the PAR reagent.[25,98] The conditions are so chosen that the alkaline earth metal replaces the equivalent amount of Zn^{2+} from the ZnEDTA, which then reacts with the PAR. As Zn^{2+} has one of the highest molar absorptivities with PAR, a much higher sensitivity for the alkaline earth metal is obtained. The equations in Scheme 6.1 show this in simplified form for Ca^{2+} determinations, ignoring lesser side reactions.

When comparing the EDTA stability constants of Zn^{2+} and the alkaline earth metal complexes it would appear that little exchange should take place.[33] However, the presence of PAR, hydroxide ion and to a lesser extent ammonia lowers the conditional constant of ZnEDTA substantially, allowing almost complete exchange of Zn^{2+} for Ca^{2+}. The displacement reaction technique not only gives much improved sensitivities for the alkaline earth metals, but should also show improvement for some other metals such as Pb^{2+}. This is because Pb^{2+} is partially hydrolysed at pH 10 and only reacts with one PAR molecule as mentioned earlier. Thus the sensitivity is potentially doubled using ZnEDTA.

Jezorek and Freiser studied the use of PAR-ZnEDTA as a post-column reagent for 14 metals.[98] As expected, when compared to PAR alone, they found huge increases in the peak response for the alkaline earth metals and a substantial increase for Pb^{2+}. Other metals gave more or less the same sensitivity except for Al^{3+}, Fe^{3+} and La^{3+}. They discussed the possibility of hydroxy products suppressing the exchange reaction, which is reasonable as these three metals are strongly hydrolysed in the alkaline region. A similar problem with Al^{3+} and Fe^{3+} was observed by Yan and Schwedt when they compared the detection ability of the PAR-ZnEDTA reagent with PAR-ZnDCTA and PAN-ZnEDTA reagents.[67] Eriochrome cyanine R and chrome azurol S based post-column reagent mixtures containing additions of 0.01% CTMAB–0.05% Triton X-100 surfactants gave much higher sensitivities. This is not too surprising as these metallochromic reagents are usually used at lower pHs where potential hydrolysis is much reduced. Interestingly, they also found that modification of the PAR-ZnEDTA reagent by using the much weaker complexing nitrilotriacetic acid (NTA) instead of EDTA, lowered the response to specific transition metals and completely suppressed the PAR response to lanthanides.[99]

The use of exchange reactions substantially increases the complexity of the PCR where 20 or more competing species will be present. Empirical optimisation will be difficult and metal speciation modelling could be of major help in future work as discussed in the previous section, but use of accurate stability constants is essential. Also, the kinetics of displacement reactions requires additional attention, especially in the presence of various complexing agents in the mobile phase. The kinetic aspects of the attainment of equilibrium for the PAR-ZnEDTA post-column reaction detection system for the determination of alkaline earth metals was investigated by Lucy and Dinh.[100]

6.4.2 Fluorescence Detection

The converse of spectrophotometric absorbance is fluorescence detection, where emission of light is measured. Usually, the analytical signal in fluorescence detection is one order of magnitude more sensitive than absorbance. Although colorimetric PCRs are by far the most commonly used for trace metal detection in IC, some fluorimetric reactions have been used with success.

Detection

As strong fluorescence will only occur in metal complexes where the metal ion has a full set of *d*-orbitals and is diamagnetic, it is a more selective technique and studies have been mainly restricted to the alkaline earths, Zn^{2+}, Cd^{2+}, Al^{3+}, Ga^{3+} and In^{3+}.[101,102] Trace Al^{3+} determinations have attracted a lot of attention with fluorimetric PCRs, 8-HQS being particularly sensitive.[101,103] A series of papers have been published studying Al speciation in natural waters using a short IC column linked to 8-HQS fluorimetric PCR detection.[104–108]

The extremely sensitive detection of aluminium has been reported for a PCR with lumogallion and fluorescent detection[83] (see details in Table 6.5). The reaction of aluminium with lumogallion is more selective than with 8-HQS, but requires increased reaction time and, hence, quite long reaction coils and reactor temperature adjusted at 70–72 °C. A further reactor temperature increase will result in the fluorescence being reduced. The optimum pH range for this reaction was found to be in the range between 6.0 and 6.2. The adsorption of aluminium–lumogallion complex on the internal walls of the PTFE reactor tubing was noted.

Displacement reactions can also be used for fluorescence detection. In this case the displaced metal reacts with a chelating agent such as 8-hydroxyquinoline sulphonate to produce a strongly fluorescent complex. Williams and Barnett used the displacement of Mg^{2+} from MgEDTA with subsequent reaction with 8-HQS for the fluorescence detection of a wide range of metals including the transition metals[109] (see Table 6.5). This is an interesting way of extending fluorescence detection to paramagnetic metal species, which do not form fluorimetric complexes by direct reaction. A similar approach was used by Ye and Lucy, who replaced EDTA by CDTA in the displacement reaction.[84] This substitution of the competing ligand allowed a decrease in the background response of the system and sensitivity improvement.

6.4.3 Chemiluminescence Post-column Reaction Detection

Chemiluminescence (CL) is potentially the most sensitive of detection techniques as light is produced by a chemical reaction against a dark background and not by absorption and/or re-emission of incident radiation. One particularly sensitive CL reaction occurs when the luminol (5-amino-2,3-dihydrophthalazine-1,4-dione) oxidation to aminophthalate by hydrogen peroxide is catalysed by a number of transition metals according to the reaction in eqn (6.1).

$$\text{luminol} + H_2O_2 \xrightarrow[\text{Base}]{\text{Metal ion}} \text{aminophthalate} + N_2 + h\nu\ (425\ \text{nm}) \quad (6.1)$$

Thus, CL is somewhat complementary to fluorescence detection and mainly involves transition elements, though again is restricted to a relatively small group of metals. The first application of chemiluminescence for the detection of metal ions separated on an ion exchange column was reported in 1974 by Neary et al.[116] Very high sensitivities are obtained for Co^{2+}, Mn^{2+}, Fe^{2+}, Cr^{3+} and Cu^2 using the luminol/peroxide CL reaction and adaptation to PCRs with IC separations has been studied in some detail by Jones and workers.[117–119] The exceptional LOD found for Co^{2+} using 200 µL injections is noteworthy, namely, 0.5 ng L^{-1} defined as three times the baseline noise level.[117] The speciation of Cr^{3+} and Cr(VI) in water was also studied, where on-line reduction of Cr(VI) was required as it is not CL active. Though Cr^{3+} is not as sensitive as Co^{2+}, LODs close to 2 ng L^{-1} range were obtained.[120] The luminol reaction has also been studied using a different oxidising agent. With peroxydisulfate oxidation, silver could be determined in simulated pressurised water reactor coolant down to 0.1 µg L^{-1}, though with 1 mL injections.[121]

Chemiluminescence determination can be based on the reactions of other organic substrates including, flavin mononucleotide, pyrogallol and lucigenin. For example, Du et al.[122] reported LODs 0.1 µg L^{-1} for the determination of Cr^{3+} and Cr(VI) using H_2SO_3 as reducing agent for Cr(VI) and lucigenin (N,N-dimethyl-9,9-diacridinium dinitrate) and KIO_4 under alkaline conditions. The use of another substrate, uranine, for the chemiluminescent detection of cobalt and copper in IC was also reported.[123] A significant enhancement of chemiluminescence intensity by the addition of ethyldimethylacetylammonium bromide (EDAB) was noted in this investigation. Nevertheless, up to now a luminol based PCR provides better sensitivity of detection in IC (Table 6.6).

Displacement reactions have also been investigated to extend the range of metals determined by CL.[119] The displacement of Co^{2+} from CoEDTA was the one chosen and indeed a much greater range of metals could be detected from the s, p, d and f block metals. However, very high background light levels gave relatively high LODs similar to colorimetric detection and the non-linearity of some calibration curves was not ideal. CL detection is very sensitive to the presence of metals ion, so there are important requirements for using very pure chemicals for the preparation of the mobile phase and of using fully plastic metal free chromatography parts. The construction of a flow cell for CL is simple and typically is a flat coil fixed to the window of a photomultiplier covered with silver foil to act as a mirror.[124]

6.5 Hyphenated Techniques

There are a number of other types of detection systems that can be used in IC. One of note involves the so-called hyphenated techniques. This is when the IC system is combined with another major analytical technique. Atomic spectroscopy methods are normally an alternative to IC for the determination of trace metals. However, they cannot distinguish between the different oxidation states of a metal and different compound types of the same metal such as non-labile organometallic compounds. This type of study is generally known as metal

Table 6.6 Chemiluminescent detection of metal ions in liquid chromatography.

Metal	PCR reagent composition	LODs ($\mu g\ L^{-1}$)	Reference
Co^{2+}, Cu^{2+}, Fe^{2+}, Cr^{3+}	5.0×10^{-5} M luminol, 5.0×10^{-3} M H_2O_2, pH 12.0 (NaOH)	Co^{2+} (0.01), Cu^{2+} (5.0)	124
Cr^{3+}, CrO_4^{2-}	Two-coil system: **(1)** 0.015 M K_2SO_3 at pH 3.0 (HNO_3); **(2)** 3.4×10^{-4} M luminol, 0.1 M H_3BO_3, 0.01 M H_2O_2, pH 11.5 (KOH)	Cr^{3+} (0.1), Cr(VI) (0.3)	118
Cr^{3+}, CrO_4^{2-}	Two-coil system: **(1)** 0.015 M K_2SO_3 at pH 3.0 (HNO_3); **(2)** 3.4×10^{-4} M luminol, 0.1 M H_3BO_3, 0.01 M H_2O_2, pH 11.5 (KOH)	0.002	120,125
Co^{2+}	1.1×10^{-3} M luminol, 0.1 M hydrogen peroxide, 0.1 M H_3BO_3, pH 12.5 (KOH)	0.0005	117
Ag^+	5.0×10^{-5} M luminol, 3.0×10^{-3} M $K_2S_2O_8$, 0.1 M H_3BO_3, pH 10.5 (KOH)	0.5	121
Mg^{2+}, Ca^{2+}, Sr^{2+}, Ba^{2+}, Fe^{2+}, Fe^{3+}, Co^{2+}, Ni^{2+}, Cu^{2+}, Zn^{2+}, Al^{3+}, Th(IV), Ga^{3+}, In^{3+}, Pb^{2+}, Bi^{3+}, lanthanides	Two-coil system: **(1)** 5.0×10^{-5} M Co-EDTA, **(2)** 1.1×10^{-3} M luminol, 0.1 M hydrogen peroxide, 0.1 M H_3BO_3, pH 12.0 (KOH)	From 2 to 100	119
Co^{2+}, Cu^{2+}	0.001 M uranine, 0.006 M EDAB, 5.0×10^{-4} M hydrogen peroxide, 0.04 M NaOH	Co^{2+} (0.01), Cu^{2+} (7.5)	123

speciation analysis. A hyphenated system is not necessarily required for metal speciation determinations as IC methods on their own can be used using the detectors already mentioned. Nevertheless, the IC method may not be selective or sensitive enough and hyphenated systems can be a better option, particularly as ng L^{-1} levels can be achieved, though at the price of higher cost and complexity of operation. Organometallic species of Sn, As and Hg have been studied quite extensively by hyphenated techniques involving LC-ICP-MS and LC-ICP-OES.[41,126,127] Non-labile aluminium complexes in water have also been studied by LC-ICP-MS and compared to simple IC methods with fluorescence detection.[104,106] One rather special example involves the determination of extremely low levels of plutonium in environmental and biological samples.[128] Although, these ng L^{-1} levels can be achieved for plutonium by high resolution ICP-MS it suffers from isobaric interference, which can only be solved using an HPCIC-ICP-MS hyphenated system. More details are given in the Applications section (Section 7.5.2.).

References

1. W. W. Buchberger and P. R. Haddad, *J. Chromatogr. A.*, 1997, **789**, 67–83.
2. W. W. Buchberger, *TrAC Trends Anal. Chem.*, 2001, **20**, 296–303.
3. P. N. Nesterenko and T. A. Bolshova, *Vestn. Mosk. Univ. Ser. 2 Khim.*, 1990, **31**, 167–169.
4. A. I. Elefterov, S. N. Nosal, P. N. Nesterenko and O. A. Shpigun, *Analyst*, 1994, **119**, 1329–1332.
5. P. N. Nesterenko, G. Z. Amirova and T. A. Bol'shova, *Anal. Chim. Acta*, 1994, **285**, 161–168.
6. P. N. Nesterenko and G. Z. Amirova, *J. Anal. Chem.*, 1994, **49**, 447–451.
7. P. N. Nesterenko, I. B. Yuferova and G. V. Kudryavtsev, *Zh. Vses. Khim. O va im D. I. Mendeleeva.*, 1989, **34**, 426–427.
8. P. N. Nesterenko, T. I. Tikhomirova, V. I. Fadeeva, I. B. Yuferova and G. V. Kudryavtsev, *J. Anal. Chem. USSR (Engl. Transl.)*, 1991, **46**, 800–806.
9. I. P. Smirnov and P. N. Nesterenko, *Fibre Chem.*, 1993, **24**, 422–424.
10. J. S. Fritz and D. T. Gjerde, *Ion Chromatography*, Wiley-VCH Verlag, Weinheim, 2009, p. 377.
11. R. M. Cassidy and B. D. Karcher, Post-column reaction detection of inorganic species, in *Reaction Detection in Liquid Chromatography*, ed. I. S. Krull, Marcel Dekker, New York, 1986, pp. 129–194.
12. G. J. de Jong, U. A. Th. Brinkman and R. W. Frei, Postchromatographic reaction detection, in *Detection-oriented Derivatization Techniques in Liquid Chromatography*, ed. H. Lingeman and W. J. M. Underber, Marcel Dekker, 1990, pp. 153–192.
13. H. Jansen and R. W. Frei, Modern post-column reaction detection in high-performance liquid chromatography, in *Selective Sample Handling and Detection in High-Performance Liquid Chromatography*, ed. K. Zech and R. W. Frei, Elsevier, Amsterdam, 1989, pp. 208–259.
14. P. K. Dasgupta, *J. Chromatogr. Sci.*, 1989, **27**, 422–448.
15. P. K. Dasgupta, Approaches to ionic chromatography, in *Ion Chromatography*, ed. J. G. Tarter, Marcel Dekker, 1987, pp. 191–368.
16. E. D. Katz and R. P. Scott, *J. Chromatogr.*, 1983, **268**, 169–175.
17. P. R. Haddad and P. E. Jackson, *Ion Chromatography: Principles and Applications*, Elsevier, Amsterdam, 1990, p. 776.
18. C. Hansch and A. Leo, *Exploring QSAR: Fundamentals and Applications in Chemistry and Biology*, American Chemical Society, Washington, 1995, p. 557.

19. R. Weinberger and R. A. Femia, Commercial aspects of post-column reaction detectors for liquid chromatography, in *Selective Sample Handling and Detection in High-performance Liquid Chromatography*, ed. R. W. Frey and K. Zech, Elsevier, Amsterdam, 1988, pp. 395–435.
20. S. Moore and W. H. Stein, *J. Biol. Chem.*, 1948, **176**, 367–388.
21. I. M. Kolthoff, E. B. Sandell, E. J. Meehan and S. Bruckenstain, *Quantitative Chemical Analysis*, Macmillan, New York, 1969, p. 1199.
22. E. B. Sandell and H. Onishi, *Colorimetric Determination of Traces of Metals in Photometric Determination of Traces of Metals. Part 1. General Aspects*, John, New York, 1978.
23. K. Ueno, T. Imamura and K. L. Cheng, *Handbook of Organic Analytical Reagents*, CRC Press, Boca Raton, 2000, p. 615.
24. J. S. Fritz and J. N. Story, *Anal. Chem.*, 1974, **46**, 825–829.
25. M. D. Arguello and J. S. Fritz, *Anal. Chem.*, 1977, **49**, 1595–1598.
26. X. J. Ding, S. F. Mou, K. N. Liu, A. Siriraks and J. Riviello, *Anal. Chim. Acta*, 2000, **407**, 319–326.
27. J. Dugay, A. Jardy and M. Doury-Berthod, *Analusis*, 1995, **23**, 196–212.
28. R. M. Cassidy, *Chem. Geol.*, 1988, **67**, 185–195.
29. V. T. Hamilton, W. Dalespall, B. F. Smith and E. J. Peterson, *J. Chromatogr.*, 1989, **469**, 369–377.
30. E. A. Gautier, R. T. Gettar, R. E. Servant and D. A. Batistoni, *J. Chromatogr. A*, 1997, **770**, 75–83.
31. J. Cowan, M. J. Shaw, E. P. Achterberg, P. Jones and P. N. Nesterenko, *Analyst*, 2000, **125**, 2157–2159.
32. A. Hulanicki, S. Glab and G. Ackermann, *Pure Appl. Chem.*, 1983, **55**, 1137–1230.
33. A. E. Martell and R. M. Smith, *NIST Standard Reference Database 46. Version 8.0.*, 2004.
34. S. Vlckova, L. Jancar, V. Kuban and J. Havel, *Collect. Czech. Chem. Commun.*, 1982, **47**, 1086–1099.
35. M. Pollak and V. Kuban, *Collect. Czech. Chem. Commun.*, 1979, **44**, 725–741.
36. P. Jones, The mysterious case of the missing lead: Improvements in detection limits for IC determination of trace metals using stability constant data. Presentation 26 at International Ion Chromatography Symposium, IICS 2001, Chicago, 2001.
37. MEDUSA and HYDRA software for chemical equilibrium calculations. Royal Institute of Technology (KTH), Stockholm, 1999.
38. S. Shibata, 2-Pyridylazo compounds in analytical chemistry, in *Chelates in Analytical Chemistry*, ed. H. A. Flashka and A. J. Barnard, Marcel Dekker Inc, New York, 1972, p. 1.
39. N. Cardellicchio, S. Cavalli, P. Ragone and J. M. Riviello, *J. Chromatogr. A*, 1999, **847**, 251–259.
40. C. Sarzanini, High-performance liquid chromatography: Trace metal determination and speciation, in *Advances in Chromatography*, ed. P. R. Brown and E. Grushka, Marcel Dekker, New York, 2001, **41**, 249–310.

41. R. Michalski, *Crit. Rev. Anal. Chem.*, 2009, **39**, 230–250.
42. D. B. Gladilovich, V. Kuban and J. Havel, *Collect. Czech. Chem. Commun.*, 1988, **53**, 526–542.
43. M. J. Shaw, P. Jones and P. N. Nesterenko, *J. Chromatogr. A.*, 2002, **953**, 141–150.
44. M. J. Shaw, J. Cowan and P. Jones, *Anal. Lett.*, 2003, **36**, 423–439.
45. D. Yan and G. Schwedt, *Fresenius Z. Anal. Chem.*, 1985, **320**, 325–329.
46. A. Mazzucotelli, A. Dadone, R. Frache and F. Baffi, *J. Chromatogr.*, 1985, **349**, 137–142.
47. H. T. Lu, X. Z. Yin, S. F. Mou and J. M. Riviello, *J. Liq. Chromatogr. Relat. Technol.*, 2000, **23**, 2033–2045.
48. T. Yamane and Y. Yamaguchi, *Anal. Chim. Acta*, 1997, **345**, 139–146.
49. K. A. Tony, S. Kartikeyan, T. P. Rao and C. S. P. Iyer, *Anal. Lett.*, 1999, **32**, 2665–2677.
50. A. Hrdlicka, J. Havel and M. Valiente, *J. High Resolut. Chromatogr.*, 1992, **15**, 423–427.
51. W. Bashir, E. Tyrrell, O. Feeney and B. Paull, *J. Chromatogr. A*, 2002, **964**, 113–122.
52. E. Sugrue, P. Nesterenko and B. Paull, *J. Sep. Sci.*, 2004, **27**, 921–930.
53. Nuryono, C. G. Huber and K. Kleboth, *Chromatographia*, 1998, **48**, 407–414.
54. R. Garcia-Valls, A. Hrdlicka, J. Perutka, J. Havel, N. V. Deorkar, L. L. Tavlarides, M. Munoz and M. Valiente, *Anal. Chim. Acta*, 2001, **439**, 247–253.
55. M. J. Shaw, P. N. Nesterenko, G. W. Dicinoski and P. R. Haddad, *Aust. J. Chem.*, 2003, **56**, 201–206.
56. P. N. Nesterenko and P. Jones, *J. Chromatogr. A.*, 1998, **804**, 223–231.
57. H. Kumagai, Y. Inoue, T. Yokoyama, T. M. Suzuki and T. Suzuki, *Anal. Chem.*, 1998, **70**, 4070–4073.
58. H. Kumagai, T. Yokoyama, T. M. Suzuki and T. Suzuki, *Analyst*, 1999, **124**, 1595–1597.
59. J. Tria, P. R. Haddad and P. N. Nesterenko, *J. Sep. Sci.*, 2008, **31**, 2231–2238.
60. M. Foltin, S. Megová, T. Prochacková and M. Steklac, *J. Radioanal. Nucl. Chem.*, 1996, **208**, 295–307.
61. P. Jones, P. J. Hobbs and L. Ebdon, *Anal. Chim. Acta*, 1983, **149**, 39–46.
62. M. J. Shaw, P. Jones and P. R. Haddad, *Analyst*, 2003, **128**, 1209–1212.
63. R. T. Gettar, E. A. Gautier, R. E. Servant and D. A. Batistoni, *J. Chromatogr. A.*, 1999, **855**, 111–119.
64. J. I. Itoh, J. H. Liu and M. Komata, *Talanta*, 2006, **69**, 61–67.
65. W. Bashir and B. Paull, *J. Chromatogr. A.*, 2001, **910**, 301–309.
66. M. J. Shaw, S. J. Hill, P. Jones and P. N. Nesterenko, *J. Chromatogr. A.*, 2000, **876**, 127–133.
67. D. Yan and G. Schwedt, *Fresenius Z. Anal. Chem.*, 1985, **320**, 252–257.
68. S. Motellier and H. Pitsch, *J. Chromatogr. A.*, 1994, **660**, 211–217.

69. P. Jones and G. Schwedt, *J. Chromatogr.*, 1989, **482**, 325–334.
70. E. H. Borai, E. A. El Sofany, A. S. Abdel-Halim and A. A. Soliman, *TrAC Trends Anal. Chem.*, 2002, **21**, 741–745.
71. G. Schwedt and P. Schneider, *Fresenius Z. Anal. Chem.*, 1986, **325**, 116–120.
72. J. N. King and J. S. Fritz, *J Chromatogr*, 1978, **153**, 507–516.
73. C. Fu and B. Zuo, *Fenxi Huaxue*, 1981, **9**, 635–639.
74. B. Divjak, M. Franko and M. Novic, *J. Chromatogr. A*, 1998, **829**, 167–174.
75. H. Huang, F. Kai, C. Uragami, M. Hirohata, H. Chikushi, H. Yanaka, M. Honda and M. Nakamura, *Bull. Chem. Soc. Jpn.*, 1991, **64**, 1982–1984.
76. E. Ohyoshi, K. Jyodoi and S. Kohata, *Nippon Kagaku Kaishi*, 1997, 190–193.
77. K. Ohshita, H. Wada and G. Nakagawa, *Anal. Chim. Acta*, 1983, **149**, 269–279.
78. T. Makino, M. Kiyonaga and K. Kina, *Clin. Chim. Acta*, 1988, **171**, 19–28.
79. T. Makino, *Clin. Chim. Acta*, 1991, **197**, 209–220.
80. P. Jones and P. Nesterenko, Developing a computer model for optimising the PAR post-column reaction for IC determination of trace metal determinations: the ammonia problem, *Abstr. of 17th Annual Ion Chromatography Symposium, IICS.* 20–23 September 2004. Trier, Germany, L-30, p. 37.
81. R. M. Cassidy and S. Elchuk, *J. Chromatogr. Sci.*, 1981, **19**, 503–507.
82. A. W. Al Shawi and R. Dahl, *Anal. Chim. Acta*, 1999, **391**, 35–42.
83. J. Tria, P. N. Nesterenko and P. R. Haddad, *Chem. Listy*, 2008, **102**, 319–323.
84. C. A. Lucy and L. W. Ye, *J. Chromatogr. A.*, 1994, **671**, 121–129.
85. H. E. Mash and Y. P. Chin, *Anal. Chem.*, 2003, **75**, 671–677.
86. R. L. Grob and E. F. Barry, (eds), *Modern Practice of Gas Chromatography*, John Wiley, Hoboken, 1985, p. 1064.
87. K. A. Rubinson and J. F. Rubinson, *Contemporary Instrumental Analysis*, Prentice Hall, New Jersey, 2000, p. 840.
88. ICH Harmonised Tripartite Guideline, Guideline on the Validation of Analytical Procedures: Q2A, *Fed. Reg.*, 1995, pp. 11260–11262.
89. J. M. Miller, in *Chromatography: Concepts and Contrasts*, Wiley-Interscience, New York, 2005, pp. 1–25.
90. R. P. W. Scott, *Technique and Practice of Chromatography*, Marcel Dekker, 1995.
91. L. R. Snyder and J. J. Kirkland, *Introduction to Modern Liquid Chromatography*, John Wiley, 1979, p. 863.
92. L. S. Ettre, *Pure Appl. Chem.*, 1993, **65**, 819–872.
93. www.jppchromatography.co.uk.
94. R. Chiarizia, E. P. Horwitz, S. D. Alexandratos and M. J. Gula, *Sep. Sci. Technol.*, 1997, **32**, 1–35.

95. C. N. Ferrarello, M. R. Fernández de la Campa and A. Sanz-Medel, *Anal. Bioanal. Chem.*, 2002, **373**, 412–421.
96. B. Paull, M. Foulkes and P. Jones, *Anal. Proc.*, 1994, **31**, 209–211.
97. P. Jones, M. Foulkes and B. Paull, *J. Chromatogr. A*, 1994, **673**, 173–179.
98. J. R. Jezorek and H. Freiser, *Anal. Chem.*, 1979, **51**, 373–376.
99. A. C. Co, A. N. Ko, L. W. Ye and C. A. Lucy, *J. Chromatogr. A*, 1997, **770**, 69–74.
100. C. A. Lucy and H. N. Dinh, *Anal. Chem.*, 1994, **66**, 793–797.
101. K. Soroka, R. S. Vithanage, D. A. Phillips, B. Walker and P. K. Dasgupta, *Anal. Chem.*, 1987, **59**, 629–636.
102. P. K. Dasgupta, K. Soroka and R. S. Vithanage, *J. Liq. Chromatogr.*, 1987, **10**, 3287–3319.
103. P. Jones, L. Ebdon and T. Williams, *Analyst*, 1988, **113**, 641–644.
104. B. Fairman, A. Sanz-Medel, P. Jones and E. H. Evans, *Analyst*, 1998, **123**, 699–703.
105. J. I. G. Alonso, A. L. Garcia, A. Sanzmedel, E. B. Gonzalez, L. Ebdon and P. Jones, *Anal. Chim. Acta*, 1989, **225**, 339–350.
106. B. Fairman, A. Sanzmedel, P. Jones and E. H. Evans, *J. Anal. Atom. Spectrom.*, 1995, **10**, 281–285.
107. P. Jones, *Int. J. Environ. Anal. Chem.*, 1991, **44**, 1–10.
108. P. Jones and B. Paull, *Anal. Proc.*, 1992, **29**, 402–404.
109. T. Williams and N. W. Barnett, *Anal. Chim. Acta*, 1992, **264**, 297–301.
110. J. Carnevale and P. E. Jackson, *J. Chromatogr. A.*, 1994, **671**, 115–120.
111. S. Yamazaki, H. Omori and O. Eon, *J. High Res. Chromatogr.*, 1986, **9**, 765–766.
112. T. Williams and N. W. Barnett, *Anal. Chim. Acta*, 1992, **259**, 19–23.
113. M. D. Prat, R. Compano, M. Granados and E. Miralles, *J. Chromatogr. A.*, 1996, **746**, 239–245.
114. C. Sarzanini, M. Aceto, G. Sacchero and E. Mentasti, *Ann. Chim. (Rome)*, 1994, **84**, 71–80.
115. T. Vortmuller and G. Wunsch, *J. Prakt. Chem./Chem.- Ztg.*, 1994, **336**, 11–15.
116. M. P. Neary, R. Seitz and D. M. Hercules, *Anal. Lett.*, 1974, **7**, 583.
117. P. Jones, T. Williams and L. Ebdon, *Anal. Chim. Acta*, 1989, **217**, 157–163.
118. T. Williams, P. Jones and L. Ebdon, *J. Chromatogr.*, 1989, **482**, 361–366.
119. P. Jones, T. Williams and L. Ebdon, *Anal. Chim. Acta*, 1990, **237**, 291–298.
120. M. Derbyshire, A. Lamberty and P. H. E. Gardiner, *Anal. Chem.*, 1999, **71**, 4203–4207.
121. P. Jones and H. G. Beere, *Anal. Proc.*, 1995, **32**, 169–171.
122. J. X. Du, Y. H. Li and R. Guan, *Microchim. Acta*, 2007, **158**, 145–150.
123. C. Lu, J. M. Lin, C. W. Huie and M. Yamada, *Anal. Sci.*, 2003, **19**, 557–561.
124. B. L. Yan and P. J. Worsfold, *Anal. Chim. Acta*, 1990, **236**, 287–292.

125. H. G. Beere and P. Jones, *Anal. Chim. Acta*, 1994, **293**, 237–243.
126. P. R. Haddad, P. N. Nesterenko and W. Buchberger, *J. Chromatogr. A*, 2008, **1184**, 456–473.
127. M. J. Shaw and P. R. Haddad, *Environ. Int.*, 2004, **30**, 403–431.
128. J. B. Truscott, P. Jones, B. E. Fairman and E. H. Evans, *J. Chromatogr. A.* 2001, **928**, 91–98.

CHAPTER 7
Practical Applications

7.1 Potential HPCIC Applications

Versatile selectivity and insensitivity to salt concentrations are two major advantages of chelating ion exchangers over standard ion exchange sorbents. These advantages can be exploited in a number of very different ways, including:

1. *Analysis of samples comprising a complex matrix.* The application of HPCIC to such samples has demonstrated the possibility of achieving direct trace elemental analysis with minimal sample pre-treatment. As a rule, these samples contain high concentration levels of alkali and alkaline-earth metals in the form of dissolved salts, and occasionally massive amounts of other metals such as iron.
2. *Trace analysis of metals* whereby a single chelating ion exchange column can be used for both simultaneous preconcentration and separation. Such applications of HPCIC have been shown involving large injection volumes of sample, injected directly onto the front of the analytical column, followed by gradient elution, to improve detection limits, for applications such as seawater analysis.
3. *Simultaneous separation of complex mixtures of metals*, where other separation techniques have failed to provide the desired selectivity for a specific group of metals, such as the lanthanides for example.
4. *Speciation of metals.* Apart from the obvious separation of metal species in different oxidation states, HPCIC can also allow differentiation of the labile and inert fractions of the metal in the sample, as well as the fraction of the metal strongly bound into various organic/inorganic forms. Although speciation studies are possible with traditional ion exchange, this can be restricted by the presence of an excess of other charged species.
5. *Analysis of biological tissues and fluids*, particularly samples where the presence of various charged bio-molecules may interfere with the

RSC Chromatography Monographs No. 14
High Performance Chelation Ion Chromatography
By Pavel N. Nesterenko, Phil Jones and Brett Paull
© The Royal Society of Chemistry 2011
Published by the Royal Society of Chemistry, www.rsc.org

determination of metals by traditional ion exchange. This also includes food products, or other samples with a similar organic matrix. Such organic substances usually necessitate wet digestion or dry ashing, which can produce a final sample matrix with a high content of salts and metals as in case (1), above.

7.2 Relative Advantages of HPCIC

In many cases the application of HPCIC is the only option for the direct analysis of complex samples. Atomic spectroscopic methods (*e.g.* AAS, AES, ICP-AES and ICP-MS) have matrix limitations when applied to the analysis of very concentrated saline solutions, which necessitate significant consecutive dilution of the samples. Where the concentration of target metals in the sample is low, such methods cannot provide sufficient sensitivity following such dilution steps. In addition, large and consecutive dilution of the sample will also seriously decrease the accuracy of the determination. Sample volume is another crucial issue. In many instances, sample volume may be too small to be analysed directly, with many atomic spectroscopic methods requiring a minimum sample volume in the order of 10 mL and upwards, although coupling with sample handling techniques such as FIA can help address this issue.

Simplicity, low cost and the small physical dimensions of instrumentation required for HPCIC is also a major advantage. Ideally, in many cases, the analysis may need to be performed in the field or shipboard, where complexity, reagent (and gas) requirements and size limitations are significant. Furthermore, remote and hazardous environments, such as oil rigs, for example, do not allow naked flames or plasmas. Obviously, in such cases the possible application of flame or plasma based atomic spectroscopic methods for elemental analysis is particularly limited.

The advent of highly efficient ion exchange chromatography in the 1970s, linked to sensitive detection methods, transformed the ability to determine ionic analytes at trace levels with a considerable reduction in analysis time. When coupled with the simultaneous development and improvement in computing integration software, this can truly be considered the beginning of modern ion exchange chromatography, now the major part of a related group of techniques, generally included within the term IC. Initially, the main driving force for instrumental development was for the rapid trace determination of the common anions. Interest in metal ions soon followed, with particular emphasis on the alkali and alkaline-earth metals, then later extending to the transition and heavy metals using PCR detection as discussed earlier. A recent major review on the applications of IC for metal ions shows the wide range of metal species that can be separated and determined using these techniques.[1] The vast majority of applications in this review still involve the analysis of waters, although there are a significant variety of other sample types included. As the majority of applications involve simple ion exchange separations, samples with high ionic strength are avoided or analysed following significant sample dilution. Chelation based separation methods are little mentioned in the review except for the Dionex CIC system, which as discussed in Chapter 1, is actually simple ion exchange separation preceded by pre-concentration with a small chelating column. However, a number of other current reviews specifically discuss HPCIC applications, the most significant of which are discussed more fully herein.[2–5]

Of course, not all metals are amenable to IC analysis and this also applies to HPCIC. The highly charged metals are the most challenging, as they hydrolyse in acid solution and hence are generally known as the acid hydrolysing metals. The high charge also leads to stronger complexation with the chelating stationary phase and also with any complexing agents within the mobile phase. This can result in slower reaction kinetics and excessively broad and tailing peaks, although in some cases increasing the column temperature can successfully overcome the peak broadening, giving reasonable efficiencies. Little work has been done with the platinum group metals and certain other heavy metals such as Nb, Ta, Mo, W, Tc and Re, which appear to be unsuitable for HPCIC separation. Also, the extreme radioactivity of some of the actinides limits their determination unless special laboratory facilities are available. Table 7.1 shows the groups of metals successfully separated by HPCIC. The alkali metals and ammonium are weakly chelated and are mainly separated by simple ion exchange, but there are a few examples where surface chelation can be used.

This chapter describes a range of applications for trace metal analysis using HPCIC techniques, including very recent work. The special nature of chelating substrates will be quite evident, involving a separation mechanism in its own right, comparable with the other major mechanisms, adsorption, partition, ion exchange and exclusion based methods. The flexibility in terms of selectivity control can also be seen, together with the ability to handle samples with the highest salt concentrations. The elution conditions are normally chosen as to make chelation the main factor controlling selectivity. A small amount of ion exchange can also be present in certain separations as explained in Chapter 2, which may also influence selectivity, but is constant for a given mobile phase system. Table 7.2 summarises the key applications considered here.

Table 7.1 Metals separated and determined by HPCIC.

Oxidation state	Metal ions	Comments
I	Li, Na, K, Rb, Cs, NH_4	More relevant to neutral complexing resins
II	Be, Mg, Ca, Sr, Mn, Fe, Co, Ni, Cu, Zn, Pb, Cd	—
III	Al, Ga, In, Fe, Bi, Y, lanthanides	—
IV	Ti, Zr, Hf, V, Th, U, Pu, Np	Present as oxocations or partly hydrolysed species
V	V	Present as oxocations
VI	U	Present as oxocations

Table 7.2 Practical application of HPCIC.

Application	Chelating phase	Column size, mm	Mobile phase	Detection	Concentration ($mg\ L^{-1}$ or $mg\ kg^{-1}$)	Reference
Seawater, saline lake water						
Cd^{2+}, Co^{2+}, Zn^{2+}, Ni^{2+}, Cu^{2+} in seawater	Table 3.3, 16	250 × 4.0	Three-step gradient: 0–10 min, $0.5\,M\ KCl$–$5 \times 10^{-4}\,M\ HNO_3$, 10–30 min. $0.08\,M$ tartaric acid and 30–50 min $0.01\,M$ picolinic acid	PCR PAR, 550 nm	—	6
Be^{2+} in seawater	Table 3.3, 16	250 × 4.0	$0.4\,M\ KNO_3$, pH 2.5	PCR, CAS, 590 nm	0.004	7
U(vi) in saline lake water	Table 3.1, 2	150 × 4.1	$0.5\,M\ KNO_3$, $0.04\,M\ HNO_3$	PCR, arsenazo III, 600 nm	0.001	8
Ca^{2+}, Cd^{2+}, Pb^{2+}, Co^{2+}, Zn^{2+}, Ni^{2+}, Cu^{2+} and Mn^{2+} in coastal seawater	Table 3.3, 9 and Table 3.3, 16	50 × 3.0 and 150 × 3.0	Two-dimensional separation: $5.0 \times 10^{-3}\,M$ oxalic acid in first dimension and $0.015\,M$ oxalic acid with $3.0 \times 10^{-3}\,M$ ethylenediamine in second	PCR PAR, 540 nm	0.05–0.005	9
REE in seawater	Table 3.2, 8	250 × 3.0	$4.0 \times 10^{-3}\,M$ oxalic acid, pH 4.0	PCR PAR, 540 nm	—	10
Mg^{2+} and Ca^{2+} in coastal sea water	Table 5.5	250 × 4.6	Tartrate buffer, pH 5.5	Fl, 525 nm	—	11
Sr^{2+} in an Antarctic saline lake water	Table 3.1, 3	150 × 4.1	$0.5\,M\ KNO_3$, $2.0 \times 10^{-4}\,M$ o-CPC, $0.02\,M$ borate buffer, pH 9.5	Vis, 575 nm	75.3	12
Al^{3+} in seawater	Table 3.1, 5	100 × 4.6	$1\,M\ KNO_3$, $0.05\,M$ lactic acid, step pH from 4.0 to 1.8 upon injection, to pH 1.1 after 10 min	PCR PCV, 580 nm	6.0	13

Table 7.2 (Continued).

Application	Chelating phase	Column size, mm	Mobile phase	Detection	Concentration ($mg\,L^{-1}$ or $mg\,kg^{-1}$)	Reference
U(VI) in seawater (NASS-4 reference sample)	Table 3.3, 16	150 × 4.0	0.25 M NaCl–0.04 M HNO_3	PCR lumogallion, Fl, 525 nm	1.0×10^{-6}	14
Mg^{2+} and Ca^{2+} in seawater	Table 3.1, 9	100 × 4.6	1 M KNO_3, 0.5 M HNO_3, 1.0×10^{-4} M dipicolinic acid	PCR, arsenazo III, 654 nm	0.05	15
Mg^{2+} and Ca^{2+} in seawater	PGC coated with PP	100 × 4.6	52% methanol–48% 4.0×10^{-4} M PP, pH 10.5.	Direct at 572 nm	370–415 Ca^{2+}, 1283–1308 Mg^{2+}	16,17
Mg^{2+} and Ca^{2+} in saline lake water	PGC coated with PP	100 × 4.6	45–58% MeOH, 4.0×10^{-4} M PP, pH 10.0–10.5	Vis, 575 nm	13.34×10^3 Mg^{2+}, 1.82×10^3 Ca^{2+}	17
Mn^{2+}, Zn^{2+}, Pb^{2+}, Ni^{2+}, Cu^{2+} in coastal seawater, reference material CASS-2	Table 3.1, 1	100 × 4.6	1 M KNO_3, 0.05 M lactic acid, step pH gradient elution	PCR PAR, 490 nm	In agreement with certified data	18
Mn^{2+}, Zn^{2+}, Pb^{2+}, Ni^{2+} and Cu^{2+} in seawater	Table 3.1, 1	100 × 4.6	1 M KNO_3, 0.05 M lactic acid, three-step pH gradient from 6 to 3.7, then to 2.0 after 1 min and 0.7 after 6 min	PCR PAR, 490 nm	7.1–15.2 Zn^{2+}, 0.7–1.3 Pb^{2+}; 0.2–0.8 Ni^{2+}; 1.2–5.0 Cu^{2+}	18,19
Cu^{2+}, Cd^{2+}, Co^{2+}, Zn^{2+} in seawater	Table 3.3, 24	100 × 4.6	0.025 M oxalate, 0.025 M $NaNO_3$ at pH 4.2	PCR, PAR, 510 nm	0.64 Cu^{2+}, 22.4 Cd^{2+}, 11.8 Co^{2+}, 1.31 Zn^{2+}	20
Mg^{2+} and Ca^{2+} in seawater	Table 3.5, 7	75 × 7.5	0.2 M KCl, 0.05 M NaH_2PO_4/Na_2HPO_4, 1.0×10^{-4} M PP, pH 5.3	Direct at 575 nm	1170–1230 Mg^{2+}; 385–436 Ca^{2+}	21

Analyte	Table	Column	Eluent	Detection	Result	Ref.
Laboratory chemicals and brines						
Mn^{2+}, Zn^{2+}, Pb^{2+}, Ni^{2+} and Cu^{2+} in laboratory chemicals (KNO_3, Na_2SO_4, CsI)	Table 3.1, 1	100 × 4.6	1 M KNO_3, 0.05 M lactic acid, 3 step pH gradient from 6 to 3.7, then to 2.0 after 1 min and 0.7 after 6 min	PCR PAR-ZnEDTA, 490 nm	from 2.0 to 60.0 in various salts	19
Mg^{2+}, Ca^{2+}, Sr^{2+}, Ba^{2+}, Mn^{2+}, Zn^{2+}, Ni^{2+}, Cu^{2+} in 2.4 M KCl and 5.1 M NaCl brines	Table 3.1, 1	100 × 4.6	1 M KNO_3, 0.05 M lactic acid, three-step pH gradient from 6 to 3.7, then to 2.0 after 1 min and 0.7 after 6 min	PCR PAR-ZnEDTA, 490 nm	0.5 Mg^{2+}; Mn^{2+}; 1.0 Zn^{2+}, Cu^{2+}; 2 Ca^{2+}, Ni^{2+}, Ba^{2+}, 4 Sr^{2+}	19
Ba^{2+} and Sr^{2+} in offshore oil-well brines	Table 3.1, 2	100 × 4.6	0.5 M KNO_3 with 0.05 M lactic acid	PCR PAR-ZnEDTA, 490 nm	1 Ba^{2+}, 8 Sr^{2+}	22
Mg^{2+} and Ca^{2+} in and laboratory grade KCl	Table 2, 2	250 × 4.0	1 M KNO_3, pH 4.9	PCR PP, 572 nm	0.21 Mg^{2+} 0.014 Ca^{2+}	23
Soils, sediments, minerals						
Be in stream sediment (standard reference sample)	Table 3.3, 22	50 × 4.6	1 M KNO_3, 0.5 M HNO_3, 0.08 M ascorbic acid	PCR, CAS, 560 nm	0.035	24
U in stream sediment (certified material GBW07311)	Table 3.1, 9	100 × 4.6	1.0×10^{-4} M dipicolinic acid, 1 M KNO_3, 0.5 M HNO_3	PCR, arsenazo III, 654 nm	8.6	15
Pu in standard reference sample NIST 4353 Rocky Flats Soil	Table 3.1, 9	100 × 4.6	1.0×10^{-4} M dipicolinic acid, 0.75 M HNO_3	ICP-MS	2.9×10^{-6}	25
Zn^{2+} in gypsum	Table 3.1, 2	150 × 4.1	0.5 M KNO_3, $2.0 \cdot 10^{-4}$ M MTB, pH 2.7	Direct UV, 600 nm	208	26

Table 7.2 (Continued).

Application	Chelating phase	Column size, mm	Mobile phase	Detection	Concentration ($mg\ L^{-1}$ or $mg\ kg^{-1}$)	Reference
			Mineral, fresh and rain waters			
Mg^{2+} and Ca^{2+} in mine process sample and deep lakes	PGC coated with PP	100×4.6	45–58% methanol, 4.0×10^{-4} M PP, pH 10.0–10.5	Vis, 575 nm	161 Mg^{2+}, 262 Ca^{2+}	17
Mn^{2+}, Cd^{2+}, Co^{2+}, Zn^{2+} in freshwater standard reference sample (NIST 1640)	Table 3.3, 16	250×4.0	0.035 M KCl, 0.065 M KNO_3, pH 2.5	PCR PAR, 550 nm	In agreement with certified data	27
Ba^{2+} and Sr^{2+} in fresh water reference material (IAEA/W4)	Table 3.1, 2	100×4.6	1 M KNO_3, 0.05 M lactic acid, step pH gradient: 0–5 min pH 8.5, then pH 3.0	PCR PAR-ZnEDTA, 490 nm	In agreement with certified data	28
U(vi) in mineral water	Table 3.1, 9	100×4.6	1 M KNO_3, 0.5 M HNO_3, 1.0×10^{-4} M dipicolinic acid	PCR, Arsenazo III, 654 nm	0.05	15
Ba^{2+} in mineral water	Table 3.1, 2	100×4.6	1 M KNO_3, 0.05 M lactic acid, step pH gradient: 0–5 min pH 8.5, then pH 3.0	PCR PAR-ZnEDTA, 490 nm	0.09–0.84	28
Be^{2+} in freshwater standard reference sample (NIST 1640)	Table 3.3, 16	250×4.0	0.4 M KNO_3, pH 2.5	PCR, CAS, 590 nm	0.02	7
Na^+, NH_4^+, K^+ in drinking water, river water, rain water	Table 3.3, 16	250×4.6	0.01 M 18-crown-6 in HNO_3, pH 2.75	Conductivity	0.05–0.564 Na^+, 0.018–0.858 K^+, 0.027–0.628 NH_4^+	29
Mg^{2+} and Ca^{2+} in mineral water	Table 3.3, 16		1.25×10^{-3} M $HClO_4$, 2.06×10^{-3} M dipicolinic acid	Conductivity	1–7	30

Practical Applications

Other samples					
Mg^{2+}, Ca^{2+} in NaCl eyewash saline solution	Table 3.3, 16	250 × 4.0	1 M KNO_3, pH 4.9	PCR o-CPC, 572 nm	6–7	23
Pb, Cd, Cu in rice flour	Table 3.1, 10	300 × 4.6	1 M KNO_3, 2.5×10^{-4} M 4-chlorodipicolinic acid, pH 1.5	PCR PAR, 520 nm	0.019 Cd, 0.73 Pb^{2+}, 2.6 Cu^{2+}	31
Sr^{2+} in milk powder	Table 3.1, 3	100 × 4.6	1 M KNO_3 with 0.05 M lactic acid, pH 10.2	PCR PAR-ZnEDTA, 490 nm	0.15	28
Mn, Fe and Zn in tomato leaves (NBS reference material)	Table 3.3, 11	250 × 4.6	0.06 M sodium oxalate, pH 4.5	PCR, PAR, 550 nm	57–730 µg g^{-1} in agreement with certified data	32
Pu in human lung (standard reference sample NIST 4251)	Table 3.1, 9	100 × 4.6	1.0×10^{-4} M dipicolinic acid, 0.75 M HNO_3	ICP-MS	0.57×10^{-6}	25
Cu and Zn in oyster tissue (NIST SRM 1566a)	Table 3.3, 24	100 × 4.6	0.025 M oxalate, 0.025 M $NaNO_3$, pH 4.2	PCR, PAR, 510 nm	64.1 Cu^{2+}, 920 Zn^{2+}	20
Fe^{2+}, Fe^{3+} and Cu^{2+} in biofuel and ethanol	Table 3.3, 16	150 × 4.0	2.5×10^{-4} M dipicolinic acid, 0.012 M HCl, 60% methanol	PCR, PAR, 520 nm	$(1-3) \times 10^{-3}$	33
Mn, Fe, Zn and Cu in vitamin tablets leaves (NBS reference material)	Table 3.3, 11	250 × 4.6	0.06 M sodium oxalate, pH 4.5	PCR, PAR, 550 nm	1.8–15 mg Mn^{2+}, Fe^{2+}, Zn^{2+} and Cu^{2+} per tablet in agreement with certified data	32
Zn^{2+} and Pb^{2+} in waste waters from a galvanic bath	Table 3.3, 16	250 × 3.0	1 mM dipicolinic acid, pH 3.0	UV, 295 nm	1.67 Zn^{2+} and 2.71 Pb^{2+}	46

7.3 Fresh and Potable Waters

The determination of trace metals in fresh and potable waters is still the most common application of IC methods involving simple ion exchange separations. For selected metal ions, HPCIC provides an alternative selectivity within such an application, together with more versatile selectivity control, particularly in samples where large amounts of matrix metals are present.

Paull *et al.* studied the separation of Mg^{2+} and Ca^{2+} on a porous graphitic carbon (PGC) stationary phase dynamically modified with *o*-cresolphthalein complexone or phthalein purple (PP) containing IDA groups.[17] The developed method was unusual in that the PP also acted as the colour forming reagent for detection, so no PCR was required. Mg^{2+} and Ca^{2+} were determined in fresh and mine process water in under 12 min. Shaw *et al.* using a similar dynamic modification approach went on to determine six transition and heavy metals in

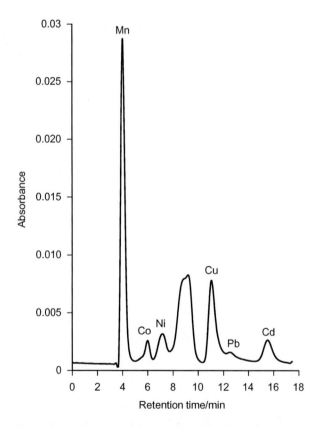

Figure 7.1 Chelation ion chromatogram of the transition metal ions in the certified soft water sample TMDA 54.2. Column: 300 × 4.6 mm i.d. packed with PRP-1 7 μm PS-DVB. Mobile phase: 1 M KNO_3–6 mM HNO_3 and 0.25 mM 4-chlorodipicolinic acid. Detection: photometric after PCR with PAR at 520 nm. Reproduced, with permission, from Shaw *et al.*[31] Copyright (2002) Elsevier.

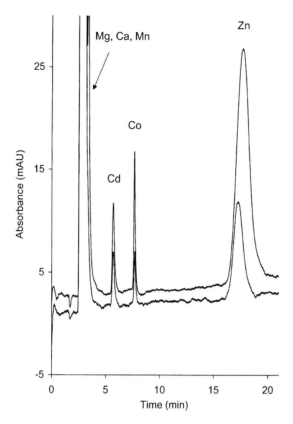

Figure 7.2 Chelation ion chromatogram of NIST 1640 freshwater sample overlaid with spiked sample. Column: Diasorb IDA, 250 × 4.0 mm i.d., 8 μm. Mobile phase conditions: 0.035 M KCl, 0.065 M KNO$_3$, pH 2.5. Sample: 0.3 mg L^{-1} for Mg^{2+} and Ca^{2+}, 0.05 mg L^{-1} for Cd^{2+}, Co^{2+} and Zn^{2+}, and 0.02 mg L^{-1} for Mn^{2+}. Detection: spectrophotometric at 495 nm after PCR with PAR. Reproduced, with permission, from Bashir and Paull.[27] Copyright (2002) Elsevier.

a certified soft water sample using a resin column dynamically modified with 4-chlorodipicolinic acid,[31] shown here as Figure 7.1.

Bashir and Paull, after a detailed study of the effect of ionic strength, pH and temperature on selectivity of an IDA bonded silica, separated and determined Mn^{2+}, Cd^{2+}, Co^{2+} and Zn^{2+} within a NIST standard reference fresh water sample using the 8 μm IDA bonded silica with a simple mobile phase containing 0.1 M potassium ions,[27] shown here as Figure 7.2. Unfortunately, under the conditions applied lead could not be determined within the reference sample.

Later, the same authors using a similar IDA bonded silica column, developed a highly selective separation of Be^{2+} from the other alkaline earths. This was achieved by carefully balancing the contribution to retention of surface chelation and ion exchange, through control of mobile phase pH and KNO$_3$ concentration.[7] The optimised conditions were used to determine Be^{2+} in

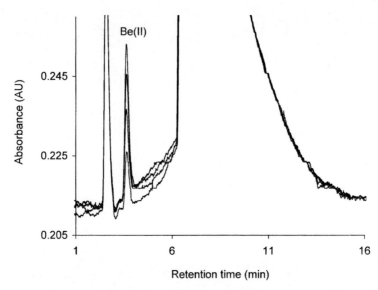

Figure 7.3 Overlaid chelation ion chromatograms of reference material NIST 1640 and the same sample spiked with 20, 40 and 60 μg L^{-1} Be^{2+}. Column: Diasorb IDA, 250 × 4.0 mm i.d., 8 μm. Mobile phase conditions, 0.4 M KNO$_3$, pH 2.5. Spectrophotometric detection at 590 nm after PCR with CAS. Reproduced, with permission, from Bashir and Paull.[7] Copyright (2001) Elsevier.

potable and wastewaters, with detection limits close to 1 ng mL^{-1} using a PCR with CAS reagent in MES buffer at pH 6.0. The resultant chelation ion chromatogram of a NIST 1640 freshwater standard is shown as Figure 7.3.

The determination of uranium in mineral water using a resin dynamically modified with dipicolinic acid was reported by Shaw et al.[15] A relatively large injection volume of 0.5 mL was used to improve sensitivity, due to the relatively broad peak shape obtained on the column, although the final detection limit was still relatively high at 20 ng mL^{-1}. The spiked mineral water sample chromatogram is shown here as Figure 7.4.

Under traditional cation exchange conditions, Sr^{2+} and particularly Ba^{2+} are strongly retained and elute last of the alkaline earth metal cations, as relatively broad peaks. This makes achieving low detection limits difficult, as well as prolonging analysis times. Within HPCIC, Ba^{2+} can be eluted first, followed by Sr^{2+}, as relatively sharp peaks as long as chelation is the only significant retention mechanism, with ion exchange interactions strongly suppressed using high salt containing mobile phases. Jones et al.[28] investigated the determination of Ba^{2+} and Sr^{2+} in mineral waters using a methylthymol blue impregnated column and found values ranging from 90 to 330 ng mL^{-1} for Ba^{2+} and 250 to 800 ng mL^{-1} for Sr^{2+}, with a detection limit of approximately 30 ng mL^{-1} for both metals.[34] Figure 7.5 shows two example chromatograms, the first a certified reference water and the second a bottled mineral water. The mineral water had a surprisingly high level of Ba^{2+}, reported as 330 ng mL^{-1}.

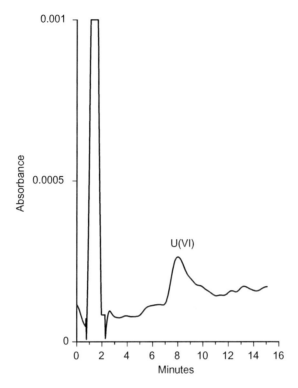

Figure 7.4 Chelation ion chromatogram of 50 mg L^{-1} U(VI) from matrix metals in a spiked mineral water. Column: 100 × 4.6 mm i.d. packed with PRP-1 7 μm PS-DVB. Mobile phase: 1 M KNO$_3$ 0.5 M HNO$_3$ and 0.1 mM dipicolinic acid. Spectrophotometric detection at 654 nm after PCR with arsenazo III. Reproduced, with permission, from Shaw et al.[15] Copyright (2000) Springer.

This example illustrates the significant advantage of the orthogonal selectivity exhibited by HPCIC columns in comparison to simple cation exchangers, where Ba^{2+} elutes first as a sharp peak, resulting in reduced detection limits, than achievable if eluted last.

In terms of efficiency, recently available commercial 5 μm particle sized silica IDA columns, are some of the best performing HPCIC columns to-date, approaching 60 000 plates per metre for some metals, thus allowing the development of more rapid HPCIC applications. For example, six transition and heavy metals can be separated isocratically in under 7 min using a complexing mobile phase on a short 5 cm column (see Chapter 4 for elution details) (separation shown here as Figure 7.6). This column has been applied to the determination of Cu^{2+} and Zn^{2+} in a potable water sample, the chromatogram for which is shown as Figure 7.7. Speciation applications are also possible using the same short column, with the rapid separation of iron as Fe^{2+} and Fe^{3+}, where, using a dilute oxalate mobile phase and highly selective PAR detection, very low detection limits can be achieved. Figure 7.8 shows the separation and

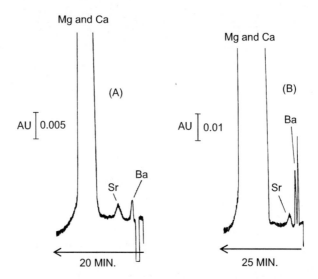

Figure 7.5 Chelation ion chromatogram showing the separation of Ba^{2+} and Sr^{2+} in (A) a certified fresh water, and (B) Highland Spring mineral water. Column, 100×4.6 mm i.d., packed with 8.8 μm PS-DVB resin impregnated with methylthymol blue. Mobile phase: 1 M KNO_3 with 0.05 M lactic acid (pH 8.5 stepped down to pH 3.0 after 5 min). Sample volume, 100 μL. Photometric detection at 495 nm after PCR with PAR-ZnEDTA. Reproduced, with permission, from Paull.[34]

ultra-trace determination of iron species in tap and mineral water in approximately 3 min.

In an application study by Tria *et al.*, involving the determination of Al in wastewaters from a paper mill, speciation capabilities were also reported using an HPCIC column, which demonstrated differing selectivity toward a number of non-labile aluminium species.[35] Figure 7.9 shows example chromatograms from this study, showing two distinct aluminium containing peaks. Peak A was considered to be 'free' *i.e.* labile Al, as confirmed by spiking experiments, with peak B a non-labile species, as confirmed by ICP-MS. The retention of peak A decreases with decreasing sample pH, however peak B remained constant. This finding was supported by a study of aluminium speciation using the IDA functionalised chelating resin, Chelex-100, reported by Pesavento *et al.*, although here not applied within a chromatographic format.[36,37]

7.4 Saline Samples

One important area in which HPCIC has found particular application, is the analysis of saline samples. This is due to the ease in which HPCIC handles samples with high ionic strengths, particularly when compared to alternative

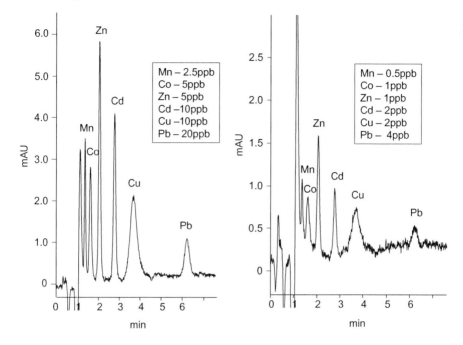

Figure 7.6 Fast chelation ion chromatography of six metals at two different concentration levels in under 7 min. Column: IDA-silica 50 × 4.0 mm i.d., 5 μm IDA. Mobile phase, 25 mM oxalic acid at pH 2.6 containing picolinic acid and chloride. Photometric detection after PCR with PAR/NH$_3$ at pH 10.5 using signal extraction noise reduction. 100 μL injection. Unpublished data, courtesy of JPP Chromatography Ltd.

standard IC methods. Even salts close to saturation can be analysed relatively easily. Ion exchange based methods obviously require high degrees of sample dilution, even for samples of modest salt concentration, which obviously considerably increases method LODs. The following applications clearly illustrate this essentially unique property of HPCIC.

7.4.1 Seawater and Estuarine Water

Seawater, with its complex mix of metals covering a huge range of concentrations in a NaCl matrix of about 0.5 M, has been, and remains a major challenge for all trace analytical techniques. Early work by Challenger and Jones, and then later by Paull and Jones on chelating dye impregnated columns clearly showed the potential of HPCIC for dealing with samples of high salt content (see Chapter 3).[19,22,28,38,39] Within the above studies, the analysis of both sea and estuarine waters using chelating dyes with either IDA or salicylic acid functionalities was demonstrated. The selectivity of the chelating phases

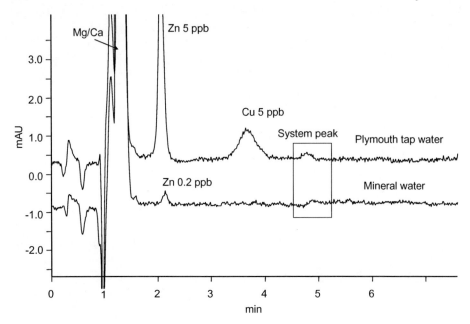

Figure 7.7 Determination of copper and zinc in drinking water using chelation ion chromatography. Column and conditions as in Figure 7.6. Unpublished data, courtesy of JPP Chromatography Ltd.

used was such that the relatively massive amounts of Ca^{2+} and Mg^{2+} within the samples eluted first. However, to ensure complete elution of these alkaline earths and to avoid overlap with later eluting metals, a step gradient approach was developed. An added advantage of the step gradients developed was the ability to combine large volume injections with minimal effect on peak efficiency. This was an important development, as concentration levels of transition metals within these environmental samples are extremely low, with large volume injections needed to achieve detection limits close to single ppb levels.

Challenger et al. determined Zn^{2+} and Cu^{2+} in coastal seawater using a simple three step pH gradient on a XO impregnated resin column, with a 20 mL injection volume.[19] Paull et al., improved on this work using purified reagents, giving better baselines and lower blanks. They determined Mn^{2+}, Ni^{2+}, Zn^{2+}, Cu^{2+} and Pb^{2+} at low ppb levels in coastal sea water and estuarine water fed by acid mine waste.[18] (see Figure 7.10)

A similar exercise was carried out on estuarine water by Nesterenko and Jones using an IDA-silica column.[47] A three-step gradient was used, but this time with complexing acids in the mobile phase, namely tartaric and picolinic acids. As expected, better peak shapes were obtained compared to the dye impregnated columns, due to the higher efficiency of the silica substrate. These separations are shown as Figure 7.11.

Toei investigated a Tosoh column containing a hydrophilic resin bonded with IDA groups for the determination of Mg^{2+} and Ca^{2+} in seawater.[21]

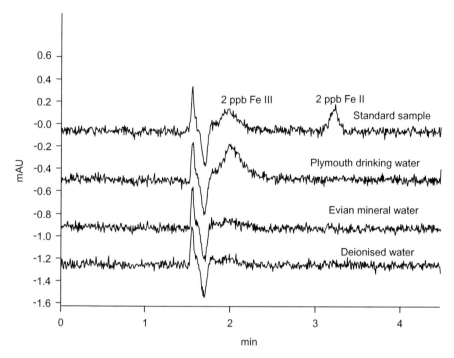

Figure 7.8 Chelation ion chromatograms of iron species in drinking waters. Column as in Figure 7.6. Mobile phase, 2.5 mM oxalic acid pH 2.6. Photometric detection after PCR with reducing agent and PAR/NH$_3$. Samples, 100 μL injection after acidification to pH 2.5. Unpublished data, courtesy of JPP Chromatography Ltd.

However, the elution order obtained did not support chelation as being the dominant retention mechanism in the separation shown. This was probably due to a number of factors, such as the low column capacity, pH and low ionic strength of the mobile phase, thus allowing cation exchange contribute to cation retention. Only 1 μL injections were used in this method, so the cation exchange sites were not 'swamped' by the salt content of the sample. Yamazaki used an ODS-5 silica column modified by coating with *N-n*-dodecyliminodiacetic acid[11] for the same application as Toei. A chromatogram showing the separation of Mg^{2+} and Ca^{2+} in seawater was shown and the retention order in this work indicated that chelation was the major mechanism involved.

In an early example of multi-dimensional HPCIC, Voloschik *et al.*[9] determined Cd^{2+} and Pd^{2+} and other transition and heavy metals in sea, natural and wastewaters. The partial separation and determination of Cu^{2+}, Ni^{2+}, Co^{2+}, Mg^{2+}, Ca^{2+}, Sr^{2+} and Fe^{2+} were achieved on a dynamically coated chelating column with conductimetric detection. After passing through the conductimetric cell, the eluate zones containing insufficiently separated components were withdrawn from the eluate flow and injected into a second

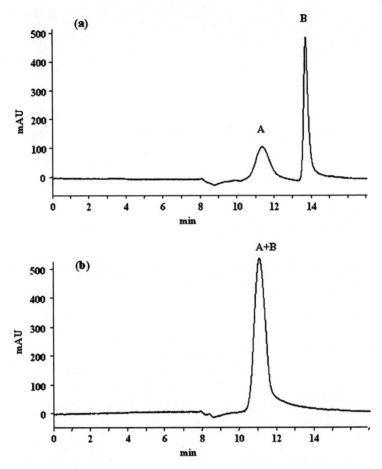

Figure 7.9 (a) Chelation ion chromatogram of Al species in mill whitewater using spectrophotometric detection at 580 nm after optimised PCR with ECR. Column: IDA-silica 200 × 4.0 mm i.d. column, 5 μm. Mobile phase: 0.25 M KCl–0.04 M HNO$_3$ delivered at 0.3 mL min^{-1}. Column and PCR temperature 71 °C. Sample was filtered and acidified to pH 4.8. Injection volume 20 μL. (b) As in (a) except sample was acidified to pH 1 with HCl. Fluorimetric detection after PCR with lumogallion. Reproduced, with permission, from Tria et al.[35] Copyright (2008) Wiley-VCH Verlag GmbH & Co. KGaA

HPCIC dimension using IDA-silica and amidoxime-silica analytical columns. Complete separation of Pb^{2+}, Zn^{2+}, Mn^{2+} and Cd^{2+} was achieved on these second dimension columns.

Other chelating groups bonded to high efficiency substrates have been applied to seawater analysis. Itaconic acid containing carboxylic acid groups,[40,41] hydroxamic acid ligands[20,42] and a cryptand type ligand[20] are worthy of note, though in these cases only seawater spiked with metal standards were analysed.

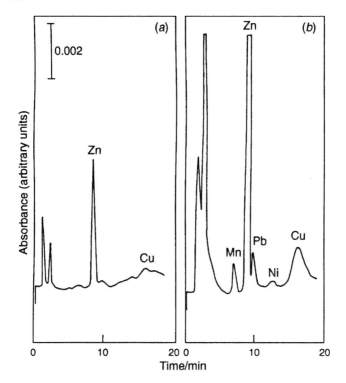

Figure 7.10 Chelation ion chromatograms showing (a) procedural blank and (b) preconcentration and separation of transition metals in coastal seawater from Plymouth Sound, UK. Column: 100 × 4.6 mm i.d. packed with 10 μm neutral PS-DVB resin and coated with XO. Preconcentration volume, 20 mL. Spectrophotometric detection at 490 nm after PCR with PAR/NH$_3$, pH 10.2. Reproduced, with permission, from Paull et al.[18] Copyright (1994) Royal Society of Chemistry.

Nevertheless, it is clear from these studies that substrates involving very different chelating groups are all unaffected by high salt containing samples.

A number of other studies on sea and estuarine waters focused on just one or two metal species using special elution conditions and/or PCR detection to increase the selectivity from other metals. For example, Paull et al. used lower pH mobile phases to separate and determine Al^{3+} and Fe^{3+} in coastal seawater,[13] for which the chromatograms obtained are shown within Figure 7.12.

The possibility of the direct determination of Al^{3+} in seawater was also studied by Tria et al.[14,35] using an IDA-silica column, but this time at the very low levels found in open ocean water. The exceptionally low detection limit of 0.39 nM (16 ng L^{-1}) were achieved by using a heated silica IDA column linked to fluorescence detection with a lumogallion based PCR.[14]

Using the same chromatographic conditions for fresh waters as discussed above Bashir and Paull demonstrated the determination of Be^{2+} in

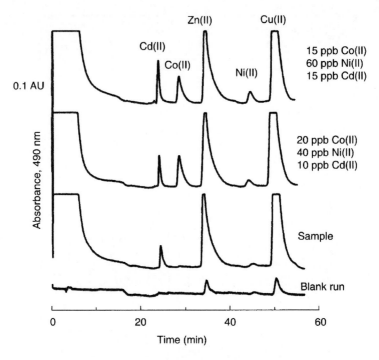

Figure 7.11 Chelation ion chromatographic determination of trace metals in estuarine seawater showing two standard additions. Column: 250 × 4 mm, 6.5 μm silica IDA. Step gradient: 0–10 min, 0.5 mM HNO_3–0.5 M KCl; 10–30 min, 80 mM tartaric acid; 30–50 min, 10 mM picolinic acid. Spectrophotometric detection at 490 nm after PCR with PAR/NH_3. Sample: 6 mL water from Carnon Estuary, UK. Reproduced with permission from Jones and Nesterenko.[47] Copyright (2004) Elsevier.

simulated seawater spiked at the 40 ppb level.[7] This chromatogram is shown as Figure 7.13.

Uranium has received special attention by a number of workers. Shaw et al. used a polymeric column dynamically modified with dipicolinic acid to separate and determine uranium in a NASS-4 open ocean reference sample spiked at the 50 ng mL^{-1} level[15] (shown as Figure 7.14). The unique selectivity allowed U(VI) be well separated from Fe(III), which elutes relatively early. Cowan determined uranium at natural seawater levels using a dipicolinic acid dynamically loaded column with a 2 mL injection volume. He found close to the expected level of 2.5 ng mL^{-1} uranium in water from the English Channel.[43] A typical chromatogram from this application is shown as Figure 7.15. The large volume injection caused a significant matrix metal peak which eluted close to the uranium peak, although this did not interfere with quantification. As far as can be ascertained, this is the first example of the direct determination of uranium in seawater at natural levels by an IC method without preconcentration.

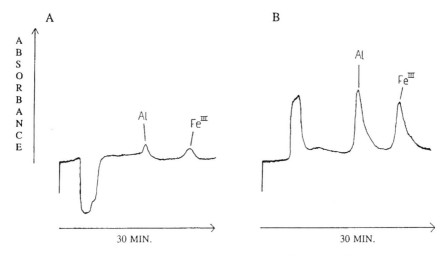

Figure 7.12 Chelation ion chromatograms showing Al^{3+} and Fe^{3+} in (A) procedural blank, (B) seawater from Plymouth Sound, UK, preconcentrated (pH 4.0) and separated (pH 1.8). Column: 100 × 4.6 mm i.d. packed with neutral 2.8 µm PS-DVB resin and coated with CAS. Preconcentration volume 20 mL. Spectrophotometric detection at 580 nm after PCR with PCV. Reproduced, with permission, from Paull and Jones.[13] Copyright (1996) Springer.

7.4.2 Highly Saline Waters

There are certain situations where saline waters can become super concentrated, either naturally, as in certain lakes, or through industrial processes, such as brines for use within the oil industry. Paull and Haddad have successfully applied HPCIC to the determination of uranium at the low ppb level in hyper saline Antarctic lake waters, naturally concentrated in NaCl up to 3 M in concentration. This analysis was achieved using large volume injections with a one step gradient elution approach on a MTB impregnated column.[8] The chromatogram obtained in this study, for a 10 ng mL^{-1} level spiked hyper saline sample is shown as Figure 7.16.

A detailed study of the retention characteristics of alkaline-earth metals on a resin column dynamically coated with PP was carried out by Paull et al.[12] The conditions were chosen such that Sr^{2+} was well separated from large amounts of Ca^{2+} and Mg^{2+}. Strontium was then determined in a similar hyper saline Antarctic lake sample, though in this example some sample dilution was required due to the extremely high level of Mg^{2+} present, approximately 0.6 M. This example chromatogram is shown as Figure 7.17.

Oil well brines or formation waters are used in the petroleum industry to maintain hydrostatic pressure as oil is removed. Salt levels can rise to 30% in these formation waters and levels of alkaline earths need to be controlled to prevent precipitation. Paull et al. once again applied HPCIC using a MTB

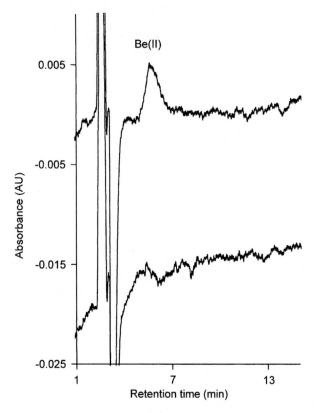

Figure 7.13 Overlaid chelation ion chromatograms of simulated seawater and simulated seawater spiked with 40 µg L^{-1} of Be^{2+}. Mobile phase conditions: 0.4 M KNO$_3$, pH 3.0. Other conditions as in Figure 7.3. Reproduced, with permission, from Bashir and Paull.[7] Copyright (2001) Elsevier.

impregnated column, to the determination of Ba^{2+}, Sr^{2+}, Mg^{2+} and Ca^{2+} in three oil well brine samples. Barium was a particular challenge as it was present at a much lower level than the others. However, as it eluted first as a sharp peak with no overlapping peak interferences, allowing a detection limit of 1 µg mL^{-1} be achieved.[22] The chromatograms of the brine samples shown with Figure 7.18 illustrate the complexity of these highly saline samples clearly.

7.4.3 Commercial Products and Fine Chemicals

Analysis of commercially produced fine chemicals for trace elemental contaminants is a major challenge for standard IC using ion exchange columns, although it presents less of a problem for HPCIC. This was demonstrated in an early study by Challenger and co-workers, who determined a range of

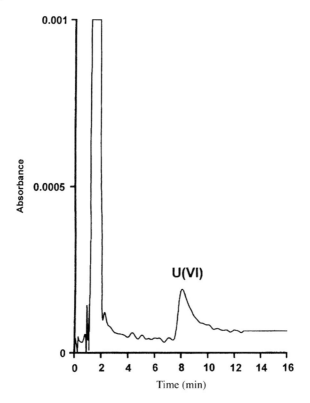

Figure 7.14 Chelation ion chromatographic separation of 50 ng mL^{-1} of U(vi) from matrix metals in a spiked seawater. Conditions as in Figure 7.4. Reproduced, with permission, from Shaw et al.[15] Copyright (2000) Springer.

transition and heavy metals in KCl, Na$_2$SO$_4$ and CsI salts from 1 to 5 M solution concentrations using HPCIC with a XO impregnated column. Large volume injections were again used with step gradient elution to reduce analysis time.[19] The chromatograms obtained for the three commercial salts are shown as Figure 7.19.

The higher efficiency commercial silica bonded IDA columns which are now available can give even lower detection limits under isocratic conditions with normal injection volumes when analysing these concentrated salt solutions. Figures 7.20 and 7.21 show the high speed trace determination of some transition and heavy metals in a two reagent grade salts using a short 50 mm silica bonded IDA column with a mixed oxalate/chloride/picolinate mobile phase, similar to that used for the chromatograms shown in Figure 7.6. Sub-ppb detection limits were easily obtained for some of the metals with a standard 100 μL injection, which compares very favourably with modern advanced atomic spectroscopic methods, without the complicating issue of salt matrix interferences.

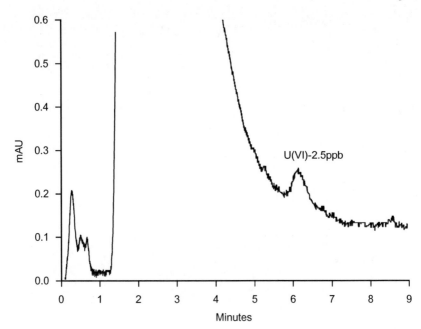

Figure 7.15 Direct chelation ion chromatographic determination of U(VI) in seawater using a 100 × 4.6 mm i.d. PS-DVB resin column dynamically modified with 0.1 mM dipicolinic acid. Mobile phase: 1 M HNO_3. Sample volume: 2 mL. Spectrophotometric detection after PCR with arsenazo III at 654 nm using signal extraction noise reduction. Reproduced, with permission, from Cowan.[43]

Aminophosphonic acid bonded silica, although not yet applied to the same degree as IDA based columns in HPCIC, also shows great potential for determining trace metals in concentrated salts. Figure 7.22 shows the determination of a mix of alkaline earths, transition and heavy metals in 30% w/v (saturated) laboratory grade KNO_3.[4]

Bashir and Paull studied the separation of the alkaline-earth metals with various high salt concentration mobile phases on a silica IDA column.[23] This application provides a good example of the effect of the presence of some ion exchange interaction on the observed selectivity, as discussed earlier within Chapters 2 to 4. With 'pure' chelation, Ba^{2+} would be expected to come out first. In fact in a 0.1 M KNO_3 mobile phase it eluted last and only eluted first when the mobile phase contained 1.5 M KNO_3, under which conditions the influence of the ion exchange groups was finally suppressed to a very low level. This finding compares well with that reported by Challenger and co-workers, who also found a reversal of Ba^{2+} retention with increasing ionic strength on dye impregnated columns containing the IDA functionality. However, the Ba^{2+} reversal occurred at a significantly lower concentration of KNO_3, almost certainly as a consequence of the much lower capacity of the dye impregnated

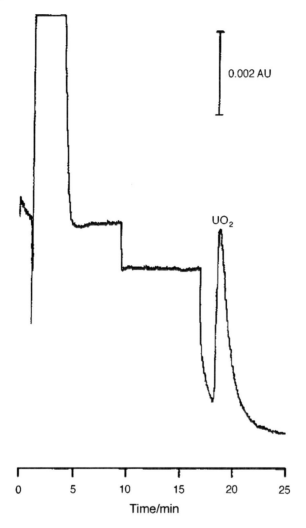

Figure 7.16 Chelation ion chromatograms obtained from the 2 mL injection of a saline lake sample spiked with $10\,\mu g\,L^{-1}$ UO_2^{2+}. Column, 150×4.1 mm i.d. packed with 5 μm PS-DVB resin coated with MTB. Mobile phase: step gradient from 6 to 60 mM HNO_3. Spectrophotometric detection after PCR with arsenazo III in 0.4 M acetic acid. Reproduced, with permission, from Paull and Haddad.[8] Copyright (1998) Royal Society of Chemistry.

columns. After optimising the separation for Ca^{2+} and Mg^{2+}, Bashir and Paull determined these cations directly in saline medical eyewash and KCl solutions,[23] the results of which can be seen in Figure 7.23.

Perhaps the most challenging HPCIC application to-date is the determination of alkaline earth metals in saturated brines used for the large scale industrial production of NaOH and chlorine. The original Castner–Kellner

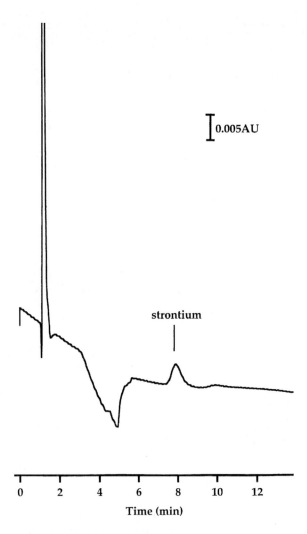

Figure 7.17 Chelation ion chromatographic determination of Sr^{2+} in an Antarctic brine sample (Deep Lake, 20 m). Column: Hamilton PRP-1 150 × 4.1 mm i.d. packed with 5 μm PS-DVB resin. Mobile phase: 0.5 M KNO_3, 0.2 mM PP, 20 mM sodium tetraborate, pH 9.5. 100 μL injection. Spectrophotometric detection at 575 nm. Reproduced, with permission, from Paull et al.[12] Copyright (1998) Elsevier.

method involving electrolyis with pools of mercury is now considered environmentally harmful and is gradually being replaced by a membrane electrolysis technique. However, the saturated NaCl feed solution needs to be highly pure, with only very low levels of trace metals, particularly Mg^{2+}, Ca^{2+} and Sr^{2+}. This is because the high pH levels found close to the membrane surface can cause precipitation of the metals as the hydroxides, which can clog the pores,

Practical Applications

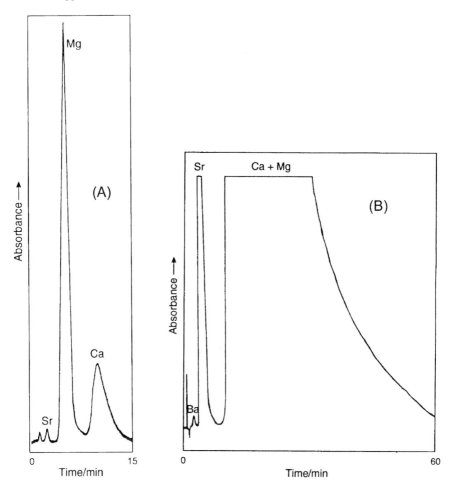

Figure 7.18 (A) Chelation ion chromatograms showing the separation of Sr^{2+}, Mg^{2+} and Ca^{2+} in an oil well brine (diluted 1:100 with deionised water) using a MTB impregnated 8.8 µm PS-DVB resin column, 100 × 4.6 mm i.d. Mobile phase: 1 M KNO_3, pH 8.0; flow rate: 1 mL min^{-1}. Spectrophotometric detection at 490 nm after PCR with PAR-ZnEDTA. (B) Chromatogram showing the separation of Ba^{2+} and Sr^{2+} from Mg^{2+} and Ca^{2+} in an oil well brine (diluted 1:1 with deionised water). Mobile phase: 1 M KNO_3, pH 9.2; flow rate: 1 mL min^{-1}. Column and detection as in (A). Reproduced, with permission, from Paull et al.[22] Copyright (1994) Royal Society of Chemistry.

substantially reducing the efficiency of electrolysis. Thus, the NaCl feed solution is purified mainly by passing through large columns containing a chelating resin, normally involving phosphonate groups, to reduce the concentration of trace metals. The cleaned-up brine then needs to be monitored to make sure the low levels are maintained during the purification cycle. The main challenge isn't

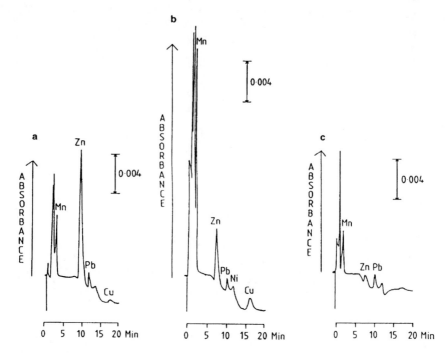

Figure 7.19 Chelation ion chromatograms showing single column preconcentration and separation of trace metal ions in laboratory chemical samples (a) 1 M KNO$_3$, (b) 1 M Na$_2$SO$_4$ and (c) 1 M CsI. Column: 100 × 4.6 mm i.d. packed with 8 μm PS-DVB resin and coated with XO. Mobile phase: 1 M KNO$_3$ containing 0.05 M lactic acid using a three step pH gradient: pH 6, pH 3.7 and finally pH 2. Flow rate: 1 mL min^{-1}. Sample volumes were 6 mL adjusted to pH 6. Spectrophotometric detection at 490 nm after PCR with PAR/NH$_3$. Reproduced, with permission, from Challenger et al.[19] Copyright (1993) Elsevier.

the high concentration of the NaCl, which has as already been seen is no problem for chelating phases used in HPCIC, it is more the very low levels of detection required for the alkaline earth metals. This challenge has been overcome by a combination of the latest developments in high efficiency chelating stationary phases, PCR detection and noise reduction. Figure 7.24A shows the separation in under 6 min of a standard mixture of Mg^{2+}, Sr^{2+} and Ca^{2+} using a 5 μm silica bonded IDA packed column. The PCR was PAR-ZnEDTA optimised for alkaline earth metal detection, combined with signal extraction noise reduction, giving detection limits of 4, 10 and 8 ng mL^{-1} for Mg^{2+}, Sr^{2+} and Ca^{2+} respectively for a 20 μL injection. The Mg^{2+} calibration curve gave a negative intercept as the sample contained less Mg^{2+} than the mobile phase itself. Figure 7.24B is an injection of an actual sample of a 3 M purified brine from an Indian chlor-alkali plant. The sample contained Mg^{2+} below the 4 ng mL^{-1} detection limit, hence the negative peak, some Sr^{2+} near the detection limit of 10 ng mL^{-1} and 34 ng mL^{-1} of Ca^{2+}. The levels before

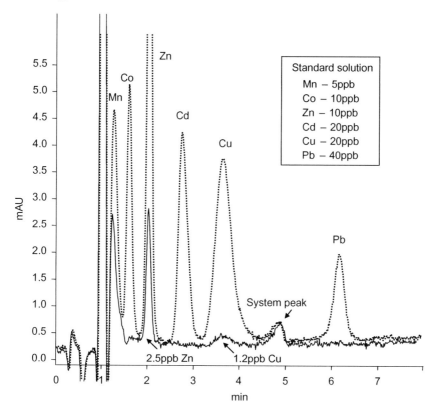

Figure 7.20 Chelation ion chromatograms of a 2 M sample of reagent grade KNO_3 with a 50 × 4 mm 5μm silica IDA column. Solid line, 100 μL injection of the sample. Dotted line, 100 μL injection of sample with standard addition of six metals with concentrations as listed in the figure. All other conditions as in Figure 7.6. Courtesy of JPP Chromatography Ltd, unpublished data.

purification were 14 ng mL^{-1} Mg^{2+}, 660 ng mL^{-1} Sr^{2+} and 3.65 μg mL^{-1} of Ca^{2+}, showing the clean-up column was performing well. The signal extraction baseline noise reduction was essential because of the very high absorbance background of the PCR reagent producing a large amount of baseline noise.

7.5 Solid Samples

The analysis of solid samples can present even further challenges when compared to water analysis. Sample treatment, including dissolution, is common to all techniques requiring special care to prevent contamination or loss. High levels of matrix metals such as Ca^{2+}, Mg^{2+} and Fe^{2+} could be present making it difficult to separate the metals of interest and avoid co-eluting peaks. HPCIC

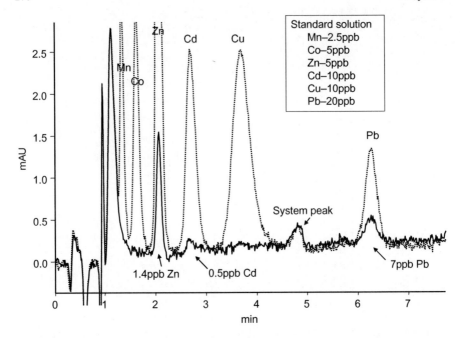

Figure 7.21 Chelation ion chromatogram of a 0.25 M sample of reagent grade KCl with a 50 × 4 mm silica IDA column. Solid line, 100 µL injection of the sample. Dotted line, 100 µL injection of sample with standard addition of six metals with concentrations as listed in the figure. All other conditions as in Figure 7.6. Courtesy of JPP Chromatography Ltd., unpublished data.

can present a useful alternative in this regard, as there is a larger range of approaches available to control selectivity, making it much more versatile than simple ion exchange procedures.

7.5.1 Sediments, Soils and Minerals

For sediments, soils and mineral type samples, leaching or complete dissolution is required to extract the trace metals. These sample treatment methods are well established and will not be detailed here. Certified reference materials are normally available for most sample types for quality control and method development. Sediment analysis is the focus of some attention as it can be important sink for trace metals, which under different conditions can release them into the fresh or saline waters. Shaw *et al.* exploited the unusual selectivity of resin columns dynamically modified with dipicolinic acid to determine trace U(VI) in a stream sediment.[15] What was unusual about the conditions chosen, was the lack of retention of the common +2 metal ions, and of Al^{3+} and Fe^{3+}, which were themselves present in relatively massive amounts. These cations all eluted within the dead volume and did not interfere

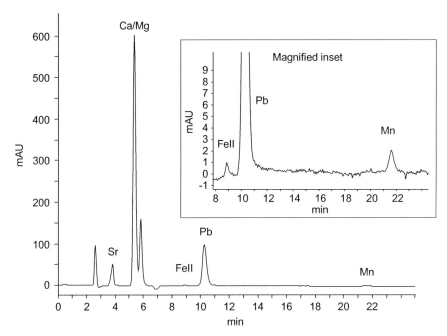

Figure 7.22 Chelation ion chromatographic analysis of a 30% (saturated) solution of ACS reagent grade KNO_3 with an APA bonded 3 μm silica column, dimensions, 150 × 4 mm. Mobile phase: 0.05 M HNO_3 in 0.8 M KNO_3 containing 3 mM ascorbic acid; flow rate: 0.7 mL min^{-1}. All iron is converted to Fe^{2+}; 100 μL injection volume. Spectrophotometric detection at 510 nm after PCR with PAR-ZnEDTA, pH 11.7 with signal extraction noise reduction. Measured concentrations: Pb^{2+} 1.2 μg mL^{-1}, Ca^{2+} 3.0 μg mL^{-1}, Mg^{2+} 420 ng mL^{-1}, Sr^{2+} 250 ng mL^{-1}, Mn^{2+} 17 ng mL^{-1} and Fe^{2+}, 33 ng mL^{-1}. Reproduced, with permission, from Nesterenko and Jones.[4] Copyright (2007) Wiley-VCH Verlag GmbH & Co. KGaA.

with the uranium peak. Zr(IV) was also well separated, but not determined, as shown in Figure 7.25. In a similar study, Cowan and co-workers also applied HPCIC to the determination of Bi^{3+} as well as U(VI) in a second type of sediment sample.[43]

Obviously silica bonded aminophosphonic acid (APA) phases have also been shown to exhibit great potential in the determination of trace metals in such complex samples. Under certain conditions the selectivity is complementary to IDA based phases. For example, for the common transition metals in simple mineral acid mobile phases, Ni^{2+} elutes first and Mn^{2+} last, the exact opposite to IDA separations. APA also shows unique selectivity for Be^{2+}, with the element being strongly retained compared to all other divalent cations. This fact was exploited by Shaw *et al.* who developed a method for determination of Be^{2+} in a stream sediment,[24] the chromatogram from which is shown as Figure 7.26. Though relatively broad, the Be^{2+} peak elutes well away from all the common divalent metal ions and Al^{3+}, which all elute close to the

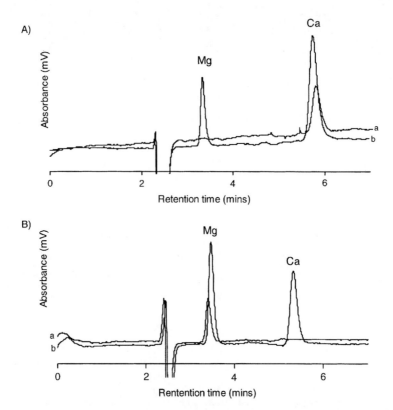

Figure 7.23 Chelation ion chromatograms showing (A) saline eyewash sample (0.9% NaCl) overlaid with spiked (200 µg L^{-1}) sample and (B) 0.5 M KCl sample solution overlaid with spiked (300 µg L^{-1}) sample. Column, 250 × 4 mm i.d., containing 8 µm IDA bonded silica. Mobile phase: (A), 1 M NaNO$_3$ (pH 4.9), (B), 1 M KNO$_3$ (pH 4.9); flow rate: 1.0 mL min^{-1}. Spectrophotometric detection at 572 nm after PCR with PP, pH 10.4. Reproduced, with permission, from Bashir and Paull.[23] Copyright (2001) Elsevier.

solvent front with the strong acid mobile phase used. Fe^{3+} was also strongly retained, but was reduced to Fe^{2+} by the use of ascorbic acid in the mobile phase. In this application, EDTA, whose complexation with Be is very weak, was added to the PCR to suppress the response of the transition metals with the CAS PCR, and so reduce the size of the solvent front peak, which could otherwise swamp the Be^{2+} peak.

Determination of trace metals in a mineral such as gypsum is a particular challenge because of the huge amount of Ca^{2+} in the dissolved sample. One reason for the interest in this mineral is for the investigation of not only the age of rocks, but also for the evaluation of the provenance of foodstuffs and wines. The determination of the isotope ratios of certain elements in wine by high resolution mass spectrometry can give an idea of the geographical area of

Practical Applications 273

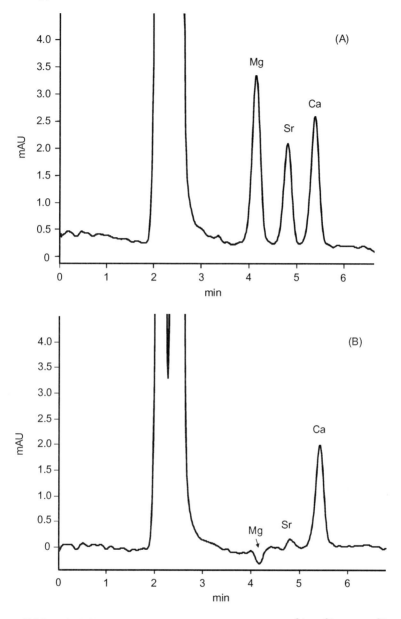

Figure 7.24 Chelation ion chromatograms showing (A) Mg^{2+}, Sr^{2+} and Ca^{2+} standards and (B) saturated brine for the chlor-alkali industry. Injection volume 20 μL. Concentrations in standards mixture, Mg^{2+} 40 ng mL^{-1}, Sr^{2+} 80 ng mL^{-1}, Ca^{2+} 40 ng mL^{-1}. Column: 150 × 4 mm i.d. column containing 5 μm IDA silica. Mobile phase: 0.6 M high purity KNO_3 containing 0.3 mM HNO_3; flow rate: 0.8 mL min^{-1}. Spectrophotometric detection after PCR with PAR-ZnEDTA/NH_3, pH 10.1. Courtesy of JPP Chromatography Ltd., unpublished data.

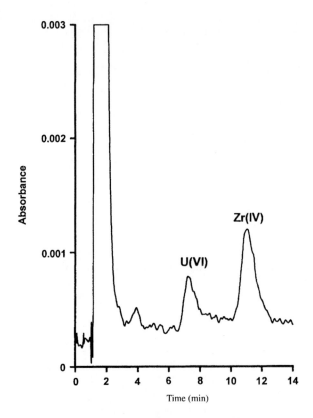

Figure 7.25 Chelation ion chromatographic separation of U(VI) from matrix metals in a sediment certified reference material GB07311. Column, 100 mm × 4.6 mm i.d., packed with 7 μm PS-DVB resin. Mobile phase: 1 M KNO_3, 0.5 M HNO_3 and 0.1 mM dipicolinic acid, flow rate 1 mL min^{-1}. Spectrophotometric detection at 654 nm after PCR with arsenazo III. Reproduced, with permission, from Shaw et al.[15] Copyright (2000) Springer.

origin.[44] This is because the isotope ratios in plants, for elements such as strontium, should reflect the ratios in the surrounding soil minerals. The ratio of ^{87}Sr to ^{86}Sr in a mineral such as gypsum depends upon the decay of ^{87}Rb in the same sample. The exact isotopic composition of strontium in the mineral will depend on the age and rubidium/strontium ratio of that mineral. The ratio found in plants and hence food and wine should be the same ratio as in the mineral in the soil and hence give evidence of provenance. To obtain reliable mass spectrometric data, the Sr^{2+} needs to be isolated not only from the Rb^+, which will give isobaric interference, but also from large amounts of other metals that could cause ionisation problems in the mass spectrometer.

Sutton used a PP impregnated hyper-cross-linked polystyrene resin (MN200, Purolite, UK) column to separate Sr^{2+} from Rb^+ and the massive amount of

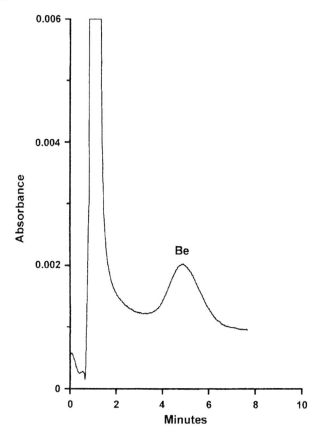

Figure 7.26 Chelation ion chromatographic separation of Be^{2+} from matrix metals in a certified reference material (GBW07311). Column, 50 mm × 4.6 mm, containing 5 μm silica bonded APA. Mobile phase: 1 M KNO_3, 0.5 M HNO_3 and 0.08 M ascorbic acid; flow rate: 1.0 mL min^{-1}. Spectrophotometric detection at 560 nm after PCR with CAS at pH 6 containing EDTA. Reproduced, with permission, from Shaw et al.[24] Copyright (2000) Elsevier.

Ca^{2+} in a sample of gypsum dissolved in HCl.[45] Figure 7.27 shows the optimised separation used, and the strontium fractions were collected from the effluent by noting the times of the start and end of the peak. With the PCR removed, the combined strontium fractions from a number of injections were concentrated by evaporation and sublimation of the ammonium salt and then subjected to high resolution mass spectrometry. Good agreement was found with Sr^{2+} fractions obtained by the standard, but much more complicated and prolonged, chemical extraction procedure.

In a different type of study, Zn^{2+} leached from gypsum was determined by Paull et al.[26] using a PS-DVB resin column dynamically modified with methylthymol blue. As in a previous study using a dynamically loaded PP

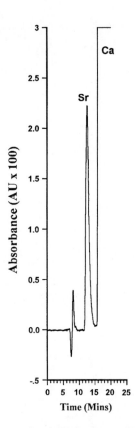

Figure 7.27 100 μL injection of 6 g L^{-1} G4 gypsum rock sample. Column, 250 mm × 4.6 mm i.d., packed with 20 μm Purolite MN200 resin impregnated with PP. Mobile phase: 4 M NH$_4$NO$_3$ at pH 10, flow rate 1.0 mL min^{-1}. Spectrophotometric detection at 495 nm after PCR with PAR-ZnEDTA. Reproduced, with permission, from Sutton.[45]

column,[16,17] no PCR detection was required as the MTB also acted as the colour forming reagent.

Concern about the possible presence of highly radioactive isotopes in the environment has led to the development of highly sensitive analytical methods for their determination. ICP-MS techniques are commonly used because of the exceptional sensitivities obtainable. However, in some instances direct analysis of the treated sample cannot be achieved successfully without additional separation methods. One such situation concerns the determination of plutonium in environmental and biological material in the presence of relatively large amounts of uranium. There is very serious isobaric interference of ^{239}Pu from ^{238}U^{1}H$^+$ produced in the plasma. Truscott *et al.* solved this problem by separating the ^{238}U and ^{239}Pu in a soil sample by HPCIC using a resin column dynamically modified with dipicolinic acid.[25] There was considerable peak overlap with the ^{235}U component of the sample, but was not important as the

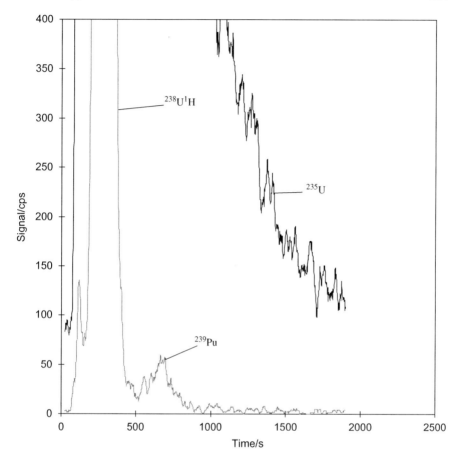

Figure 7.28 Chelation ion chromatographic seperation of ^{239}Pu from uranium and subsequently the ^{238}U^{1}H^{+} interference for NIST 4353 Rocky Flats Soil (No. 1) reference material. Column, 100 mm × 4.6 mm i.d., packed with 7 μm PS-DVB resin. Mobile phase: 0.75 M HNO$_3$ containing 0.1 mM dipicolinic acid, flow rate 0.5 mL min^{-1}. (^{239}Pu elutes a little earlier than usual due to the strong acidity of the sample compared to the mobile phase). Detection, ICP-MS. Reproduced, with permission, from Truscott et al.[25] Copyright (2001) Elsevier.

mass spectrometer could easily differentiate it from the ^{239}Pu, as shown in Figure 7.28.

7.5.2 Biological Materials and Foodstuffs

Biological materials require unique sample treatment procedures for trace metal analysis, mainly wet or dry ashing to remove the organic material. Sometimes leaching can be used depending on the ease of metal removal. As seen previously in relation to water samples, Sr^{2+} is readily separated from

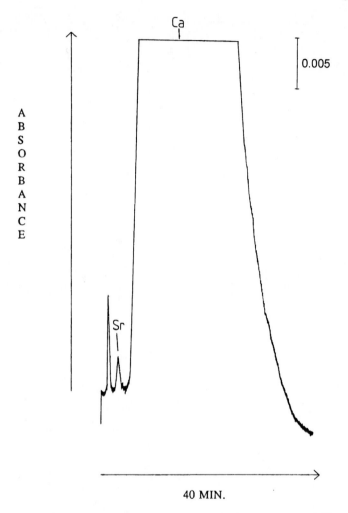

Figure 7.29 Chelation ion chromatogram showing the separation of Sr^{2+} from Ca^{2+} in milk powder digest on a MTB impregnated 8.8 μm PS-DVB resin 100 × 4.6 mm i.d. column. Mobile phase: 1 M KNO_3 with 0.05 M lactic acid, pH 10.2; flow rate: 1.0 mL min^{-1}. Photometric detection at 490 nm after PCR with PAR-ZnEDTA. Reproduced, with permission, from Jones et al.[28] Copyright (1994) Elsevier.

large amounts of Ca^{2+} as it elutes prior to it on most chelating ion exchange columns. After dry ashing and dissolution in dilute HNO_3, Jones et al. determined Sr^{2+} in milk powder using a resin column impregnated with MTB.[28] Good agreement was obtained with results for the same sample obtained using from ICP-OES. The milk power sample chromatogram is shown as Figure 7.29.

In a similar food product application of HPCIC, Shaw et al. determined Cd, Pb and Cu in a certified rice flour.[31] Through exploiting the control of selectivity through mobile phase pH, Shaw et al. used a resin column dynamically

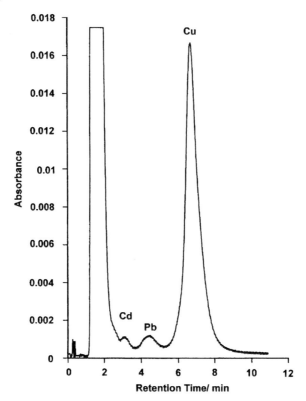

Figure 7.30 Chelation ion chromatographic separation of Cd^{2+}, Pb^{2+} and Cu^{2+} from matrix interferences in the certified rice flour GBW08502 on the 100 × 4.6 mm i.d. PRP-1 7 μm PS-DVB column. Mobile phase: 1 M KNO_3, 0.03 M HNO_3 and 0.25 mM 4-chlorodipicolinic acid, flow rate 1.0 mL min^{-1}. Photometric detection at 520 nm after PCR with PAR using signal extraction noise reduction system. Reproduced, with permission, from Shaw et al.[31] Copyright (2002) Elsevier.

loaded with 4-chlorodipicolinic acid for the analysis, with samples digested within a closed vessel microwave digester and subsequent dissolution. A typical chromatogram from this application is shown as Figure 7.30.

Hsu et al. used HPCIC to determine Cu and Zn in a certified oyster tissue using a silica bonded macrocyclic cryptand type column followed with PAR PCR detection. Results from the analysis agreed well with certified values.[20] Simonzadeh et al. used a phenylhydrazone bonded silica phase column for the investigation of trace metals in tomato leaves and vitamin tablets.[32] Peaks were relatively broad and tailing, but there was sufficient resolution to determine Mn^{2+}, Fe^{2+}, Zn^{2+} and Cu^{2+} within the sample. Truscott et al. using the same HPCIC conditions as for the Pu in sediment application discussed earlier, also determined Pu in certified human lung tissue.[25] The sensitivity was notable, giving a result of 570 fg g^{-1} Pu, well within the certified range. This sample chromatogram is shown as Figure 7.31.

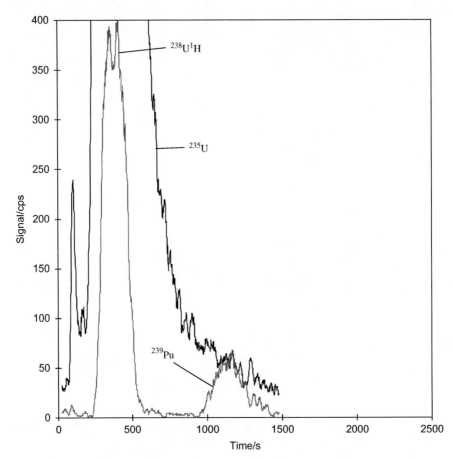

Figure 7.31 Chelation ion chromatographic separation of ^{239}Pu from uranium and subsequently the ^{238}U^{1}H$^+$ interference in NIST 4351 Human Lung reference material. Column, 100×4.6 mm i.d., packed with 7 µm PS-DVB resin. Mobile phase: 0.75 M HNO$_3$ containing 0.1 mM dipicolinic acid, flow rate 0.5 mL min^{-1}. Detection, ICP-MS. Reproduced, with permission, from Truscott et al.[25] Copyright (2001) Elsevier.

7.5.3 Miscellaneous

There are several other applications which have some unique aspects worthy of note compared to the applications discussed so far. The untypical nature of the sample itself necessitates the use of the versatile HPCIC method for the analysis. An example which stands out in this regard is the determination of trace elements in non-aqueous solvents.

Although the vast majority of IC methods for trace metals, including HPCIC, involve totally aqueous solutions, occasionally, small amounts of polar organic solvents are added to improve peak shape or modify selectivity.

However, when the sample containing trace metals is completely non-aqueous this produces a rather interesting situation. Ethanol produced from fermented sucrose is increasingly being used as a biofuel to reduce petroleum consumption. Trace metals such as copper and iron in the ethanol biofuel need to be strictly controlled and therefore continuously monitored during manufacture. As ethanol is completely miscible in aqueous solutions, IC shows potential for determining the trace metals in the solvent, particularly if the mobile phase contains some alcohol to improve compatibility. Nevertheless, unexpected results are likely to be found since the metal complexation reactions will likely differ from that in totally aqueous mobile phases.

Nesterenko and co-workers have investigated the determination of trace metals in ethanol using HPCIC with IDA silica columns and a dipicolinic acid containing mobile phase, also containing varying amounts of ethanol or methanol. When optimised they obtained a separation of 9 metals spiked in an ethanol sample using a mobile phase consisting of 60% methanol containing dipicolinic acid and hydrochloric acid.[33] This final separation is shown here as Figure 7.32.

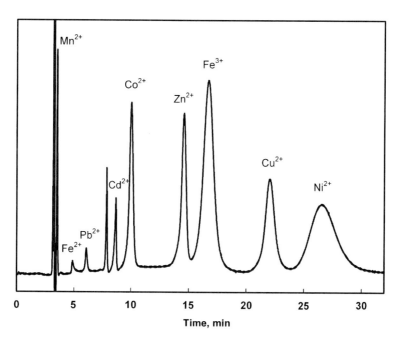

Figure 7.32 Chelation ion chromatogram of a model mixture of metals in bioethanol. Column: IDA-silica 150×4.0 mm i.d., 5 µm. Mobile phase: 60% methanol–40% of 2.5 mM DPA–0.01 M HCl, flow rate 0.8 mL min^{-1}. Sample volume, 100 µL. Spectrophotometric detection at 510 nm after PCR with PAR. Adopted from Dias et al.[33] Copyright (2010) Royal Society of Chemistry.

References

1. R. Michalski, *Crit. Rev. Anal. Chem.*, 2009, **39**, 230–250.
2. M. J. Shaw and P. R. Haddad, *Environ. Int.*, 2004, **30**, 403–431.
3. P. Jones and P. N. Nesterenko, *J. Chromatogr A.*, 1997, **789**, 413–435.
4. P. N. Nesterenko and P. Jones, *J. Sep. Sci.*, 2007, **30**, 1773–1793.
5. B. Paull and P. R. Haddad, *TrAC Trend. Anal. Chem.*, 1999, **18**, 107–114.
6. P. N. Nesterenko and P. Jones, *J. Chromatogr. A.*, 1997, **770**, 129–135.
7. W. Bashir and B. Paull, *J. Chromatogr. A.*, 2001, **910**, 301–309.
8. B. Paull and P. R. Haddad, *Anal. Comm.*, 1998, **35**, 13–16.
9. I. N. Voloschik, M. L. Litvina and B. A. Rudenko, *J. Chromatogr. A.*, 1994, **671**, 205–209.
10. C. Y. Liu, N. M. Lee and T. H. Wang, *Anal. Chim. Acta*, 1997, **337**, 173–182.
11. S. Yamazaki, H. Omori and O. Eon, *J. High Res. Chromatogr.*, 1986, **9**, 765–766.
12. B. Paull, M. Clow and P. R. Haddad, *J. Chromatogr. A.*, 1998, **804**, 95–103.
13. B. Paull and P. Jones, *Chromatographia*, 1996, **42**, 528–538.
14. J. Tria, P. N. Nesterenko and P. R. Haddad, *Chem. Listy*, 2008, **102**, 319–323.
15. M. J. Shaw, S. J. Hill, P. Jones and P. N. Nesterenko, *Chromatographia*, 2000, **51**, 695–700.
16. B. Paull, P. A. Fagan and P. R. Haddad, *Anal. Comm.*, 1996, **33**, 193–196.
17. B. Paull, M. Macka and P. R. Haddad, *J. Chromatogr. A.*, 1997, **789**, 329–337.
18. B. Paull, M. Foulkes and P. Jones, *Analyst*, 1994, **119**, 937–941.
19. O. J. Challenger, S. J. Hill and P. Jones, *J. Chromatogr.*, 1993, **639**, 197–205.
20. J. C. Hsu, C. H. Chang and C. Y. Liu, *Fresenius J. Anal. Chem.*, 1998, **362**, 514–521.
21. J. Toei, *Fresenius Z. Anal. Chem.*, 1988, **331**, 735–739.
22. B. Paull, M. Foulkes and P. Jones, *Anal. Proc.*, 1994, **31**, 209–211.
23. W. Bashir and B. Paull, *J. Chromatogr. A.*, 2001, **907**, 191–200.
24. M. J. Shaw, S. J. Hill, P. Jones and P. N. Nesterenko, *J. Chromatogr. A.*, 2000, **876**, 127–133.
25. J. B. Truscott, P. Jones, B. E. Fairman and E. H. Evans, *J. Chromatogr. A.*, 2001, **928**, 91–98.
26. B. Paull, P. Nesterenko, M. Nurdin and P. R. Haddad, *Anal. Comm.*, 1998, **35**, 17–20.
27. W. Bashir and B. Paull, *J. Chromatogr. A.*, 2001, **942**, 73–82.
28. P. Jones, M. Foulkes and B. Paull, *J. Chromatogr. A.*, 1994, **673**, 173–179.
29. A. Haidekker and C. G. Huber, *J. Chromatogr. A.*, 2001, **921**, 217–226.
30. A. I. Elefterov, P. N. Nesterenko and O. A. Shpigun, *J. Anal. Chem. (Engl. Transl.)*, 1996, **51**, 887–891.
31. M. J. Shaw, P. Jones and P. N. Nesterenko, *J. Chromatogr. A.*, 2002, **953**, 141–150.

32. N. Simonzadeh and A. A. Schilt, *Talanta.*, 1988, **35**, 187–190.
33. J. C. Dias, L. T. Kubota, P. N. Nesterenko, G. W. Dicinoski, and P. R. Haddad, *Anal. Methods*, 2010, **2**, in press.
34. B. Paull, Investigation of chelating dye impregnated resins for the ion chromatographic determination of trace metals, PhD thesis, University of Plymouth, UK, 1994.
35. J. Tria, P. R. Haddad and P. N. Nesterenko, *J. Sep. Sci.*, 2008, **31**, 2231–2238.
36. G. Alberti, G. D'Agostino, G. Palazzo, R. Biesuz and M. Pesavento, *J. Inorg. Biochem.*, 2005, **99**, 1779–1787.
37. M. Pesavento, R. Biesuz and C. Palet, *Analyst.*, 1998, **123**, 1295–1301.
38. O. J. Challenger, S. J. Hill, P. Jones and N. W. Barnett, *Anal. Proc.*, 1992, **29**, 91–93.
39. P. Jones, O. J. Challenger, S. J. Hill and N. W. Barnett, *Analyst*, 1992, **117**, 1447–1450.
40. W. Bashir, E. Tyrrell, O. Feeney and B. Paull, *J. Chromatogr. A.*, 2002, **964**, 113–122.
41. P. N. Nesterenko and P. Jones, in *Abstracts of International. Ion Chromatography Symposium, IICS-2000*, 11 September 2000, Nice, France, No. 65.
42. M. J. Chen, J. D. Fang and C. Y. Liu, *J. Chin. Chem. Soc.*, 1992, **39**, 124–128.
43. J. Cowan, The development and study of chelating substrates for the separation of metal ions in complex matrices, PhD thesis, University of Plymouth, UK, 2002.
44. P. Horn, P. Schaaf, B. Holbach, S. Holzl and H. Eschnauer, *Z. Lebensmittel-Untersuch. Fors.*, 1993, **196**, 407–409.
45. R. M. C. Sutton, Investigation of chelating dye impregnated resins for the selective adsorption and separation of trace metals from aqueous solutions, PhD thesis, University of Plymouth, UK, 1996.
46. A. I. Elefterov, S. N. Nosal, P. N. Nesterenko and O. A. Shpigun, *Analyst*, 1994, **119**, 1329–1332.
47. P. Jones and P. N. Nesterenko, Ion Exchange. Chelation Ion Chromatography, in *Encyclopedia of Analytical Science*, Eds. P. Worsfold, A. Townsend and C. Poole, Elsevier, 2005, **4**, 467–481.

Subject Index

Note: Figures and Tables are indicated by *italic page numbers*

acetate buffer (for post-column reactions) 221
acetonitrile, as mobile phase additive 31–2, 133
acid–base properties, chelating ion exchanger ligands 39, 41–2
acid hydrolysing metals 244
acidic functional groups, protonation of 72–3
actinides (actinoids)
 detection of 205, 206, 214, 221
 separation of
  by counter-current chromatography 164, 169–70, 173, 174
  by dynamic HPCIC 105, 260, 264, 276, 277
  by extraction chromatography 184–5, *185*, 187, *189*, *190*
adsorption–complexing chromatography 7–8, 11
adsorption of metal ions, properties required 36–7
air conditioning units, baseline fluctuations due to 229, *231*
Alberon *see* chrom(e)azurol S
alkaline-earth metal ions
 detection of 205, 213, 214, 217, 218, 233, 235
 formation constants of complexes *120*, *121*, 144, 145

 retention order affected by mobile phase parameters 117, *118*, 119, *123*, *129*, 264
 separation of
  in brines 268–9, *273*
  commercial available chelating ion exchangers for 64, 68–9
  by counter-current chromatography *171*, *173*
  with dynamically modified phases 99, *100*, 250–1
  effect of organic solvent additives in mobile phase 31–2, 133
  by extraction chromatography *189*, *190*
  with pre-impregnated phases 80, 86, 87, *87*, 92, *93*, *100*, *101*, 252, 254
  temperature effects 124, *125*
alkylamidoxime functionalised silica 57
 preparation of 54, 62
alpha coefficients 136
 calculations 143
alpha ray spectrometry, detection using *173*, *174*
aluminium
 separation efficiency affected by temperature 124, *126*
 speciation analysis for 233, 254

Subject Index

aluminium(III) ion
  detection of *205, 214, 215, 217, 218,* 221, 229, 232, 233, *235*
  determination of
    in seawater 259, *261*
    in wastewater 254, *258*
  formation constants of complexes *121, 145*
Aluminon *see* aurin tricarboxylic acid (ATA)
americium, separation from other actinides 170, *173, 174,* 180, *185, 189, 190*
amidoxime functionalised silica 39, *40,* 64, 141
amidoxime type group 38
α-amino acid type group 38
aminocarboxylic chelating agents, role of ligand upon complexation kinetics 19–20
γ-aminohydroxamate resin 48, *50*
aminophosphonic acid functionalised silica *see* APA-silica
3-aminopropylsilica
  coverage affected by porous structure 72
  hydrolytic stability 63
  reactions 54
  surface complexation 16
ammonia
  as buffer for post-column reactions 221
  effect on complex formation 208–10
  with PAR, as PCR reagent *255, 256, 257, 259, 260, 268*
ampholytic ion exchangers
  influence of solution ionic strength on Cu(II) sorption capacity 22–3
  sorption enthalpy values 28
anion exchangers 119
  influence of solution ionic strength on Cu(II) sorption capacity 22
  sorption enthalpy values 27–8

Antarctic saline lake waters *245,* 261, *265, 266*
APA-silica 39, *41, 60*
  applications
    in fine chemicals analysis 264, *271*
    in sediment analysis 271, *275*
  preparation of 54, 62
argentation chromatography 9
aromatic interactions 2
arsenazo I
  chemical structure *202*
  as PCR reagent 207, *214*
arsenazo III
  chemical structure *202*
  data for formation of coloured complexes with various metal ions *205–6*
  as mobile phase ligand 98
  as PCR reagent *12, 122, 130, 164, 169, 171, 172, 173,* 180, *184, 186, 187, 188, 189,* 207, *214,* 221, *225, 245, 246, 247, 248, 253, 264, 265, 274*
ascorbic acid, as mobile phase additive *247, 271, 272, 275*
atomic emission spectrometry (AES), detection using *167, 168, 171*
atomic spectroscopic methods, limitations 243, 276
aurin tricarboxylic acid (ATA)
  chemical structure and physical data *83*
  impregnated phases using 89, *90*
azo type metallochromic ligands, selectivity 92, *93*

barium ion
  detection of *213,* 217, 230, *235*
  determination of 87, 99–100, *101,* 252–3, *254*
    in mineral water *248,* 252–3, *254*
    in oil well brines *247,* 262, *267*
  formation constants of complexes *120, 145*

baseline noise
  reduction of 226–30
  types 224–6
baseline wander 225
basic chromatographic principles 1–3
beryllium ion
  detection of 205, 215, 221
  determination of
    in fresh/potable water 251–2
    in seawater 259–60, 262
    in sediments 271–2, 275
  formation constants of
    complexes 121, 145
  separation from other alkaline
    earths 251–2
bioethanol 281
biofuels
  determination of trace metals
    in 249, 281
  separation of metal ions in 133, 281
biological materials, analysis
  of 242–3, 276, 277, 279, 280
bis(2-ethylhexyl)phosphonic acid
  (BEHPA), as stationary phase in
  HSCCC 162, 171
bis(2-ethylhexyl)succinamic acid
  (BEHSA), as stationary phase in
  HPEC 181, 185, 186
bis(2-hydroxyethyl)-dithiocarbamate
  (BHEDTC), as PCR reagent 216
bismuth(III) ion
  detection of 205, 214, 235
  determination in sediments 271
bis(1,1,3,3-tetramethylbutyl)phos-
  phinic acid (HMBP), as stationary
  phase in HPEC 178, 180, 182, 188
borate buffer (for post-column
  reactions) 221, 245
brines (industrial/oil-well), analysis
  of 13, 86, 87, 247, 261–2, 265–9, 273
5-Br-PADAP
  data for formation of coloured
    complexes with various metal
    ions 205–6
  as PCR reagent 201, 202, 213, 219,
    222

CA12, as stationary phase in
  HSCCC 162, 172
cadmium(II) ion
  detection of 200, 204, 205, 210, 212,
    213, 214, 216, 217, 218, 220, 233
  determination of
    in foodstuffs 249, 278, 279
    in fresh and potable water 248,
      251
  formation constants of
    complexes 120, 145
  temperature effect on separation
    efficiency 124, 125
caesium iodide, analysis of 263, 268
calcium ion
  detection of 213, 214, 216, 217,
    218, 230, 235
  determination of, in
    seawater 256–7
  formation constants of
    complexes 120, 145
calmagite (CAL)
  chemical structure and physical
    data 84
  impregnated phases using 91, 92,
    93
  as PCR reagent 85, 215
carboxylic acid type cation
  exchangers 64, 65–6
CAS see chromazurol S
Castner–Kellner process 265–6
cation exchangers
  commercially available 65–6
  influence of solution ionic strength
    on Cu(II) sorption capacity 22
  sorption enthalpy values 28
chelate, meaning of term 6
chelating ion exchange ligands
  effects of immobilisation 67, 70
  properties 36–42
    acid–base properties 39, 41–2
    adsorption of metal ions 36–7
    functional selectivity 37–9, 40
    ligand chemical stability 37
    suitable coordinating sites 37
    types 37, 38

Subject Index

chelating ion exchangers
  advantages compared with ion exchange sobents 242–3
  commercially available 64–7, *68–9*
  complexation kinetics at surface of 17–21
  coordinating groups used 37, *38*
  with covalently bonded chelating groups 42–67
  equilibria within 21–7
  exploitation of advantages 242–3
  first developed 10
  historical development of 10–11
  monolithic 77–9
  polymer-based 46–53
  silica-based 54–64
  surface complexes 16–17
    two-step formation of 17–18
  synthetic methods 42–6
    direct immobilisation (single-step grafting) approach 42–3
    oxirane chemistry approach 45–6, 54
    surface assembly method 43–4
    two-step reaction process 45–6, 54
  types 42
chelating stationary phases 35–115
  dynamically modified phases 79, 95–108
  evolution of 9–11
  homogeneous distribution of functional groups in 36
  hydrolytic stability requirements 35
  impregnated 79–80, 80–95
  mass-transfer requirements 36
  matrix effects 67, 70–9
    monolithic materials 77–9
    particle porosity effects 73–4
    phase effects 73–4
    surface distribution of ligands 70–3
  mechanical stability requirements 35
  mixed mode retention on 4

  surface distribution of covalently bound ligands 70–3
    homogeneous coverage 71, *71*
    'island-like' coverage 70, *71*
    random coverage 70, *71*
  thermal stability requirements 36
chelation
  meaning of term 3
  separation selectivity affected by 3–4, 6
'chelation effect' 3, 4, 72
chelation ion chromatography (CIC) 5–6
chelation ion exchange 4
  compared with ion exchange 46, 92
Chelex-100 resin 46, 72, 254
chiral phase chromatography (CPC) 4, 67
chlor-alkali industry brine, alkaline-earth metal determination of 268, *273*
chloride ion
  formation constants of complexes with various metal ions *120–1, 145*
  as mobile phase ligand 144, *145, 146*
4-chlorodipicolinic acid
  chemical structure and physical data *106*
  dynamic modification of stationary phase by 141, 278–9
  as mobile phase ligand 107–8, *107*, 141, *249, 250*, 251
chlorophosphonazo III
  chemical structure *202*
  as PCR reagent 52, 98, 207, *214*
chlorosulfonazo III, as mobile phase ligand 98
chromatofocusing of proteins and peptides 67, 151–2
chrom(e)azurol S (CAS)
  chemical structure and physical data *82, 202*

chrom(e)azurol S (CAS) (*continued*)
  impregnated phases using 80, 85, *91*, 92, *93*, *261*
  as PCR reagent 207, *215*, 222, 232, *245, 247, 248, 275*
chromium ions, detection 205, *215*, *235*
chromotropic acid, bisazo derivatives 207
chromotropic acid type group *38*
citric acid
  in direct photometric detection 141, 194
  as mobile phase ligand 211
cobalt(II) ion
  detection of *200, 204, 205, 210, 213, 214, 215, 216, 217, 220, 235*
  determination of, in fresh and potable water *248*, 251
  formation constants of complexes *120, 145*
commercially available chelating ion exchangers 64–7, 68–9
competitive complexation
  of ligands in mobile phase 138–40, 166–7, 208–12
  of metal ions in mobile phase 146–7
complex mixtures of metal, simultaneous separation of 242
complex samples, analysis of 242
complexation, separation selectivity affected by 3
complexation ion chromatography 5, 9
concentration gradient elution 147–8
  example(s) *149, 177, 245*
conditional formation constants 136, 137, 143–4
conductimetric detection 179, *248*, 257
coordinate bonding 2
copper(II) ion
  detection of *204, 205, 212, 213, 214, 215, 216, 217, 218, 235*

determination of
  in drinking water 253, *256*
  in estaurine and sea water 256, *258, 259*
  in foodstuffs *249, 278, 279, 279*
  in seawater *246, 256, 259*
formation constants of complexes *121, 145*
counter-current chromatography (CCC) 8–9, 158–70
  applications 169–70, *171–3*
  classification of methods 159
  see also high-performance centrifugal partition chromatography; high-speed counter-current chromatography
CPC surfactant 222
o-cresolphthalein complexone (*o*-CPC)
  chemical structure and physical data *82, 202*
  as mobile phase ligand 245
  as PCR reagent *249*, 250
  see also phthalein purple
m-cresolphthalexon S *see* xylenol orange
o-cresolphthalexon S *see* xylenol orange
CTAB surfactant 222
Cyanex 272, as stationary phase in HSCCC *162, 171*
Cyanex 301, as stationary phase in HSCCC *162, 172*

dative covalent bonding 2
desorption of metal ions 21
detection
  by alpha ray spectrometry *173, 174*
  by atomic emission spectrometry *167, 168, 171*
  conductimetric 179, *248*, 257
  fluorescent detection, reagents 207, *217–18*

Subject Index

in hyphenated techniques 234–6
  by ICP-MS *173*, *185*, *189*, *190*,
    236, *247*, *249*, *277*, *280*
  photometric detection
    after post-column change of
      pH *102*, *103*, *104*
    after post-column reactions 13,
      195–222
    mobile phase ligands
      facilitating 98, 99, *100*, *101*,
      141, 194–5
  by scintillation counter *172*, *173*
  see also photometric detection;
    post-column reactions
detector noise
  electronic methods for reduction
    of 226–7
  measurement of 224
  types 224–6
dialkylphosphinic acids, as
  stationary phases in HSCCC
  162, 166
dialkylphosphonic acids, as stationary
  phases in HSCCC *162*, 166
dialkylphosphoric acids, as stationary
  phases in HSCCC 161, *162*, 163,
  166
dichlorophenyldithiophosphinic acid
  (DCPDTPI), as stationary phase in
  HSCCC *162*, *172*
dicyclohexano-18-crown-6
  (DCH18C6), as stationary phase in
  HSCCC *163*, *173*
N,N'-diethylcarbamoylmethylene-
  phosphonic acid di-n-hexyl ester
  (DHDECMP), as stationary phase
  in HSCCC *162*, *173*
di-(2-ethylhexyl)phosphoric acid
  (DEHPA)
  as stationary phase in HPEC *182*,
    186, *187*, *188*
  as stationary phase in
    HSCCC 161, *162*, 163, *164*, 166,
    167, *168*, *171*
diffusion coefficient, mobile-phase
  solutes 29

diglycolic acid, as complexing
  agent *120–1*, 140
β-diketonate functionalised silica *56*
  preparation of 54, 62
β-diketonate type group *38*
N,N'-dimethyl-N,N'-dibutyldode-
  cyloxyethylmalonamide
  (DMDBDDEMA), as stationary
  phase in HSCCC *162*, *174*
N,N'-dimethyl-N,N'-
  dioctylhexylethoxymalonamide
  (DMDOHEMA), as stationary
  phase in HSCCC *162*, *173*
di-2-methylnonylphosphoric acid
  (DMNPA), as stationary phase in
  HSCCC *162*, *171*
Dionex CIC system 5–6, 244
Dionex PCR systems 201
Dionex resins and columns 51, 64, *68*,
  76, 90, *91*
1,5-diphenylcarbazide, as PCR
  reagent *215*
Diphonix resin 229
dipicolinic acid
  chemical structure and physical
    data *106*
  as direct photometric detection
    reagent 141, 194
  dynamic modification of stationary
    phase by 105, *246*, 252, *253*, 260,
    263, *264*, 270, *274*, 276, *277*
  effect on complex formation 211
  formation constants of complexes
    with various metal ions *120–1*,
    145
  as mobile phase ligand 132, *134*,
    140, 141, 210, *212*, *225*, *247*, *248*,
    249
dipole–dipole interactions 2, 4
direct current plasma atomic emission
  spectrometry (DCP-AES),
  detection using *167*, *168*
dispersion (London) forces 2
distribution ratio
  of chelating ligands, in extraction
    chromatography 177–8

distribution ratio (*continued*)
  of metal ions 23, 158
    in competitive complexation elution 139
    in counter-current chromatography 163
    in mixed mode mechanism 23–4
    in substrate-only complexation elution 137
dithiocarbamate functionalised silica *61*
  preparation of 54, 62
dithiocarbamates, as PCR reagents *202*, 207, *216*
dithizones, as PCR reagents *202*, 203, 207, *214–15*, 222
dodecyl crown ethers, in stationary phase for HPEC *183*, *189*
drift (of baseline) *224*, 225
drinking water, analysis of 85, *248*
droplet centrifugal partition chromatography (DCPC) 159
droplet counter-current chromatography (DCCC) 159
dynamically modified stationary phases 79, 95–108
  advantages of use 98–9, 100
  applications
    actinides separation 105, 260, *264*, 276, *277*
    alkaline-earth separation 99, *100*, 250–1, 261, *266*
    transition-metal separation *53*, 98, 99, 101–4, 105–8, 250–1, 253, *255*, 257
    aromatic and heterocyclic acids used 104–8, 141, *246*, 252, *253*, 260, *263*, *264*, 270, *274*, 276, *277*, 278–9
  compared with pre-impregnated phases *100*, *101*
  metallochromic ligands used 97–104, 141, *245*, 248, 250, 261, *266*, 275–6
  theory 95–7

ECR *see* eriochrome cyanine R
EDTA
  as PCR additive 272, *275*
  see also zinc EDTA
effective (chelating) capacity of stationary phases 71
  3-aminopropylsilica 72
  IDA-silicas *74*
electrostatic interactions 2, 4, 16
  suppression of 25
    effect of mobile phase ionic strength 21–3, 24, 26, 35, 116–19
elution 116–54
  see also mobile phase
elution modes 135–54
  gradient elution 147–54
    pH step gradient mode *85*, 86, 89, *90*, 149–54
  isocratic elution 135–47
English Channel, uranium levels in 260
equilibrium stability constants, IDA–metal complexes 25
eriochrome black T, as PCR reagent *215*, 222
eriochrome cyanine R (ECR), as PCR reagent *215*, 232, *258*
estuarine seawater, analysis of 148, 256, *260*
ethanol (as biofuel) 281
  trace metals in *249*, 281
ethylenediamine, as mobile phase ligand *188*, 245
ethylenediamine type group *38*
2-ethylhexylphosphonic acid (EHPA)
  as stationary phase in HPEC *188*
  as stationary phase in HSCCC *162*, *165*, 166
ethylphosphonic acid, complexes with various metal ions *120–1*
Evian mineral water *257*
extraction chromatography 8, 9, 11, 170, 174–90
  applications 180, 184–7
    examples listed *188–90*

chelating reagents used 178–80, *181–3*
compared with HPCIC 174
stationary phase supports 175–6
eyewash saline solution, analysis of *249*, 265, *272*

fast centrifugal partition chromatography (FCPC) 159
fine chemicals, analysis for trace metals 263–4, *269–71*
flow-injection analysis (FIA), post-column reactions used 196
fluorescent detection, PCR reagents used 207, *217–18*, 232–3, *258*, 259
foodstuffs, analysis of *249*, 278–9
formation constant(s) 24, 136
(logarithms) listed for various metal complexes with common ligands *120–1*, 148
plots 143–4, *145*
fresh water
determination of trace metals in *248*, 250–4
speciation study for aluminium 233
functional groups, homogeneous distribution in chelating stationary phases 36

gallium(III) ion, detection of *217*, 233, *235*
galvanic bath waste waters, analysis of *249*
glutamic acid bonded silica 41, *59*, 133
3-glycidoxypropyl groups, silica activated by 46
glycine cresol red (GCR)
chemical structure and physical data *83*
impregnated phases using *91*, 92, *93*
gradient elution 147–54
concentration gradients 147–8, *149*

pH gradients 149–54
examples *85*, 86, 89, *90*, *171*, *172*, *173*
gypsum, determination of trace metals in *247*, 272, 274–6, *276*

hafnium(IV) ion, detection of *205*, *214*
heavy metal ions
detection of 200, *201*, *204*, *205*, *212*, *213*, *214*, *216*, *217*
elution order affected by mobile phase ionic strength 117–18
separation of 8, *49*, *50*, *51*, 53, 86, *87*, 89, *90*, 92, *93*
heavy metals, isolation and separation of 8
high-performance centrifugal partition chromatography (HPCPC) 159
commercial equipment 161
instrumental designs *160*
high-performance chelation ion chromatography (HPCIC)
advantages 168, 243–4, 254–5
comparison with HSCCC 161, 168
current status 13
detection methods 194–235
elution modes 135–54
first example 12
limitations 244
meaning of term 5
metals successfully separated by *244*
mobile phase parameters 116–34
practical applications *245–9*, 250–81
stationary phases 35–115
temperature effects 27–9, 124–30
high-performance extraction chromatography 170, 174–90
applications 180, 184–7
examples listed *188–90*
extracting/chelating ligands used 178–80, *181–3*
retention mechanism influenced by ligand loading stability 177–8
stationary phase supports 175–6

high-performance ion exchange chromatography (HPIEC) 5
high-performance liquid chromatography (HPLC)
  interactions involved 1–3
  normal-phase (NP-HPLC) 1, *2*
  reversed-phase (RP-HPLC) 1, *2*
high-speed counter-current chromatography (HSCCC) 159
  advantages 168–9
  applications 169–70, *171–3*
  commercial equipment 161
  comparison with HPCIC 161, 168
  efficiency of metal separations using 167–9
  extracting ligands 161–3
    (listed) *162–3*
  separation efficiencies 161, 167–9
  separation selectivity 163–7
Highland Spring mineral water *254*
historical developments 6–9
HMBP *see* bis(1,1,3,3-tetramethylbutyl)phosphinic acid
8-HQ *see* 8-hydroxyquinoline
8-HQS *see* 8-hydroxyquinoline-5-sulfonic acid
human lung tissue, determination of plutonium in *249*, 279, *280*
hydration sphere, of metal cations, factors affecting disruption of 19
hydrodynamic counter-current chromatography 159
  instrumentation for *160*
  see also high-performance centrifugal partition chromatography; high-speed counter-current chromatography
hydrogen bonding 2
hydrolytic stability of stationary phases 35, 62–4
hydrophilic interaction liquid chromatography (HILIC) 67
hydrophobic complexing ligands
  in counter-current chromatography 158
  in dynamic HPCIC 95, 104, 105, 141
  in extraction chromatography 8, 170, 178–80, *181–3*
hydrophobic stationary phases 47, 48, 91, *93*, *96*, 99, 133, 141, 175
hydrostatic counter-current chromatography 159
  instrumentation for *160*
hydroxamic acid type group *38*
4-hydroxy-N,N-dihexylbutyramide (4HHBA), as stationary phase in HPEC 179, *182*, *184*, *189*
α-hydroxyisobutyric acid (α-HIBA), in mobile phase 105, 178, *184*, *186*, *187*, *188*, *189*
hydroxylamine, as buffer for post-column reactions 221
8-hydroxyquinoline
  chemical structure *202*
  as direct photometric detection reagent 141
  early use in HPCIC *18*, 19, 54, 141
  as PCR reagent *216*
  in precipitation chromatography 6, 7
8-hydroxyquinoline bonded silica, preparation of 43–5
8-hydroxyquinoline type group *38*
8-hydroxyquinoline-5-sulfonic acid (8-HQS)
  chemical structure *202*
  with MgCDTA, as PCR reagent *217*, 233
  with MgEDTA, as PCR reagent *217*, 233
  as mobile phase ligand in IEC 98
  as PCR reagent *189*, 207, *217*, 233
8-hydroxyquinolinium ion
  as chelating reagent 119
  as counter-ion 119
Hypersil C_{18} column, in extraction chromatography *184*, *186*, *188*, *189*
hyphenated techniques, detection in 234–6

Subject Index

ICP-MS *see* inductively coupled plasma mass spectrometry
IDA functionalised phases, effect of stationary phase structure on separation selectivity 76
IDA functionalised resin 46, 72, 254
IDAA SAMMS chelating ion exchanger 72
IDA-silica 58–9
  applications
    commercial brines 268, *273*
    fine chemicals analysis 263, 264–5, *269, 270*
    seawater analysis 256, 258, *260*
    water analysis 251–2, 253–4, *255, 256, 257*
  commercially available 64, *68*
  dependence of effective capacity on particle porosity *74*
  elution order of metal ions affected by mobile phase parameters 117–19, *123, 132, 134*
  hydrolytic stability 63
  pH dependence for metal ion retention 38–9, *40*
  separation selectivity, factors affecting 27, 38–9, *40*, 75, 76, 124, *126, 130, 200, 201*
iminodiacetic acid (IDA)
  complexes with various metal ions 25, *120–1*
    stability in water–methanol solutions 31, *32*
  see also poly-IDA
iminodiacetic acid (IDA) functional groups *18, 19*, 38
  ion-pair formation with divalent metal cations *18*, 19
iminodiacetic acid (IDA) functionalised resins 10–11
  see also IDA-silica
immobilised artificial membrane chromatography (IAMC) 4

immobilised metal ion affinity chromatography (IMAC) 3, 9, 67, 77
impregnated stationary phases 79, 80–95
  disadvantages of use 97–8
  effect of mobile phase pH 31
  effects of resin structure upon ligand impregnation 89–95
  metallochromic ligands used 80, *81–4*, 85–8, *91, 92, 93*, 252, *254*, 256, *259*, 261, *261*, 263, *265, 267, 268*, 274, *276*, 278
  production of 88–9
  stability 88–9
indium(III) ion
  detection of *205, 217*, 233, *235*
  formation constants of complexes *121, 145*
induction (polarisation) forces 2
inductively coupled plasma mass spectrometry (ICP-MS)
  detection using *173, 185, 189, 190*, 236, *247, 249*, 254, *277, 280*
  determination of radioisotopes by 236, 276
  in hyphenated systems 180, 235, 236
  limitations 243, 276
inner-sphere complexes 19
  dissociation of 21
ion chromatography (IC) 4, 5, 243
  limitations 244
ion–dipole interactions 2, 4
ion exchange, definition 5
ion exchange chromatography (IEC) 4, 243
  compared with chelation-based methods 46
  electrostatic interactions in 2, 4
  temperature-responsive selectivity 28
ion exchange resins, first discovered 10
ion-induced dipole interactions 2
ion–ion interactions 2, 4

IonPac CG12A/CS12A column 31–2, 64, *68*, 124, 127, 128, *129*, *131*, 134
IonPac CS5 column 210, *212*
IonPac CS15 column 64, *65*, *68*, 127
ion-pairs, formation in surface complexation 18–19
iron, speciation analysis in drinking waters 253–4, *257*
iron(II) ion
  detection of *205*, *213*, *215*, *216*, *217*, *235*
  determination of, in foodstuffs *249*, 279
  formation constants of complexes *120*, *145*
  separation from Fe(III) *145*, 167, *168*, 253–4, *257*
iron(III) ion
  detection of *205*, *212*, *213*, *216*, 221, 232, *235*
  determination of, in seawater 259, *261*
  formation constants of complexes *121*, *145*
  separation from Fe(II) *145*, 167, *168*, 253–4, *257*
Irving–Williams order (of metal complexes) 144
isocratic elution 135–47
  competing ligands in 140–2
  examples *41*, *104*, *107*, *130*, *168*
  prediction of retention and selectivity relationships 142–3
  speciation plots 143
  types
    competitive complexation of ligands in mobile phase 138–40
    competitive complexation of metal ions in mobile phase 146–7
    substrate-only complexation 137–8
isotope ratio analysis 272, 274
itaconic acid functionalised column 64, *68*, 124, *125*, 127–8, 258

Kelex-100 impregnated ODS column 230
kinetic theory of chromatography 28
knitted loop reactors (for post-column reactions) 198, *199*
Kryptofix 222 178

laboratory chemicals, analysis for trace metals *247*, 262–4, *268–71*
lactic acid, as mobile phase additive *85*, *87*, 141, 149, 150, *151*, *245*, *246*, *247*, *248*, *249*, *254*, *268*, *278*
lanthanides (lanthanoids)
  detection of *205*, *213*, *214*, *215*, *217*, 221, *235*
  retention affected by mobile phase parameters 128, *129*, *131*, 134
  separation of
    by counter-current chromatography 163, *164*, 169, *171*, *172*, *173*
    by extraction chromatography 8, 178, 179, *180*, *184*, 186–7, *187*, *188*, *189*
    by HPCIC *50*, *51*, 52, 103, *104*, 118–19, *122*, 128, *130*
    by paper precipitation chromatography 7
  stability constants for chelates 118, *122*
lead(II) ion
  detection of *201*, *204*, *205*, *212*, *213*, *214*, *216*, *217*, *235*
  determination of
    in foodstuffs *249*, 278, *279*
    in seawater *246*, 256, *259*
  formation constants of complexes *121*, *145*
LiChrospher C-18 column, in extraction chromatography *187*, *188*
ligand-exchange liquid chromatography 3, 9, 64
ligand-impregnated stationary phases 79–80, 80–95

Subject Index

effect of mobile phase pH 31
see also impregnated stationary phases
limit of detection (LOD)
  calculation using signal-to-noise (S/N) measurements 223, 224
  factors affecting 223
  methods for improving 222–30
  PCR reagents 212
    (listed) for chemiluminescent detection 234, *235*
    (listed) for fluorescent detection *217–18*
    (listed) for photometric detection *213–16*
  see also baseline noise
limit of qualification (LOQ) 223
liquid–liquid chromatographic methods
  counter-current chromatography 158–70
  extraction chromatography 170–90
LIX 63, as stationary phase in HSCCC *163, 173*
London (dispersion) forces 2
long-term noise *224*, 225
luminol, chemiluminescent detection using 234, *235*
lumogallion
  chemical structure *202*
  as PCR reagent *126, 218*, 229, 233, *246, 258*, 259
Lysofon *see* aurin tricarboxylic acid (ATA)

macro-reticular polystyrene based resins, PAR-impregnated 92, 94
magnesium ion
  detection of *213, 214, 215, 217, 218*, 230, *235*
  determination of, in seawater 256–7
  formation constants of complexes *120, 145*
maleic acid, as complexing agent *120–1*, 140

mandelic acid
  chemical structure and physical data *106*
  as mobile phase ligand 105, 141, *189*
manganese(II) ion
  detection of *200, 204, 205, 210, 212, 213, 215, 216, 217, 218, 220*
  determination of
    in foodstuffs *249*, 279
    in fresh and potable water *248*, 251
    in seawater *246*, 256, *259*
  formation constants of complexes *120, 145*
mass-transfer requirements 36
mechanical stability requirements 35
MEDUSA speciation plotting software 143, 208
2-mercaptopropionic acid, as PCR reagent *216*
mercury(II) ion
  detection *205, 213, 214, 215*
  formation constants of complexes *121, 145*
metal-ion chelation chromatography (MICC) 5
metal speciation analysis 234–5
metal speciation plots 143, 208, *209*
metallochromic ligands
  chemical structures and physical data *81–4*
  dynamical modification of stationary phases by 97–104, 141, *245*, 248, 250, *261, 266*, 275
  impregnated (pre-coated) stationary phases using 76, 80–95, 252, *254*, 256, *259*, 261, *261*, 265, *267*, 274, *276*
methanol–water mixture, as mobile phase 31, *32*, 99, 133, *134, 189, 246*, 248, 249, 281
N-methyl P-13 resin 48, *50*
methylthymol blue (MTB)
  chemical structure and physical data *81*

methylthymol blue (MTB) (*continued*)
  as direct photometric detection reagent 275–6
  impregnated phases with 86–8, *91, 92, 93,* 252, *254,* 261, *265, 267,* 278
  as mobile phase ligand 99, 100–4, 141, *247*
  as PCR reagent 207
Metrohm PCR system 201, 203
milk powder, analysis of *249,* 278
Milligat type pump 196, 203
mine process water, analysis of *248,* 250
mineral water, analysis of 170, *248,* 252–3, *253, 254, 256, 257*
minerals, determination of trace metals in *247,* 272, 274–6, *276*
mobile phase
  chelating ligand(s) in 95–108
    see also dynamically modified stationary phases
  competing complexing ligands in 140–2
  electrolyte additives 119, *123*
  non-complexing, selectivity in 24–6
  on-line purification of 229–30
  organic solvent additives in 31–2, 132–4
  oxidising and reducing agents in 134
  parameters influencing separation performance 116–34
    ionic strength 19, 21–3, 24, 26, 35, 116–23
    organic solvent additives 31–2, 132–4
    oxidising and reducing agents 134
    pH 19, 29–31, 41–2, 130–2
    temperature effects 124–30
  secondary equilbria within 26–7
molar extinction coefficient 219
  complexes with various PCR reagents *205–6,* 219, 220

molybdenum(VI), detection of *216*
monolithic chelating ion exchangers 77–9
moving-average smoothing algorithm 227, *229*
multidimensional HPCIC, seawater analysis using *245,* 257–8
multiple wavelength approach, noise suppression using 227–9

neptunium, separation from other actinides 180, *185, 189, 190*
neptunium(IV) ion, detection of *206*
nickel(II) ion
  detection of *204, 205, 212, 213, 214, 216, 217, 235*
  determination of, in seawater *246, 256, 259*
  formation constants of complexes *120, 145*
nitro-PAPS
  chemical structure *202*
  as PCR reagent 200, 207, *213,* 219, *220,* 221
noise reduction methods 226–30
  electronic methods 226–7
  mechanical methods 226
  multiple wavelength approach 227–9
  software methods 227
non-aqueous solvents, determination of trace metals in 280–1
nuclear industry applications, liquid–liquid chromatographic methods 169–70, *174,* 180, 186, 187, *187*

octadecylsilica (ODS) phases
  in dynamic HPCIC 101–4, 105
  in extraction chromatography 175–6, *177*
offshore oil well brines, analysis of 87, *247,* 261–2, *267*
oligoethyleneamine functionalised silicas 152

Subject Index

oligoethylenediamines, protonation constants 73
open-tubular reactors (for post-column reactions) 198, *199*
orange II (OII), impregnated phases using *91*, 92, *93*
organic solvents, as mobile phase additives 31–2, 132–4
outer-sphere complexes
   conversion to inner-sphere arrangement 19
   dissociation of 21
   factors affecting formation of 18
oxalic acid
   formation constants of complexes with various metal ions *120–1*, *145*
   as mobile phase ligand 105, 132, 140, 144, *146*, 211, *245*, *246*, *249*, *255*
oxidising agents, as mobile phase additives 134
oxirane chemistry 45–6, 54, 62
3-oxo-pentanedioicacid bis-[bis-(2-ethylhexyl)-amide] (OPAEHA), as extracting ligand in HPEC *183*, 185
3-oxo-pentanedioicacid bis-diisobutylamide (OPAIBA), as extracting ligand in HPEC *183*, 185
oyster tissue, determination of trace metals in *249*, 279

π–π interactions 2, 104–5
P-13 resin 48, *50*
packed-bed reactors (for post-column reactions) 197–8
PAN, as PCR reagent 204, *213*
PAOOP *see* 2-(2-pyridylazo)-4-(octyloxy)phenol
paper mill wastewater, determination of aluminium in 254, *258*
paper precipitation chromatography 7
PAR *see* 4-(2-pyridylazo)-resorcinol
PC-88A
   as mobile phase in HPEC *188*
   as stationary phase in HPEC *188*
   as stationary phase in HSCCC *162*, 166, *169*, 170, *171*, *172*
   see also 2-ethylhexylphosphonic acid (EHPA)
PCR *see* post-column reaction
PCV *see* pyrocatechol violet
Pearson acid–base concept 37
peristaltic pumps (in post-column reactors) 196–7
   short-term noise/pulsations due to 226
pH step gradient elution 149–54
   examples 85, 86, 89, *90*, *164*, *165*, *171*, *172*, *173*, *245*, *246*, *247*, *248*, *268*
phase volume ratio 23
1,10-phenanthroline, as PCR reagent 201, *216*
phenanthroline type group *38*
m-phenylenediglycine-based ion exchange resins 10
phenylhydrazone bonded silica column 279
1-phenyl-3-methyl-4-benzoylpyrazol-5-one (HPMBP), as stationary phase in HSCCC *162*, *173*
1-phenyl-3-methyl-4-capryloylpyrazol-5-one (HPMCP), as stationary phase in HSCCC *162*, *172*
photometric detection
   after post-column change of pH *102*, *103*, *104*
   direct detection reagents 141, 194–5, 261, *266*, 275–6
   mobile phase ligands facilitating 98, 99, *100*, *101*, 141, 194–5
   post-column reagents used 12, 13, 203–19
   (listed) *202*, *213–16*
   see also post-column reaction (PCR) reagents
phthalein purple (PP)

chemical structure and physical data *82*
as direct photometric detection reagent 141, *245*, *248*, 250, 261, *266*
dynamic modification of stationary phase by 99, 141, *245*, *248*, 250, 261, *266*
impregnated phases using 75, 76, 80, 85, *91*, 92, *93*, 99, *100*, *101*, 274, *276*
as mobile phase ligand 99–100, *100*, *101*, 141, *246*, *248*
as PCR reagent 78, *171*, *190*, 207, *214*
picolinic acid
chemical structure and physical data *106*
as direct photometric detection reagent 194–5
formation constants of complexes with various metal ions *120–1*, *145*
as mobile phase ligand 105, 140, 144, *145*, *146*, *150*, *245*, *255*, 256, *260*
plutonium
determination in environmental and biological samples 236, 276, *277*, 279, *280*
separation from other actinides 170, *174*, 180, *185*, *189*, *190*, 276, *277*, 279, *280*
poly(acrylonitrile–DVB) resins 48, 52
polyampholyte containing mobile phases 140–1
polyampholyte ion exchangers 22–3
polyampholyte polybuffer mobile phases 152–3
chemical structures 153, *154*
poly(aspartic acid) functionalised silica 67, *68*
polybuffer ion exchanger 151–2
poly(butadiene–maleic acid) (PBDMA) resins 64, *65*

poly(glycidyl methacrylate) (PGMA) 49, 52
poly(glycidyl methacrylate-*co*-ethylene dimethacrylate-*co*-ethyl methacrylate) copolymer 69
poly-IDA functionalised resins 52–3, *68*, 76, 125
poly(itaconic acid) functionalised resins 64, *68*, 124, *125*
polymer-based chelating ion exchangers 46–53
listed *49–51*
polymethacrylates 48
commercially available *65*
polystyrene resin, crosslinked (MN200) 274, *276*
poly(styrene-divinylbenzene) resins *see* PS–DVB resins
poly(vinyl alcohol)-based resins 66
IDA functionalised 76
porous graphitic carbon (PGC) column 99, 141, *248*, 250
post-column reaction (PCR) approach
chemiluminescence detection 233–4, *235*
displacement reactions used 230–2
fluorescence detection 232–3
photometric detection 195–222
aims 195
buffers for 221–2
effect of surfactants 222
factors influencing 195–6
post-column reactors 196–201, 203
reagents for 203–19
recent developments in high-sensitivity reagents 219–22
post-column reaction (PCR) reagents 12, 13, 203–19
arsenazo I 207, *214*
arsenazo III *12*, *122*, *130*, *164*, *169*, *171*, *172*, *173*, *180*, *184*, *186*, *187*, *188*, *189*, 207, *214*, 221, *225*, *245*, *246*, *247*, *248*, *253*, *264*, *265*, 274
BHEDTC *216*

Subject Index

5-Br-PADAP 204, *213*, 219, 222
calmagite *85*, *215*
CAS 207, *215*, 222, 232, *245*, *247*, *248*, 275
chemical structures *202*
chlorophosphonazo III *52*, 98, 207, *214*
o-CPC *249*, 250
data for formation of coloured complexes with various metal ions *205–6*
1,5-diphenylcarbazide *215*
dithiocarbamates *202*, 207, *216*
dithizones *202*, 203, 207, *214–15*, 222
eriochrome black T *215*, 222
eriochrome cyanine R *215*, 232, *258*
8-HQ (8-hydroxyquinoline) *216*
8-HQS *189*, 207, *217*, 233
8-HQS–MgCDTA *217*, 233
8-HQS–MgEDTA *217*, 233
limits of detection 212
  (listed) *213–16*
listed
  for chemiluminescent detection *235*
  for fluorescent detection *217–18*
  for photometric detection *213–16*
lumogallion *126*, *202*, *218*, 229, 233, *246*, *258*, 259
MTB 207
nitro-PAPs 207, *213*, 219, *220*, 221
PAN 204, *213*
PAR *53*, *90*, *106*, *107*, *125*, *146*, *150*, *165*, 170, *171*, *172*, *188*, 198, *200*, *201*, 204, 207, *210*, *212*, *213*, 219, *220–1*, *220*, *230*, *231*, *245*, *246*, *248*, *250*, *251*, *279*, *279*, *281*
PAR/NH$_3$ *255*, *256*, *257*, *259*, *260*, *268*, *273*
PAR–ZnEDTA *41*, *87*, *213*, *231–2*, *247*, *248*, *249*, *254*, *267*, *268*, *271*, *273*, *276*, *278*
PBI–ZnEDTA *218*

PCV 207, *214*, 221, 222, *245*, *261*
1,10-phenanthroline *216*
PP *78*, *190*, 207, *214*, *247*
requirements 203
stability constants for 208, 210, 212
TAR *179*, *188*, 204
tiron 207, *215*
TTMAPP *215*
XO 86, 207, *213–14*, 222
ZnEDTA *151*
post-column reactors 196–201, 203
  commercial systems 201, 203
  effect of tubing materialon peak shapes 198, 200–1
  flow reaction chambers 197–8, *199*
  mixers 197
  reagent delivery systems 196–7, 203
potable water, analysis of 85
potassium chloride brine, trace-metal analysis of 86, *247*
potassium chloride solution, alkaline earths in 265, *272*
potassium nitrate, analysis of *268*, *269*, *271*
precipitation chromatography 6–7, 11
ProPac IMAC-10 column 52–3, *68*, 76, 125
proteins, chromatofocusing of 67, 151–2
provenance of foods and wines, evaluation of 272, 274
PRP-X800 column *68*, 124, *125*, 127–8
PS–DVB resins
  amino-functionalised 11, 28
  commercially available 46, *65*
  disadvantages 47–8
  dynamically modified *100*, *101*, 104–8, *253*, *264*, *266*, *274*, 275–6, *277*, *279*
  grafted with propylenediaminetetraacetic acid 12
  IDA-functionalised 11, 46

PS–DVB resins (*continued*)
  metallochromic ligand
    impregnated 75, 76, 80, 85–7, 88–95, 99, *100, 101*, 149, *151, 254, 259, 267, 278*
  sulfonated 28
PTFE tubing (in post-column reactors) 198
  effect on peak shapes 133, 198, 200, *200, 201*
pulse dampeners (for post-column reactor pumps) 226
PUREX (plutonium uranium redox extraction) 169, 187
2-(2-pyridylazo)-4-(octyloxy)phenol (PAOOP), as stationary phase in HPEC *179, 181, 188*
4-(2-pyridylazo)-resorcinol (PAR)
  with ammonia, as PCR reagent *255, 256, 257, 259, 260, 268*
  chemical structure and physical data *84, 202*
  data for formation of coloured complexes with various metal ions *205–6*
  impregnated phases using *91, 92, 93*, 94
  as PCR reagent 53, *90, 106, 107, 125, 146, 150, 165*, 170, *171, 172, 188*, 198, *200, 201*, 204, 207, *210, 212, 213*, 219, *220–1, 220, 230, 231, 245, 246, 248, 250, 251, 279, 279, 281*
  with ZnEDTA, as PCR reagent *41, 87, 213*, 231–2, *247, 248, 249, 254, 267*, 268, *271, 273, 276, 278*
2,2′-pyridylbenzimidazole (PBI)–ZnEDTA, as PCR reagent *218*
pyrocatechol type group 38
pyrocatechol violet (PCV)
  chemical structure *202*
  data for formation of coloured complexes with various metal ions *205–6*
  as PCR reagent 207, *214*, 221, 222, *245, 261*

quinaldic acid
  chemical structure and physical data *106*
  as direct photometric detection reagent 194
  as mobile phase ligand 105, *106*, 107, 133, 141

radionuclides, isolation and separation of 8, 170, *174*, 180, *185, 189, 190*, 276, *277, 279, 280*
rain water, analysis of *248*
rare earth elements *see* lanthanides (lanthanoids)
reagent grade chemicals, analysis for trace metals 263–4, *269–71*
reciprocating pumps (in post-column reactors) 196
  short-term noise/pulsations due to 226
reducing agents, as mobile phase additives 134
REEs *see* lanthanides (lanthanoids)
retention factor 23, 24
  effects of mobile phase parameters 117–18, 128, *129*, 131–2
  effects of stationary phase structure 74–6
  at H_{50} 39
  prediction under isocratic elution 142–3
retention order
  alkaline-earth metals 117, *118, 129*, 144, 166
  lanthanides (lanthanoids) *104, 122, 129, 130, 131, 164, 180*
  transition metals *129*, 144, *149, 150, 151*
rice flour, analysis of *249*, 278–9
river water, analysis of *248*

salicylic acid type group 38
saline lake water 99, *245–6*, 261, *265, 266*
saline samples, analysis of 254–69

Subject Index

Savitzky–Golay algorithm 227
scandium ion, formation constants of complexes *121*, *145*
scintillation counter, detection using *172*, *173*
SDS surfactant 222
seawater analysis 86, 141, 148, 242, 245–6, 255–60, *261–4*
sediments, analysis of *247*, 270–2, *274*, *275*
selectivity
 appropriateness 36
 functional 37–9, *40*
selectivity ratio 23
semi-xylenol orange (SXO), impregnated phases using *91*, *92*, *93*
separation efficiency
 in counter-current chromatography 161, 167–9
 factors affecting 1, 20, 29, 48, 124–7
separation selectivity
 changes using multi-component mobile phases 144, *146*
 in counter-current chromatography 158, 163–7
 factors affecting 1, 3–4, 27, 28, 31, 74–6, 127–30, 140
separation selectivity ratio 26
short-term noise *224*, 225
signal extraction, noise suppression using 228–9, *230*, *231*, *264*, 268, *271*, *279*
silica, cation exchange properties 63
silica-based chelating ion exchangers 54–64
 commercially available *65*
 disadvantages 64
 listed *55–61*
 preparation of 43–6, 54, 62
silver ion, detection of *235*
silver ion chromatography 9
Skogseid's potassium-selective resin 10
sodium chloride brine, trace-metal analysis of 86, *247*

sodium chloride feed solution (for NaOH process) 266–8
sodium sulfate, analysis of 263, *268*
soils, analysis of *247*, 276–7
solid samples, analysis of 269–81
solochrome brilliant blue B *see* chrom(e)azurol S
solution complexation kinetics 17
sorption enthalpies, various chelating ion exchangers 27–8, 127
speciation analysis
 aluminium 233, 254
 iron 253–4, *257*
speciation of metals 234–5, 242
speciation plots 143, 208, *209*
specific ion interaction theory (SIT) 24
stability constants, metal complexes *120–1*, 208, 210, 212, 219
stationary phases *see* chelating stationary phases
stream sediments, analysis of *247*, 270–2, *274*, *275*
strontium ion
 detection of *213*, *217*, 230, *235*
 determination of 87, 99–100, *101*
  in gypsum 274–5, *276*
  in milk powder *249*, 278
  in mineral water *248*, 252, *254*
  in oil well brines *247*, 262, *267*
  in saline lake water 261, *266*
 formation constants of complexes *120*, *145*
 isotopic composition 274
substrate-only complexation model of elution 137–8, *147*, 148, 152
2-(3-sulfobenzoyl)-pyridine-2-pyridylhydrazone (SPPH)
 chemical structure and physical data *84*
 impregnated phases using *91*, *92*, *93*
surface complexation 16–21
 kinetics 17–21
surfactants, post-column reactions affected by 222

TAR, as PCR reagent 179, 188
tartaric acid
  metal complexes with 120–1
  as mobile phase complexing agent 140, 150, 166–7, 245, 256, 260
temperature effects in HPCIC 27–9, 124–30
  kinetic effects 28–9, 124–7
  thermodynamic effects 27–8, 127–30
terminology 4–6
tetraethylenepentamine-silica
  pH dependence for metal ion retention 39, 40
  pH gradient elution used 152, 153
meso-tetrakis(4-N-trimethylaminopenyl)-porphine (TTMAPP), as PCR reagent 215, 220
N,N,N',N'-tetraoctyldiglycolamide (TODGA), as extracting ligand in HPEC 183, 186, 187
tetraphenylmethylenediphosphine dioxide (TPMDPD), as stationary phase in HSCCC 163, 173
thermal stability requirements 36
thorium, separation from other actinides 180, 185, 186, 189, 190
thorium(IV) ion, detection of 205, 207, 214, 235
tin(IV) ion, detection of 205, 214
tiron
  chemical structure 202
  as PCR reagent 207, 215
titanium(IV) ion, detection of 205
tomato leaves, analysis of 249, 279
trace analysis of metals 236, 242
  advantages of HPCIC 236, 242
transition metal ions
  detection of 200, 204, 205, 206, 208, 210, 212, 213, 214, 215, 216, 217, 218
  formation constants of complexes 120–1, 145

retention order affected by mobile phase parameters 117–18, 128, 129
separation of
  commercial available chelating ion exchangers for 68–9
  by counter-current chromatography 165, 171, 172
  with dynamically modified phases 53, 98, 99, 101–4, 105–8, 250–1, 253, 255, 257, 270
  by extraction chromatography 178, 179, 188
  by gradient elution 148, 149, 150, 151
  by ion exchange chromatography 28
  polymer-based chelating ion exchangers for 49, 50, 51
  with pre-impregnated phases 86, 87, 89, 90, 92, 93, 256, 259
  silica-based chelating ion exchangers for 55–61
tributylphosphate (TBP) 187
trifluoroacetylacetone, as direct photometric detection reagent 141
triphenylmethane type ligands
  effective capacities 91–2
  selectivity 92, 93
  see also aurin tricarboxylic acid; chromazurol S; glycine cresol red; methylthymol blue; phthalein purple; pyrocatechol violet; xylenol orange
Triton X-100 surfactant 222
TRU resin (for extraction chromatography) 180, 182, 185, 189, 190
TSK-Gel Chelate 5PW column 69, 75, 76, 86, 141
TTMAPP see tetrakis(4-N-trimethylaminopenyl)-porphine

Subject Index

Universal Cation column *69*, 127
uranine, chemiluminescent detection using 234, *235*
uranium, separation from other actinides 170, *174*, 180, *185*, *186*, *189*, *190*, 260, 276–7, *277*
uranium(VI) ion
    detection of *206*, 207, *214*
    determination of
        in mineral water 252, *253*
        in seawater 260, *263*
        in sediments 270–1, *274*
uranyl ion
    determination of 87, 105
        in saline lake water 261, *265*
    formation constants of complexes *121*, *145*

van Deemter equation 28–9
van der Waals forces 2
vanadium ions, detection of *206*, *214*
vanadyl ion, formation constants of complexes *121*, *145*
van't Hoff equation 28
vitamin tablets, determination of trace metals in 229, *249*

waste waters, analysis of *249*
water elimination, as rate-determining step in complex formation 18, 19, 20
waters *see* estaurine water; fresh water; potable water; saline water
Wofatit C cation exchange resin 10

xylenol orange (XO) 86
    chemical structure and physical data *81*, 150, *202*
    formation constants of complexes with metal ions 148
    impregnated stationary phases using 85, 86, 88–9, 89–90, *91*, 92, *93*, 149, *151*, 256, *259*, 263, *268*
    as mobile phase ligand 98, 99
    as PCR reagent 86, 207, *213–14*, 222

zinc EDTA
    with PAR, as PCR reagent *41*, 87, *213*, 231–2, *247*, *248*, *249*, *254*, *267*, 268, *271*, *273*, *276*, *278*
    as PCR reagent *151*
zinc(II) ion
    detection of *200*, *204*, *205*, *210*, *212*, *213*, *214*, *215*, *216*, *217*, *218*, *220*, 233, *235*
        in estaurine and seawater 256, *258*, *259*
        in gypsum leachate 275–6
    determination of
        in foodstuffs *249*, 279
        in fresh and potable water *248*, 251, 253, *256*
        in seawater *246*, 256, *259*
    formation constants of complexes *120*, *145*
zirconium(IV) ion, detection of *206*, *214*, 221